Building the Snowflake Data Cloud

Monetizing and Democratizing Your Data

Andrew Carruthers

Apress®

Building the Snowflake Data Cloud: Monetizing and Democratizing Your Data

Andrew Carruthers
Birmingham, UK

ISBN-13 (pbk): 978-1-4842-8592-3   ISBN-13 (electronic): 978-1-4842-8593-0
https://doi.org/10.1007/978-1-4842-8593-0

Managing Director, Apress Media LLC: Welmoed Spahr
Acquisitions Editor: Jonathan Gennick
Development Editor: Laura Berendson
Coordinating Editor: Jill Balzano
Copyeditor: Kim Burton

Cover photo by Fallon Michael on Unsplash

Distributed to the book trade worldwide by Springer Science+Business Media LLC, 1 New York Plaza, Suite 4600, New York, NY 10004. Phone 1-800-SPRINGER, fax (201) 348-4505, e-mail orders-ny@springer-sbm. com, or visit www.springeronline.com. Apress Media, LLC is a California LLC and the sole member (owner) is Springer Science + Business Media Finance Inc (SSBM Finance Inc). SSBM Finance Inc is a **Delaware** corporation.

For information on translations, please e-mail booktranslations@springernature.com; for reprint, paperback, or audio rights, please e-mail bookpermissions@springernature.com.

Apress titles may be purchased in bulk for academic, corporate, or promotional use. eBook versions and licenses are also available for most titles. For more information, reference our Print and eBook Bulk Sales web page at http://www.apress.com/bulk-sales.

Any source code or other supplementary material referenced by the author in this book is available to readers on GitHub.

Printed on acid-free paper

To Esther, Verity, and Diane

Table of Contents

About the Author

Andrew Carruthers is the Director of the Snowflake Corporate Data Cloud at the London Stock Exchange Group. He oversees accounts supporting both ingestion data lakes and the consumption analytics hub. The Corporate Data Cloud services is a growing customer base of over 7,000 end users. He also leads the Centre for Enablement, developing tooling, best practices, and training, and the Snowflake Landing Zone, provisioning Snowflake accounts conforming to internal standards and best practices.

Andrew has almost 30 years of hands-on relational database design, development, and implementation experience, starting with Oracle in 1993. Before joining the London Stock Exchange Group, Andrew operated as an independent IT consultant. His experience has been gained through predominantly working at major European financial institutions. Andrew is considered a thought leader within his domain with a tight focus on delivery. Successfully bridging the gap between Snowflake's technological capability and business usage of technology, Andrew often develops proof of concepts to showcase benefits leading to successful business outcomes.

Since 2020 Andrew has immersed himself in Snowflake. He is considered a subject matter expert, recently becoming CorePro certified, contributing to online forums, and speaking at Snowflake events on behalf of the London Stock Exchange Group. In recognition of his contribution to implementing Snowflake at the London Stock Exchange Group, Andrew recently received the Snowflake Data Hero of the Year award. This category recognizes a technology trailblazer who has pioneered the data cloud in their organization.

Andrew has two daughters, 19 and 21, both elite figure skaters. He has a passion for Jaguar cars, having both designed and implemented modifications, and has published articles for *Jaguar Enthusiast Magazine* and The Jaguar Drivers' Club. Andrew enjoys 3D printing and has a mechanical engineering workshop with a lathe, milling machine, and TIG welder, to name a few tools, and he enjoys developing his workshop skills.

About the Technical Reviewer

 **Mike Gangler** is a senior database specialist and architect. He's also an Oracle ACE with expertise in database reliability and architecture and a strong background in operations, DevOps, consulting, and data analysis.

Acknowledgments

Thanks are not enough to the Apress team for the opportunity to deliver this book—specifically, to Jonathan Gennick. I hope my 0.01% contribution remains worth your investment. This book is all the richer for your patient guidance, help, and assistance. Mike Gangler, you have enriched this work more than you know while teaching me American grammar and spelling. As George Bernard Shaw said: "Two people separated by a common language." Jill Balzano, thank you for your project management. Who knew writing a book would be so easy? And those unknown to me—editors, reviewers, and production staff, please take a bow. You are the unsung heroes who make things happen.

To my friends at Snowflake who patiently showed me the "art of the possible," who first encouraged, advised, and guided, then became friends, mentors, and partners. Jonathan Nicholson, Andy McCann, Will Riley, Cillian Bane, and James Hunt—quite a journey we have been on, and in no small part, this book is possible only because of your steadfast support. I owe you all a beer or three!

To my colleagues at London Stock Exchange Group who had the vision and foresight to investigate Snowflake, Mike Newton, this is for you. Also, step forward my inspiring colleagues—Nitin Rane, Srinivas Venkata, Matt Willis, Dhiraj Saxena, Gowthami Bandla, Arabella Peake, and Sahir Ahmed. You all played a bigger part in this book than you know, and I will be forever grateful to you all.

To my daughters, Esther and Verity, who knew Dad would write something like this! And finally, to Diane, you were there every day throughout the six months this book took to write. You have the patience of a saint and are more lovely than I can say.

PART I

Context

CHAPTER 1

The Snowflake Data Cloud

The Snowflake Data Cloud is a means by which data can be shared and consumed in near real time by willing participants across geographical borders—seamlessly and securely. Snowflake offers solutions to consume, reformat, categorize, and publish a wide variety of data formats and leverage siloed legacy data sets while opening new market opportunities to monetize data assets.

We must first understand the nature of the problems we currently face in our organizations. We cannot address problems if we do not acknowledge them.

I use the experience of more than 30 years working in widely varying organizations, each having its own challenges. I broadly distill my experience into several core themes discussed in this book. No one-size solution fits all, and there are always edge cases, but with some lateral thinking and innovative approaches, resolution may be at hand.

By developing an understanding of how the Snowflake Data Cloud can solve the problems we face, and providing a platform, we deliver a springboard for future success.

The journey isn't easy, and convincing our peers is equally hard, but the rewards as we progress, both in personal terms of our own development and organizations' returns for the faith they place in us, are huge.

Though leadership places a great burden upon us, identifying opportunities and having the courage and conviction to pursue them are equally important. The only way to be certain of failure is to not try. This book not only seeks to short-circuit your timeline to success but provides tools to participate in the Snowflake Data Cloud revolution far sooner than you imagine.

This chapter overviews Snowflake's core capabilities, including Secure Direct Data Share, Data Exchange, and Data Marketplace. With an understanding of each component, we then explain why we should integrate with the Snowflake Data Cloud.

© Andrew Carruthers 2022
A. Carruthers, *Building the Snowflake Data Cloud*, https://doi.org/10.1007/978-1-4842-8593-0_1

Let's approach the Snowflake Data Cloud by addressing the why, how, and what. In a TEDx Talk, Simon Sinek unpacked "How great leaders inspire action" (`www.youtube.com/watch?v=u4ZoJKF_VuA`). Please take 18 minutes to watch this video, which I found transformational. This book adopts Sinek's pattern, with *why* being the most important question.

Snowflake documentation is at `https://docs.snowflake.com/en/`. One legitimate reason for reading this book is to short-circuit your learning curve. However, there are times when there is no substitute for reading official documentation (which is rather good). We come to some of this later, but for now, at least you know where the documentation is.

Setting the Scene

Snowflake helps organizations meet the increasing challenges posed by integrating data from different systems. Snowflake helps organizations keep that data secure while providing the means to analyze the data for trends and reporting purposes. Snowflake is key to making data-driven decisions that drive future profitability for many organizations.

Almost without exception, large corporations have grown through mergers and acquisitions, resulting in diverse systems and tooling, all largely siloed around specific functions such as human resources, finance, sales, risk, or market aligned, such as equities, derivatives, and commodities. The organizational estate is further complicated by internal operating companies with less than 100% ownership and complex international structures subject to different regulatory regimes.

Growth by acquisition results in duplicate systems and data, often from different vendors using their own way of representing information. For example, Oracle Financials and SAP compete in the same space but have very different tooling, terminology, chart of accounts, mechanisms to input adjustments, and so forth.

At the macro and micro levels, inconsistency abounds in each line of business, functional domain, and lowest grain operating unit. We have armies of bright, intelligent; articulate people reconciling, mapping, adjusting, and correcting data daily, tying up talent and preventing our organizations from progressing.

Increasingly, our organizations are reliant upon data. For some, data is both their life-blood and energy. Recent mergers and acquisitions demonstrate data provides a competitive advantage, increased market share, and opportunity to grow revenue streams. One example is the London Stock Exchange Group's (LSEG) $27 billion acquisition of data and analytics company Refinitiv (`www.lseg.com/refinitiv-acquisition`) to become a major financial data provider. And there are other interesting acquisitions in the data visualization space. Google recently acquired Looker, and Salesforce acquired Tableau. Wherever we look, progressive, forward-looking companies are positioning themselves to both take advantage of and monetize their data, some with Snowflake at the center of their strategy.

These are not knee-jerk reactions. All three leading cloud providers are under pressure to provide integration pathways and tooling to enable low-friction data transfer and seamless data access, all with good reasons. Data is increasingly powering our economies, and the volume, velocity, and variety of data constantly challenge our infrastructure and capabilities to consume and make sense of it. We return to these themes repeatedly in this book and offer solutions on our journey to understanding the Snowflake Data Cloud.

The Published Research

If you need further convincing, a cursory examination of published research provides startling insight. Some studies provide insight into why Snowflake Data Cloud is important.

International Data Corporation (IDC) (`www.idc.com/getdoc.jsp?containerId=prUS46286020`) stated that in 2020, more than 59 ZB (59 trillion gigabytes) of data was created, captured, copied, and consumed. Most of these are copies of data (or worse, copies of copies), with the ratio of replicated data (copied and consumed) to unique data (created and captured) being approximately 9:1, trending toward less unique and more replicated data. Soon, the number of replicated copies is expected to be 10:1, and this growth forecast is set to continue through 2024 with a five-year compound annual growth rate (CAGR) of 26%.

The Economist Intelligence Unit, in its Snowflake-sponsored report, "Data Evolution in the Cloud: The Lynchpin of Competitive Advantage" (`www.snowflake.com/economist-research/data-evolution-in-the-cloud-the-lynchpin-of-competitive-advantage/`), states that 87% of 914 executives surveyed agree that data is the most important differentiator in the business landscape today, where 50% of respondents frequently share data with third parties to drive innovation in products and services. But in contrast, 64% admit their organization struggles to integrate data from varied sources. The 50% of organizations that frequently share data must be doing something right. How are they leveraging their infrastructure and tooling to share their data? Where did they draw their inspiration from? We return to these themes later, but an immediate conclusion we can draw is that someone, or some groups, must have changed their thinking to embrace a new paradigm and found themselves in an environment where they could try something new. Likewise, the 64% who struggle to integrate data from varied sources are at least acknowledging they have a problem and hopefully looking for solutions. If your organization is in the 64%, keep reading because out of the same sample, 83% agree their organizations will routinely use AI to process data in the next five years. Without integrating more disparate data, less complete, potentially inaccurate results will arise.

The Global Perspective

Our organizations are a microcosm of the global perspective. Whether we know or acknowledge our position in the digital economy, our organizations are subject to external market forces that force change and adaptation to new working methods. We need a global perspective to underpin our organization's success, and the Snowflake Data Cloud does just that. A single platform is delivering tooling out of the box enabling frictionless, seamless, and rapid data integration with the opportunity to monetize our digital assets.

In every organization, whether immediately obvious or not, there are data silos. Those repositories which we think may hold value but are inaccessible; data sets we are not allowed to see or use, and those data sets hidden in plain sight, the obvious and the not so obvious. Data silos are discussed in Chapter 2, where we start to challenge received wisdom in tightly segregating our data. Continuing with our theme of asking *why*, there are typically two reasons why we don't get answers to questions. The first is because knowledge is power, and the second is because the answer is unknown.

The volume, velocity, and variety of data are ever-increasing. I touched upon these earlier and later discussed the same themes in Chapter 10. As a sneak preview, we also explain how to address some of the less obvious data silos, and for the avoidance of doubt, this was my favorite chapter to write, the one with the most scope to unlock the potential of our organization's data, bear with me, as there is much to understand before we get there, oh alright, you skipped straight there.

Unavoidable in today's increasingly complex, international, and multi-jurisdictional organizations, data governance policies, processes, and procedures oversee and increasingly control every aspect of our organization's software development lifecycle (SDLC). How policies and procedures impact what, where, and how we can use our data are covered in Chapter 11. It's not all doom and gloom, there are very good reasons for data governance, and we lift the lid on some hidden reasons why we must comply.

We all know software is increasingly more complex. Anyone using a desktop PC without a deep knowledge of the underlying operating system and tools to keep both process and memory hogs at bay looks to upgrade hardware far more often than might otherwise be needed. Subtle software interactions and edge cases cause unforeseen scenarios, and we don't always have the tools to identify and remediate issues. Chapter 13 offers some ideas and concepts to remediate these challenges; not an easy subject to resolve, but we can do something to help ourselves.

Having read this, I must be clear: organizations cannot fund constant hardware, storage, network, and associated infrastructure upgrades. Increasingly costs are associated with software patching, products to help us manage our infrastructure, and improvements to our security posture dealing with both current (what we know) threats and emerging threats (those we don't yet know about). All of which is without considering up-skilling and increasing our staffing levels while the regulatory environment tightens and imposes greater controls and reporting obligations.

Every day, we focus on what we need to do to keep the lights on. Our operational staff faces more demand from ever-growing data volumes but with the same people and infrastructure, constantly reacting and firefighting to the same challenges daily. If we are fortunate, we have robust documentation explaining how we do our day-to-day jobs. But documentation only remains relevant if we devote sufficient skill, time, and energy to its maintenance. We all know the first thing to suffer in highly stretched teams is the documentation, offering a litmus test in and of itself.

If we are ever to break this never-ending cycle, we need to think differently, challenge the status quo, and quickly pivot to focus on why. Embracing automation, low-code tooling, reducing dependencies upon key people, and implementing appropriate controls and alerting mechanisms reduce the daily noise and allow us time to understand why.

We have the means at our fingertips to resolve some of these seemingly intractable problems. But we must start now because the longer we delay, the big issues compound, become harder to understand, and head toward an inflection point beyond which we will never recover. A friend described this as a diode (a discrete component that only conducts current in one direction) at some unknowable point on our upward trajectory of increasing complexity. Once passed, we can never go back because the complexity, inertia, complacency, and all other reasons make it far too hard, and the only way from this point onward is to start again.

Why Consider the Snowflake Data Cloud?

If we want a cloud-first strategy, we must look at the core capabilities required to re-homing our data. Central to answering this question is identifying a platform that supports much of what we already know and use daily, a SQL-based engine of some sort because we use SQL as the lingua franca of our databases which underpins most of our current systems. We cannot leave behind our rich legacy and skill sets but must leverage them while adopting a new paradigm.

Much has been made of data lakes using Hadoop, HDFS, map-reduce, and other tools. In certain cases, this remains valid. But largely, after more than five years of trying and showing initial promise, and in some cases, huge financial investment, the results are generally disappointing. Many independent comparisons show Snowflake to be superior across a range of capabilities.

Naturally, no single offering will ever encompass all aspects of a cloud-first strategy which must consider more than a single product and have wider utility; for example, implementing a data fabric. But there are good reasons for putting Snowflake Data Cloud at the center of our cloud-first strategy, first and foremost. It is the only one built for a cloud SaaS data warehouse with security baked in by design. In a book titled *Building the Snowflake Data Cloud*, you would expect Snowflake to be the first component in implementing our cloud-first vision, but not without sound reasons.

Having decided to investigate Snowflake, we should satisfy ourselves that the marketing hype lives up to expectation and that claims are verified. In our organization, a *proof of concept* (POC) was stood up from a standing start, where a lone developer with zero prior knowledge of Snowflake (but with significant Oracle experience) was able, over seven weeks, to use representative sample data sets, able to deliver a robust, extensible application subsequently used as the basis for production rollout. The claim may be unbelievable to some, but I was the developer, and two years later am now responsible for running the department and production rollout.

For those still unsure and thinking maybe the author got lucky or the POC was flawed, challenge yourself to imagine "what if" the same could be repeated in your organization, dare to try Snowflake for yourself, and establish your own POC and measurement criteria. I don't ask anyone to trust my word. Find out for yourself because facts and evidence are impossible to argue against. Anyone with a good understanding of relational database techniques and a firm grasp of SQL will find Snowflake readily accessible, and the learning curve is not as sharp as imagined. Moreover, Snowflake offers a free 30-day trial with sufficient credits to run a credible POC. There is no excuse for not trying.

This book challenges established thinking. Our current tooling perpetuates long-established, entrenched thinking, which does not work in the new data-driven paradigm. Something must give—we continue to spend vast sums of money supporting the status quo, or we find another way. The same rationale must underpin our decision to investigate the Snowflake Data Cloud.

Benefits of Cloud Integration

I have already discussed the broader themes leading to the inevitable conclusion that we must change our thinking and approach to take advantage of the data-driven paradigm before us. This section identifies tangible benefits of cloud integration, drawing parallels between cloud service providers and current on-premise implementations. I offer this perspective as the outcomes underpin the adoption of the Snowflake Data Cloud.

Following the theme of why, how, and what, I first offer deeper insight into why we should move to a cloud provider. Naturally, these themes cover both how and what. According to an International Data Group (IDG) research report (sponsored by Accenture AWS Business Group), benefits accruing from cloud integration include a 45% infrastructure and storage costs reduction, an organization's typical operational efficiency is increased by 53% with 43% better utilization of data. This is significant, bearing in mind our previous observation of 9:1 copies of data to original data sets. Perhaps not surprising, 45% greater customer satisfaction is also recorded.

Let's now look at some more tangible benefits the Snowflake Data Cloud offers.

Hardware and Software

The most obvious difference between cloud and on-premise (on-prem) is hardware provisioning, with software provisioning a close second. Most organizations have two or more data centers filled with hardware with high bandwidth network connections facilitating both application failover and disaster recovery in the event of server or, worst case, data center outage.

Hardware, air conditioning, physical security, staff, UPS, halon protection, miles of networking cables, switches, routers, racking, test equipment, servicing, repair, renew and replace. The list is almost endless, and each item has a fixed cost; ever depreciating, the upgrade cycle is endless and increasingly demanding.

With the cloud, organization-owned data centers are no longer required. Cloud service providers (CSPs) provision everything, available on demand, 24×7×365, infinitely scalable, fault tolerant, secure, maintained, and patched. While we pay for service provision, we do not have the headache of managing the infrastructure. Moreover, apart from storage costs, we (mostly) only pay for what we consume. Storage costs are very low. Snowflake simply passes through storage charges, and depending upon location, at the time of writing, costs vary from $23 to $24.50 per terabyte.

Figure 1-1 shows the relative costs of retaining fixed on-prem infrastructure compared to moving to cloud-based elastic provisioning.

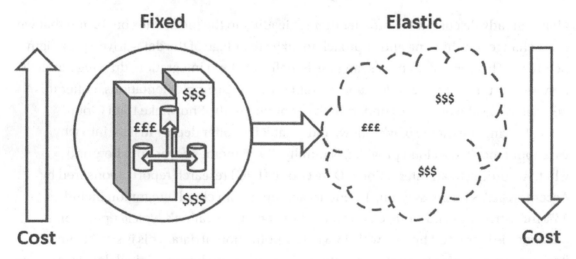

Figure 1-1. *Data cloud adoption trends*

Performance

On-Prem dedicated hardware is always limited to the physical machines, CPU, memory, and disk. Regardless of whether the servers are in use, we continue to pay for their provision, yet servers are only utilized for about 5% of the time. The same is true of network bandwidth; every organization has a limit to the amount of data carried on their internal network with utilization peaks and troughs.

With cloud-based applications, instant performance scaling and per-second costing provide performance when required, for only as long as required, with predictable costs, and then scale back. We pay for what we consume, not what we provision, and have both elastic performance and storage at our fingertips. We don't incur network bandwidth charges for cloud-to-cloud connectivity as data transits the public Internet; however, we may pay cloud provider egress charges.

When hardware failures occur, they are invisible to us. We benefit from always-on, infinite compute, infinite storage elastically provisioned on demand. We need to pay attention to authentication and network transport layer security, all discussed later.

Staffing

The highest costs to an organization are its people. With on-premise hardware, we incur security guards, physical installation costs, specialists to perform operating system and application patching, servicing, installation costs, PATS testing, and management to run teams.

With the cloud, costs are significantly reduced, but not to zero. We need more cloud security specialists instead. However, the cost savings are significant when compared to on-prem equivalents.

The COVID-19 pandemic has shown that our organizations can operate with at least the same or higher efficiency than otherwise thought possible, with reduced office capacity needed to support staff that can work remotely and connects to cloud services natively.

Control

Each organization has exclusive control of hardware and data in its own domain with on-prem implementations. Total control is physically expressed in the infrastructure provisioned.

Cloud implementations have a different perspective. Absolute control is only as good as the security posture implemented, and in a shared environment, such as may be provisioned for managed services. Across our estate, both on-prem and cloud, the security boundaries must remain impenetrable, and this is where our cybersecurity team is most important.

We also realize benefits from seamless patching and product updates that are applied "behind the scenes" are invisible to us. Vendors release updates without downtime, or outages lead to higher uptime and availability. While we give up some administrative oversight, we benefit from integrated, joined-up automated delivery throughout all our environments.

Another benefit is the feature-rich cloud environment offering built-in tooling and virtual hosting for third-party tools and applications. Driven by the need to increase market share, cloud providers continually improve their tools and integration pathways, enabling developers to innovate, reduce time to market, and find solutions to difficult business problems.

Data Security

Data security is of paramount importance to cloud-based solutions. Snowflake has a comprehensive suite of security controls—an interesting subject, which is discussed later.

Typically, each business owner determines the business data classification of each attribute in a data set with specific protections mandated for each business data classification. For on-prem data which never leaves the internal network boundaries, security considerations do not arise to the same extent as they do for cloud, where data in transit is expected to be encrypted at the connection level (conform to TLS1.2, for example) or be encrypted at rest (Bring Your Own Key, SHA256, for example). We will discuss data security in more detail, but for now, it is enough to note the distinction between on-prem and cloud-hosted solutions.

Compliance

For on-premise solutions, most regulators are content with knowing data is held in their own jurisdiction. But for cloud-based solutions, some regulators apply additional scrutiny on a per data set basis before allowing data onto a cloud platform. Additional regulation is not identical between regulators; therefore, an individual approach must be adopted to satisfy each regulator's appetite.

Additional governance may apply before approval is granted to move artifacts to the cloud, particularly where data is hosted in a location outside of the governing authority's jurisdiction. This is covered later, though it is important to note data governance and controls follow the data, not the cloud, ensuring we protect the data while maintaining the highest levels of security and compliance.

Data Volumes

You have seen from the research I identified that the typical number of data copies to original data is currently around 9:1 and expected to increase to 10:1 soon. These alarming figures hide many underlying issues. Replicated data sets as pure copies (or copies of copies) result in the inability to reconcile across data sets due to different extract times and SQL predicates.

Unreconcilable data leads to management mistrust and a desire to build siloed capabilities as the central system must be wrong. You see the problem emerging, particularly where manual out-of-cycle adjustments are made, further calling data provenance and authenticity into question.

Other direct consequences arise with increased data storage requirements and higher network bandwidth utilization as we shuffle data to and fro while requiring more powerful CPUs and memory to process data sets. There are others, but you get the idea.

With alarming regularity, new data formats, types, and challenges appear. And the velocity, volume, and variety of information are increasing exponentially. With the Snowflake Data Cloud, many of these challenges disappear. I will show you the *how* and the *what*.

Re-platforming to Cloud

If we simply want to reduce our hardware, infrastructure, and associated costs, can we simply "lift and shift" our technical footprint onto a cloud platform?

In short, yes, we can port our hardware and software to a cloud platform, but this misses the point. While superficially attractive by allowing the decommissioning of on-premise hardware, performing "lift and shift" onto the cloud only delays the inevitable.

We still require specialist support staff and cannot fully take advantage of the benefits of re-platforming.

For example, imagine porting an existing data warehouse application to the cloud using "lift and shift," we can decommission the underlying hardware and gain the advantage of both elastic storage and CPU/memory. However, typically we must bounce our applications when making configuration changes, require the same support teams but introduce the added complication of cybersecurity to ensure our new implementation is secure.

In contrast, if we were to re-platform our data warehouse to Snowflake, we would immediately reduce our platform support requirements almost to zero. Elastic storage is automatically available; no configuration is required. CPU/memory configuration changes no longer require system outages, allowing us to break open our data silos. And we can still decommission our on-premises hardware. Both approaches require cybersecurity to ensure our configurations are safe. I show you how in Chapter 4.

In summary, re-platforming to the cloud buys us little at the expense of introducing cloud security and effectively reinforces existing silos because our data remains locked up in the same platforms.

Where Is the Snowflake Data Cloud?

The Snowflake Data Cloud is not a single physical place. Rather, it is a collection of end-points on the Internet—hidden in plain sight among various CSPs.

The chances are you have accessed many cloud platforms already just by browsing the Internet. Snowflake has chosen the top three CSPs to build the Snowflake Data Cloud.

- Amazon Web Services (AWS)

- Microsoft Azure (Azure)

- Google Compute Platform (GCP)

Figure 1-2 shows where Snowflake has established its presence. Note that locations are added from time to time.

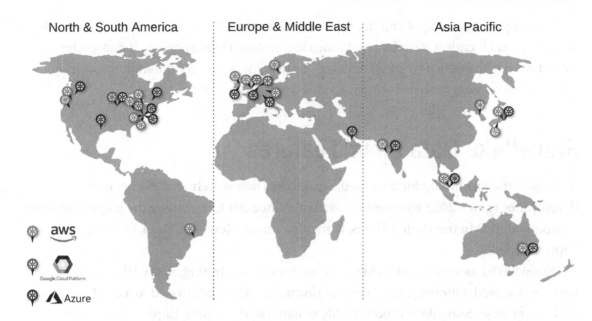

Figure 1-2. Snowflake Data Cloud locations

Several other cloud providers exist, typically focused on specific applications or technologies, whereas the preceding three clouds are relatively agnostic in their approach.

In addition to the cloud providers, we describe our legacy infrastructure in our datacenters as on-premise (or *on-prem* for short). Along with that term, the following is some additional terminology that you should know.

- Connectivity between any cloud location and on-premise is North/ South. North is in the cloud, and South is on-prem.

- Connectivity between any two cloud locations (regardless of platform) is East/West.

The distinction in connectivity terminology becomes important for several reasons.

- Establishing East/West connectivity is harder than it first appears for various reasons discussed later.

- East/West bandwidth consumption is largely irrelevant, except for egress charges, discussed later.

- North/South connectivity is usually easier than East/West and is usually established via closed connectivity (AWS Direct Connect).

The implication of applications deployed to the cloud from the end-user perspective is the physical location is irrelevant. From a technologist's perspective, it matters less than it used to, but the new considerations for the cloud are very different than for on-prem. What matters more is security, a subject we return to later.

Snowflake Data Cloud Features

Built from the ground up for the cloud, Snowflake Data Warehouse has been in development since 2012 and publicly available since 2014, becoming the largest software company to IPO in the United States, launching on the New York Stock Exchange in September 2020

Snowflake has security at its heart. From the very beginning, Snowflake has been security-focused. Offering exceptional performance when compared to a traditional data warehouse, Snowflake is both highly scalable and resilient. Implemented on all three major cloud platforms (AWS, Azure, GCP) and equally interoperable across all three, support is provided centrally. There are no installation disks, periodic patching, operating system maintenance, or highly specialized database administration tasks. These are all performed seamlessly behind the scenes, leaving us to focus on delivering business benefits and monetizing our data, which is what this book is all about.

Snowflake supports ANSI standard Structured Query Language (SQL), user-defined functions in both Java and Scala with Python soon to follow, and both JavaScript and SQL stored procedures. Almost everything you have learned about SQL, data warehousing, tools, tips, and techniques on other platforms translates to Snowflake. If you are a database expert, you have landed well here.

Snowflake is used ubiquitously across many industry sectors due to its excellent performance. New uses are emerging as the velocity, volume, and variety of information increases exponentially, and businesses search for competitive advantage and opportunities to monetize their data.

Business leaders looking to improve their decision-making process by faster data collection and information delivery should consider Snowflake. Improving knowledge delivery and increasing wisdom enables decision makers to set the trend instead of reacting to the trend. Imagine what you would do if you had knowledge at your fingertips five minutes before your competitors; Snowflake offers you the opportunity to turn this into reality. I'll show you how later.

Snowflake is the only data platform built for the cloud. Natively supporting structured, semi-structured, and (soon) unstructured data, with a plethora of built-in tooling to rapidly ingest, transform and manipulate data, Snowflake delivers.

Having identified many differences between cloud and on-premises implementations and the benefits of moving to the cloud, let's now discuss how the Snowflake Data Cloud benefits your organization.

The Snowflake Data Cloud enables secure, unified data across your organization and those you choose to collaborate with, resulting in a global ecosystem where participants choose their collaboration partners and effortlessly both publish and consume data sets and data services of choice. Huge quantities and ever-increasing data can easily and rapidly be connected, accessed, consumed, and value extracted.

When properly implemented, your organization realizes the benefits of broken-down data silos and experiences greater agility enabling faster innovation. Extracting value from your data becomes far easier leading to business transformation and opportunities to monetize your data.

There is no mention of hardware, platforms, operating systems, storage, or other limiting factors. Just pause for a moment and think about the implications. The vision is extraordinary, representing the holy grail every organization has searched for in one form or another since before the advent of the Internet age. How to seamlessly access cross-domain data sets, unlock value, find new markets, and monetize our data assets.

The Snowflake Data Cloud is the place with secure unified data, seamlessly connected, available where and when we need it. If we get our implementation right, by using Snowflake Data Cloud, we have a single source of the truth at our fingertips with the ability to go back in time up to 90 days. Seamlessly, out of the box.

Business Lifecycle

Evidence for Snowflake being at the bottom of the growth stage is not hard to find. Look at Figure 1-3, which illustrates the standard product lifecycle showing the various stages every product goes through. Supporting my assertion, financial statements are at `www.macrotrends.net/stocks/charts/SNOW/snowflake/financial-statements`. Snowflake is a Unicorn by any standard; big institutional investors at IPO include Berkshire Hathaway and Salesforce.

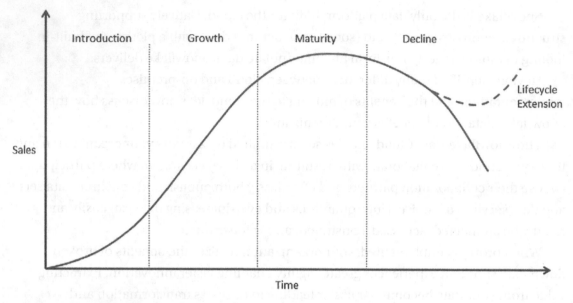

Figure 1-3. *Business lifecycle*

The six-year pre-IPO period might be regarded as the introduction stage in the diagram with low sales, high costs, and low profits. The growth stage started post seed funding with increased sales, reduced costs, and profits. Snowflake is certainly not into the maturity phase; the sheer volume of new features and product enhancements demonstrate Snowflake is ever-growing in capability and scope.

Snowflake published the numbers shown in Figure 1-4. Draw your own comparisons. The latest figures are at `https://investors.snowflake.com/overview/default.aspx`.

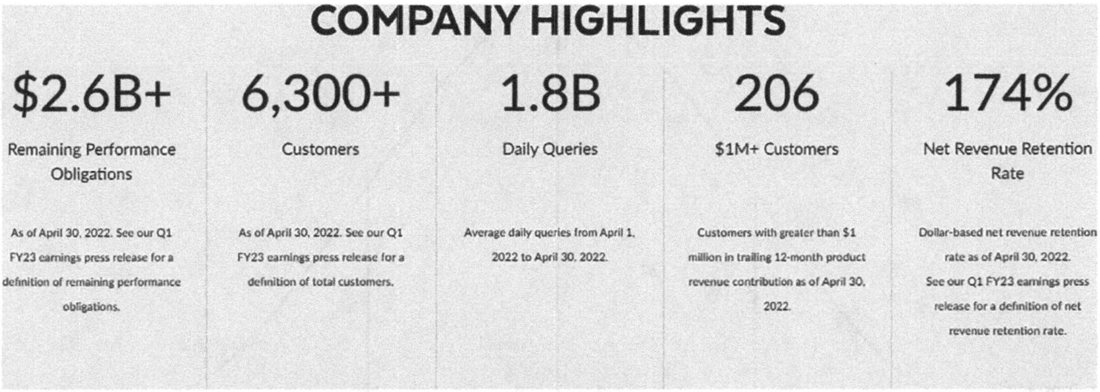

COMPANY HIGHLIGHTS

$2.6B+	6,300+	1.8B	206	174%
Remaining Performance Obligations	Customers	Daily Queries	$1M+ Customers	Net Revenue Retention Rate
As of April 30, 2022. See our Q1 FY23 earnings press release for a definition of remaining performance obligations.	As of April 30, 2022. See our Q1 FY23 earnings press release for a definition of total customers.	Average daily queries from April 1, 2022 to April 30, 2022.	Customers with greater than $1 million in trailing 12-month product revenue contribution as of April 30, 2022.	Dollar-based net revenue retention rate as of April 30, 2022. See our Q1 FY23 earnings press release for a definition of net revenue retention rate.

Figure 1-4. *Snowflake published company highlights*

Diffusion of Innovations

Snowflake's innovative design separating storage from computing is a paradigm shift from old thinking, representing a clean break from the past. Snowflake architecture is explained in Chapter 3.

When presented with new ideas, we are often challenged to accept new thinking. The Diffusion of Innovations theory provides some answers by breaking down the population into five distinct segments, each with its own propensity to adopt or resist a specific innovation. Figure 1-5 presents the bell curve corresponding to adoption or resistance to innovation. I leave it to you to determine where your organization falls.

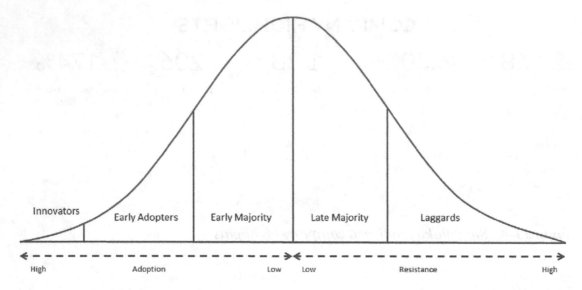

Figure 1-5. *Diffusion of Innovations*

Snowflake adoption is no different. In Figure 1-5, Snowflake is somewhere to the left of the early adopter's profile. This is where those of us looking to make a strategic leap forward find ourselves, an opportunity to embrace a rapidly maturing product, enabling the Snowflake Data Cloud and providing another opportunity to monetize our data.

On the left, the innovators, 2.5% of the population, have already adopted Snowflake, endured the challenges of dealing with an immature product, unknowingly conducted user acceptance testing, and suffered sleepless nights creating workarounds for previously broken features—the list goes on.

The next 13.5% are early adopters. If you have read this far, you are likely one of them. We take advantage of the hard work the innovators have put into Snowflake and seize the day, grasp what is before us, and move forward.

At our fingertips is Snowflake, a mature product of which the next 34%—the early majority—are just becoming aware of. It offers a window of opportunity to create the greatest commercial advantage and develop into new markets to monetize our data.

Let's pause our discussion on Diffusion of Innovations here; few want to be in the late majority, and less still in the laggards, but this is where 50% of the population find themselves, whether they know it or not.

Better to be early than just on time, never to be late.

Future State Enabled

Our world is often changing in unpredictable ways. For example, face recognition is now commonplace but unthinkable ten years ago with a consequential rise not only in volume but the velocity of data too. By the time you read this book, Snowflake will have released support for unstructured data, a subject for later in this book, which mainstream databases struggle to deal with.

And new challenges arise. If you want to extract billing information from a photographed invoice, Snowflake has the tools to do just that. If you want to save paragraphs from a Microsoft Word document into a structured data format, Snowflake can do that too. I will show you how to do both in Chapter 9.

This represents microcosms of data sets requiring particular tools and techniques to process, and Snowflake has the answers. New features arrive all the time. As I wrote this chapter, object tagging became available; see Chapter 11 for details.

Emerging themes on future data growth (all discussed later) such as the Internet of Things (IoT) generating sensor data, imaging, semi-structured records, and more, big data and analytics showcasing how we can make sense of these huge volumes of data, machine learning decision-making affecting our everyday lives.

First-Mover Advantage

Combining the Diffusion of Innovations theory with the business lifecycle gives us the knowledge to understand we have a window of opportunity and first-mover advantage. We can identify and create new markets, be the first entrants into the market, develop brand recognition and loyalty, gain competitive advantage, and consequently, a marketer's dream.

But time is not on our side. Your organization's competitors have already done their research and begun executing their strategies. Some are reading this book just as you are.

Your Journey

Your journey to the cloud is just that—a journey with some lessons to learn. Some are outside of the scope of this book, but many are illustrated here.

Not only are there significant cost savings, performance benefits, and flexibility to be gained, but also wider security considerations to address. And for those who enjoy technology, Snowflake has plenty to offer with an ever-expanding feature set delivered faster than most of us can assimilate change.

How Snowflake Implements the Data Cloud

For all approaches, it is important to understand the data owner always remains in full control. Role-based access control (RBAC) is discussed in Chapter 5, but for a sneak preview, everything in Snowflake is an object subject to RBAC, including secure direct data shares and the objects contained therein.

In other words, Snowflake is inherently highly secure from the ground up. Some of these features are discussed in Chapter 4. I will later show you how to configure Snowflake security monitoring, exceeding Snowflake recommendations and exposing some underlying features.

Figure 1-6 illustrates the options available to us, along with some background information. Each option is briefly explained next.

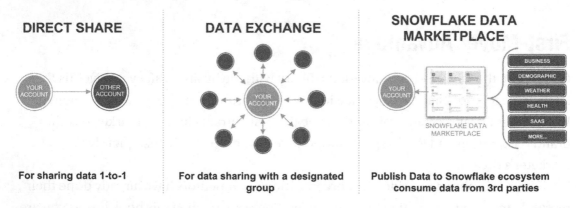

Figure 1-6. *Snowflake Data Cloud options*

Global Data Mesh

The Snowflake infrastructure underpinning the Snowflake Data Cloud is its global data mesh. Snowflake addresses the challenges of building data platforms in unique ways which readily facilitate the delivery of robust data architectures.

With some prior thought on how RBAC enables data ownership and access, adherence to published standards for data ownership, and clarity of thought on logically and physically separating data sets, delivery of the Snowflake Data Cloud is readily achievable. But only if we do things the right way from the outset, or more probable for organizations with an existing data lake, develop a plan to bring structure and governance together along with clear direction to remediate the past.

Engagement Criteria

To successfully create our contribution to the Snowflake Data Cloud, we must adopt a domain-based approach to data ownership. In practice, we must know "who owns what" and enforce data ownership as a toll-gate before allowing data onto our platform. Not as easy as one might think, our organizations are becoming more tightly regulated, and while attractive to "just throw the data in," any audit or governance function wants to know "who owns what." Better to set the standard at the outset rather than suffer the consequences later.

With multiple data domains in our data cloud, new opportunities exist for cross-domain reporting. Subject to correctly segregating our data using RBAC, we simply need to grant roles to users to enable data usage. Simple, right? No, not simple in practice because we may not want to grant access to all data in the domain, so we must restrict data to the minimal subset required to satisfy the business usage. And Snowflake has the tools to do this. Finer-grained RBAC, row-level security, object tagging, and the creation of custom views are some techniques we use.

As I will unpack shortly, data discovery and self-service are central to Snowflake Data Cloud capability.

Secure Direct Data Share

Secure Direct Data Share is a built-in, highly secure core Snowflake capability reliant upon cloud storage to share data with other Snowflake customers; that is, consumers both inside and outside your immediate organization. We unpack Secure Direct Data Share in Chapter 14 with hands-on examples.

With Secure Direct Data Share, the data owners decide the data sets to share, and technologists provide the implementation. At all times, full control is maintained over which external customers have read-only access. Addressing "who can see what,"

Snowflake is extending its monitoring capabilities to provide metrics on consumption. Chapter 14 provides a template.

Data is maintained in real time. As we update the source object data, our customers see the same changes in real time as they occur. Data sharing occurs behind the scenes. Once we declare the objects and entitle customers to access the share, our clients consume it; it is as simple as that. No more SFTP, data dumps, process management, and other cumbersome data interchange scheduling and tooling.

Costs differ according to CSP and actions performed, noting as Snowflake product capabilities evolve, the costing model may also change. For further up-to-date information, see the documentation at `https://docs.snowflake.com/en/user-guide/ billing-data-transfer.html#understanding-snowflake-data-transfer-billing`.

Suitable for all applications, Secure Direct Data Share explicitly controls which customers have access to shared data and are a great way to securely and safely monetize your data while retaining full control.

Many organizations suffer from incomplete, incorrect, divergent reference data where inconsistencies lead to rejected records, unmatched attributes lead to missing content, and over time, slowly lead to mistrust of the system.

If we can address this one seemingly simple (it is not simple) issue across the many hundreds or thousands of applications in our organizations by sourcing our reference data once and distributing it across all our applications seamlessly, we would soon enjoy the benefits of corrected, conformed data which can more easily be joined as many more attribute values will match.

And there are more opportunities; the story just gets better.

Data Exchange

Utilizing Snowflake's built-in data sharing capability, Data Exchange is the organization's internal place where data is published to consumers, that is, people and organizations both in and outside the publishing organization boundary. Data Exchange is also where organizations and individuals discover available data sets. We unpack Data Exchange with worked hands-on examples in Chapter 14.

Data sets controlled by business owners are published for consumption by others, either by approval-based subscription or freely available to all. Truly the beginnings of data democratization, data publishing and data discovery allow everyone to fully participate. They are both able and entitled to access data sets. Moreover, the silos begin to crumble, allowing data discovery, enrichments, and utilization in previously unthinkable ways.

Marking a significant step forward in an organization's capability to seamlessly interact, Data Exchange quickly allows subscribers to access their entitled data sets. Time to market is slashed by hours, or at most days, to implement data integration between organizations—no more SFTP, authentication, firewalls, or handshaking, and previous impediments disappear.

Central to the concept of data sharing is *security*. With Data Exchange, the data owner retains complete control of their own data sets and can curate, publish, manage, and remove entitlement at will. Naturally, one year retained immutable full audit trail is available; this is discussed in more detail later.

Snowflake Marketplace

Organizations use Replication and Secure Direct Data Sharing capabilities to create listings on the Snowflake Marketplace, the single Internet location for seamless data interchange.

Snowflake Marketplace is available globally to all Snowflake accounts (except the VPS account, a special case) across all cloud providers.

Because Snowflake Marketplace uses Secure Data Sharing, capability, security, and real-time features remain the same. The integration pattern differs as replication, not share, is used as the integration mechanism.

Snowflake Marketplace supports three types of listings.

- Standard: As soon as the data is published, all consumers have immediate access to the data.

- Purchase with a free sample: Upgrade to the full data set upon payment.

- Personalized: Consumers must request access to the data for subsequent approval, or data is shared with a subset of consumer accounts.

However, there is a hand-shake between subscriber and provider as information may be required to subscribe, and data sets may require pre-filtering, resulting in custom data presentation.

Once complete, the source database is replicated, becoming available to import into the subscribing account. Both options require local configuration to import the database, which becomes accessible similarly to every other database in the account.

Who Is Using the Snowflake Data Cloud?

With a very low barrier to entry, several organizations, including Refinitiv, FactSet, S&P Global, and Knoema, already use Snowflake to make their data available and are monetizing their assets, each offering a free trial enabling future clients to "try before you buy." this is data democratization at its best, enabling both individuals and organizations to decide for themselves which data sets best suit their needs with a very low barrier to entry.

Addressing the COVID-19 pandemic, many vendors offer relevant data sets enabling rapid dissemination and accrual of differing perspectives. Combined with machine learning and AI, unprecedented opportunities abound.

The variety of organizations using Snowflake Marketplace is growing rapidly, with free participation.

Beginning Your Snowflake Journey

The process of migrating to the cloud does not happen overnight. What you have read is thought-provoking, groundbreaking, and significantly impacts your organization.

You may find your organization is unprepared for the journey, and you may need to showcase features as they are developed to expose the benefits.

The Snowflake Data Cloud is not a single destination. It is a journey. Features are constantly being developed, improved, and enhanced. Our job is to show others the way and lead using the examples found in this book while building out proof of concepts showcasing evolving Snowflake capabilities.

Summary

This chapter began by setting the scene, outlining problems every organization has in scaling to meet the torrent of data, and offering ways to mitigate and reduce cost while expanding capability.

You looked at the emerging data landscape, changing the nature of data and increasing storage costs inherent in holding multiple copies of data.

Introducing the Snowflake Data Cloud as the answer to many issues we face in our organizations set us on our journey to realize the tangible benefits not just in cost savings but also in future proofing our organization by "right-platforming" for the future.

Through practical experience, the importance and value of showcasing technical capability and conducting a "hearts and minds" campaign to our business colleagues cannot be underestimated. Remember our discussion on Diffusion of Innovations: we must endeavor to "shift our thinking to the left" to help our colleagues embrace a new and unfamiliar paradigm.

And having established the right mindset in preparation for looking deeper into our data silos, let's open the door to Chapter 2.

CHAPTER 2

Breaking Data Silos

Every organization has data silos—those segregated, highly controlled IT systems and data repositories controlled by a limited, select group of individuals who closely guard their data. Traditionally, each silo operates as a self-contained micro-business, with minimal overlap and essential data interchange. Furthermore, in larger organizations, data silos have coalesced around common industry-leading applications, each with its own embedded data model leading to further challenges to data integration. And the model is self-perpetuating, with niche skillsets required to support each domain or line of business.

We might not recognize our everyday work as contributing to data silos. It is hard to see the forest for the trees when standing in the middle of the woods, but rest assured, data silos exist all around us. We just need to take a step back and think about the invisible boundaries that guard our data, whether they are physical in terms of machinery and location, logical in terms of network access, firewalls, and access groups, or otherwise. But these data silos have locked up value and prevent monetization of our most valuable asset: data.

Some examples of data silos, even for small companies, are in human resources owning employees, work location, diversity information, finance-owning business hierarchies, cost centers, and revenue by product line. There are many more (some surprising) benefits to organizations willing to break open their data silos, where cross-domain data sharing underpins a wide variety of benefits; some unexpected.

I am not advocating a free-for-all nor recommending any breach of proper security safeguards. I am simply proposing to democratize our data by freeing up essential elements for cross-domain reporting and public interchange (for a price), all in a highly secure Snowflake Data Cloud environment where the data owner retains full control.

But first, we must ingest data into Snowflake, and this is where our challenges begin.

© Andrew Carruthers 2022
A. Carruthers, *Building the Snowflake Data Cloud*, https://doi.org/10.1007/978-1-4842-8593-0_2

Why Break Down Silos

Chapter 1 discussed how 50% of respondents frequently share data with third parties to drive innovation in products and services. But in contrast, 64% admit their organization struggles to integrate data from varied sources. Are these examples of companies that have broken down their data silos or companies that have retained the status quo? It is too simplistic to argue one way or the other without context, but a clear perspective, attitude, and approach are important. Challenging all three allows you to break the status quo and realize the value held in your data.

Data in its correct context provides information. This is more clearly stated using the data, information, knowledge, wisdom (DIKW) pyramid. Typically, information is defined in terms of data, knowledge in terms of information, and wisdom in terms of knowledge. Figure 2-1 illustrates the relationship between each layer of the pyramid identifying the value gained by each layer.

Figure 2-1. *DIKW pyramid*

You must understand the questions you want to answer and, from the underlying data, build the context to provide the wisdom.

Where does my organization generate revenue from? This question is easily enough answered; you turn to your finance system and run a report. But the report is only available to nominate and suitably entitled finance staff, one of whom must be available to execute the request.

What is the revenue impact if system X fails? This complex example is not as easy to answer because the upstream system may not lie in finance; it probably lies in another data silo, such as sales, marketing, or risk. These are the kind of questions breaking down data silos answers, and using Snowflake Data Cloud, we have the platform to collate, categorize, store, and protect our data, putting information and knowledge at the fingertips of those who need it—when they need it, and how they need it.

Where Are the Silos?

Data silos exist all around us, the generally inaccessible logical containers we may think of in terms of the application name, vendor product, or regularly but invisibly used repositories.

As a starting point, we could map out our organization's operating model, lines of business, and functional units to identify likely sources of information. At this point, we are analyzing the "art of the possible," indulging ourselves by creatively imagining outcomes if only we had access to the data, in one place, at the same time...

Moltke the Elder famously stated, "No plan survives contact with the enemy." There is a ring of truth to this quote when navigating our organizations because, for many data owners, knowledge is power. Data control locks up information, knowledge, and wisdom, in and of itself, a data silo to be broken down. But you are reading this book, I encourage you to embrace new thinking, without which the status quo never changes, and we forever remain bound by the past. The Diffusion of Innovation theory was discussed in Chapter 1. The same logic applies. Do you want to be an innovator/early adopter in changing your thinking too?

Knowledge locked up in data silos quickly goes stale. For example, marketing data degrades at 25% per quarter, and technical knowledge—the data silos in our heads of how to do things—also degrades as new tools and techniques arise. Up-to-date documentation helps, and sharing knowledge is one sure way to stay current; self-education is another way.

Data silos are not just the boundaries between business functions or lines of business. They also occur between systems in the same domain. For example, one large European bank in HR alone has 146 systems, some cloud-based, some on-prem. And there is a propensity for each domain to build their data warehouse to solve their problems, often using disparate technologies perpetuating the very problem we seek to solve, rapidly heading toward complexity and the diode point mentioned in Chapter 1, after which recovery is impossible.

We also see data silos and cottage industry proliferation catering to weaknesses in vendor-supplied strategic platforms. The product may not support the workflow required, the overworked support team cannot implement required changes quickly enough, or the demand is for an answer "right now," and a weeklong delay is unacceptable. Propagation of tactical solutions increases technical debt, the burden which must be paid for at some point, unpicked, and reconciled back to the source. Another suite of data silos loosely coupled to sources, the presence of a system just to catalog these End-User Developed Applications (EUDAs) should give us all cause for concern. But, if someone had the foresight to put a central registry in place to record these EUDAs, we would have a decent starting point to identify and take out systematically.

For those of us totally engaged, utterly immersed, and consumed with passion for their work, how do we keep the lightning in the bottle? How do we retain motivation to over-perform, over-deliver, and continually push boundaries?

Some of the answers lie in our working environment. Research has shown people who work in cognitively demanding environments more readily share their skill, knowledge, and expertise, but not when there are high demands from co-workers and/or management expectations to knowledge share, bringing others up to the same standard where reliance upon a time with constant interruptions often result in withdrawal and lack of engagement, exactly the opposite kind of behavior we want to encourage.

My personal view is to keep encouraging both your visionaries and thought leaders to innovate and push boundaries. These are your 11 out of 10 performers weekly, the ones who can't wait to get to work, always have something up their sleeve to pull out and amaze, and the self-starters who see opportunity long before others. It is uncomfortable giving freedom to people who think differently. But these people find ways to break down data silos (often unconventional) and allow others to bring their skill, knowledge, and expertise to bear. They see those things hidden in plain sight. Their lateral thinking and ability to connect the dots often result in a competitive advantage.

Forgive them their sins when they overstep boundaries and create noise, provide head-cover, let them get on with innovating and ultimately delivering value to our organizations because if we constrain our rocket scientists into conformance to norms and insist on and noise reduction because management likes things quiet and peaceful, the lightning is sure to escape the bottle. Once the energy has gone, it has gone forever, and we have robbed our organization of the very essence crucial for continued success. Iron sharpens iron, and smart people spark greatness wherever they land and unlock huge potential in others, often without knowing their presence or few words of wisdom. They have a massively positive ripple effect reaching out to far corners of our organizations.

We have few thought leaders and even fewer visionaries. Let's encourage and retain those we have. There is always an option to work elsewhere with more enlightened teams and organizations, our visionaries and thought leaders' organizational contribution and value far exceed the noise they create. Put another way, noise is evidence that something is happening.

Note   The lessons are clear: facilitate, equip, enable, and release, don't constrain, overburden, and eventually kill our visionaries and thought leaders. Our organizations cannot survive without them.

Returning to discuss more tangible data silos, we might consider Word documents as data silos. Consider system documentation containing a table of contents (TOC) with information for each heading. If we can access our system documentation centrally from Snowflake, with unstructured document support, the TOC can be extracted and programmatically used to interrogate content according to each heading, storing results into a table.

Imagine a template document with common TOC. Using unstructured document support, we can bring all identically named sections together and interrogate using SQL. Alternatively, imagine scanned invoices where key-value pairs such as invoice number and amount can be extracted programmatically. Who needs expensive OCR software now? Chapter 10 explains how to implement tooling to extract text from images and provides a practical example for you to implement and extend.

Once we have captured data from unstructured documents, how will our business use the data? How will the data inform our decision-making? Why is it important to know? Only you can begin to know the answers to these questions, and naturally, this book can only go so far in exposing tooling to assuage your thirst for wisdom. But, in your hands, you have a starting point and new thinking to apply, hopefully changing perspectives.

We must not think of data silos as being limited to a subset of our organization's estate or a suite of physical servers and disk drives. Rather, we should approach data silos from the perspective of identifying where the value lies, then identify an access path to unlock the value. The data content defines whether the container is a silo or not.

Data silos are all around us, we just need the eyes to see and the imagination to unlock what is before us.

To quote Saurin Shah, Snowflake developer, "It doesn't matter how we store the data. It matters how data is shared and consumed." There is an implicit assumption in Saurin's comment; we must be able to unlock the data from its storage container, then make the data accessible to allow consumption.

Saurin's revelation must challenge our thinking on what constitutes a data silo because today's definition of what a data silo looks like will be very different in 10 years with new, currently unimaginable formats and media available. As an extreme example, who would have thought our own DNA would be such a rich (and complex) source of information? And who can foresee how the information contained in our DNA can be used for the future benefit of mankind?

Regardless of where we find data of interest or value, we must develop and maintain access patterns to unlock desired content for several reasons. Emerging data formats require new custom handlers. Existing handlers require upgrades as product and code versions change. New platforms necessitate handler refactoring. For some, this is a ready-made business idea. Plug-ins for Snowflake deliver data from hard to get at sources.

The lesson is clear. Even if we don't have a starting point today, a suite of tooling ready to deploy, or even minimal capability to execute, we must be ready. The future is rushing toward us ever faster, and now is the time to begin. A journey of a thousand miles starts with a single step.

How to Break Open Silos

By now, we have an idea of the types of data we are interested in, their likely business locations, data owners, and some thoughts on how to interact with data silos. Naturally, proper governance determines the entities and attributes permissible for access.

You must also consider your organization's network topology, as accessing data isn't simple. To bring data together, we must consider the end-to-end connectivity from source to destination, which allows our data to flow. For those of us fortunate enough to have clean running water, I like the analogy of a water tap. When I turn the tap on, I get water, but what facilitates the water arriving? Logically, there has to be pipework from the source to the destination.

Likewise, we can separate our data sourcing strategy into two parts: the plumbing, which provides the end-to-end connectivity, and the tap, which determines how much data flows through the pipework subject to the maximum capacity of the pipework or network bandwidth in our case. While related, these two components should be considered as separate deliverables. In larger organizations, an architectural governance body is mostly concerned with the plumbing. Any system-to-system interconnectivity should require approval to prevent a free-for-all situation. However, the data which flows through the plumbing are governed by a different authority. When breaking down data silos, it is important to remember the distinction and to engage each body separately, notwithstanding our cybersecurity colleagues who rightly insist on data security from connection through transit and final storage in Snowflake. And the hidden point behind this message is the high degree of human interaction and collaboration required to begin the journey of accessing silos.

Each data silo has format, structure, and entitlement (prerequisites to allow access). Some commonly used examples include SharePoint, Excel spreadsheets, business applications, archives, log files, databases, catalogs, active directories, and shared network drives. The list is not exhaustive, and each may contain one or more object types with data of interest, requiring a custom connector to interact with and derive value from.

We must ensure our organization is prepared, equipped, and ready to accelerate. We best set ourselves up for success by having the right attitude and approach, which requires investment into presentations, showcases, internal education, and continual positive reinforcement by evangelizing our approach and product. If we are not enthusiastically endorsing our own work, why should anyone else believe and have confidence in us? Technology is not the issue. People and processes are.

Note   Showcasing capability and conducting "hearts and minds" are critical to our success.

How Data Is Evolving

No conversation on data silos would be complete without discussing how data is evolving. We want to avoid repeating the mistakes of the past and enable data democratization for the future. Previously mentioned themes on future data growth included the Internet of Things (IoT) generating sensor data, imaging, and semi-structured records. But we must also be mindful of medical data and the possibility of personalized gene therapy, satellite imagery, data from different spectra, and climate data. The list goes on.

Our ambition to consume ever more complex and challenging data to process pushes conventional database technology's boundaries. We don't always have the tools to handle everything at our fingertips, but we can develop them over time. Consumption is not an end-state in and of itself. Rather, consumption facilitates convergence with existing data sets leading to new insights and opportunities from which our organizations benefit.

We are experiencing exponential growth in volume, velocity, and variety of data, broadly categorized into three themes: structured data, our (hopefully) well-understood third normal form OLTP databases, data vaults, data warehouses, or star schemas. More recently, semi-structured data such as JSON, XML, AVRO, Parquet, and ORC formats are all supported and managed by Snowflake tooling. And latterly, unstructured data, Microsoft Word documents, images, audio, video, sensor data, and so on. Considering Snowflake Data Cloud, we might view the three themes as evolutions of data, and the management thereof, as shown in Figure 2-2.

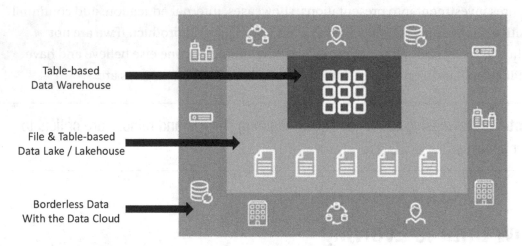

Figure 2-2. *Evolution of data management*

In Figure 2-2, there are no hard borders between each theme, and with some effort, we can develop frictionless transitions between themes, though we must invest some time and effort to develop each transition. Snowflake supplies the tools and sufficient examples to work out our integration patterns and approach to implementing our Snowflake Data Cloud. We also examine homegrown tooling in Chapter 10, specifically for semi-structured and unstructured data, with hands-on examples delivering the capability to extract textual content from image files.

The most important takeaway from the evolving data landscape is to enable data interoperability. That is, merging data across all three themes into readily usable data sets for everyone to consume. We must hide complexity while retaining full data lineage, that is, full traceability of data from the point of ingestion through to the point of consumption, and provide the means for self-service wherever possible. Snowflake has the tools to do all of this.

Common Reference Data

Surprisingly, or shocking to some, most organizations do not have a single source of standard reference data. There are exceptions, but each domain typically owns its reference data. All is well and good if the reference data matches, and in truth, countries, and currencies rarely change, but they do. Good data governance should address the issue of common reference data across an organization. But it is not always easy as our large corporates frequently purchase subsidiaries and divest other parts of our organizations. This theme is further discussed in Chapter 12.

One immediate benefit we can gain from implementing Snowflake Data Cloud is to share approved golden source reference data, managed and maintained centrally, immediately available to all.

Snowflake and OLTP

You are probably wondering why we don't use Snowflake as our primary data store and replace the databases supplied by vendors who host their data on any platform. This is a great question, and one answered in Chapter 3, but for now, let's consider the market segment in which Snowflake operates.

Designed from the ground up, Snowflake is aimed at the data warehouse market, where huge volumes of data are accessed, sorted, and filtered, answering complex business questions. The fundamental storage pattern is very different in a data warehouse than in traditional OLTP systems where individual record management occurs. It is a matter of scale. Snowflake is optimized for huge data volumes, and OLTP systems are optimized for low data volumes, albeit with the capability to implement data warehouse functionality, just not on the same scale as Snowflake.

Accessing Silos

The IT software industry has matured over many years. Numerous organizations have delivered market-leading platforms in specific market segments and industry sectors; examples include Oracle Financials, ServiceNow, Salesforce, and Workday. There are plenty more established vendors and a never-ending list of new entrants eager for your business.

Naturally, each vendor seeks to protect their intellectual property by implementing difficult (or impossible) human-readable database schemas, proprietary compression algorithms for uploaded attachments, inextricably linked data and logic held in large binary objects, and many more fiendishly difficult approaches to challenge the unwary. All of these, and many more, reinforce data silos preventing data interchange except through pre-defined and often limited interfaces, raising barriers to protect vendor interests but rarely serving the data owner's interests.

Each large-scale vendor application has three core components resulting in three primary patterns for accessing application data, as shown in Figure 2-3. This section examines each core component and the pros and cons of using each pattern. This is not an exhaustive treatise. And, yes, there are some out-of-the-box solutions for some of the issues raised in specific domains. But we are talking in general terms here to expose the underlying challenges to a wide audience.

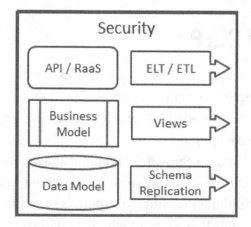

Figure 2-3. *Silo landscape and access paths*

Let's also discuss a few options to solve domain-specific issues.

Security Architecture

Data security must be "front and center" of everything we do with our data ecosystems, whether on-prem or in the cloud. However, the considerations for the cloud are far more stringent. In truth, this section belongs in Chapter 3. However, a brief mention here because we are discussing data moving between containers, and we must keep our data secure.

Before we consider the data sets to move, we must consider how to securely connect to a data source, data protection in transit (particularly over the public Internet), and how Snowflake protects inbound data (see Figure 2-4).

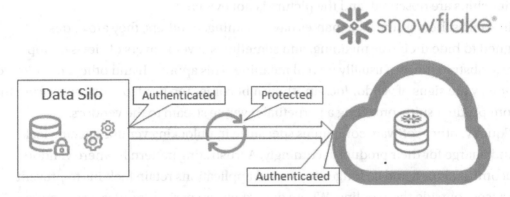

Figure 2-4. *Data in transit*

Every data source application provides at least one, and hopefully several, secure authentication mechanisms to establish a connection. We must choose a connection mechanism supported and approved by our cybersecurity architects. We must also ensure the pipework through which our data flows are protected. Our data must be encrypted in transit. Finally, our connection to Snowflake must likewise be secure. We know Snowflake does not allow third parties to create their own drivers. Instead, it develops and maintains its drivers, guaranteeing both authentication and data flow into Snowflake are secure.

Data Model

Every application has a data model. The fundamental data structures enable the application to hold and manage its version of the truth. In database terms, we refer to this as a schema. Of course, there is usually more than one schema in a database, but for this explanation, we stick to a single schema.

Think of any database as a set of Russian dolls. The analogy breaks down very quickly, but simply put, a schema is one of the inner dolls—a logical container into which we put multiple other Russian dolls, or more precisely, tables and views. A table is like a spreadsheet tab and contains data. A view is analogous to formulae applied to cells in the table, an overlay adding value if you like.

Tables can have relationships with other tables in a database schema, linking "like" data and allowing navigation between tables. The result can be likened to a tapestry because the relationships display a picture of how the data is related. However, in a badly structured schema, the opposite side of the tapestry is more representative, with threads hanging all over the place and great difficulty navigating between tables because the relationships are obscured, and the picture is not evident.

In some schemas, the table names have meaning. In others, they are codes designed to hide their true meaning, and sometimes several layers of views on top provide abstraction, but usually no real meaning. This approach and others mentioned previously are signs of vendor lock-in—how it becomes too hard to move away from their custom product suite, providing a perpetual revenue stream to the vendors.

Equally, other software companies specialize in unlocking vendor-specific lock-ins and charge for their product accordingly. A frustrating pattern is where vendors of decommissioned and no longer supported applications retain lock-ins to prevent data access outside their tooling. Where data retention periods of 10 years or more are common, a lot of on-prem hardware and disk storage are needlessly consumed to provide occasional data access. I will discuss how to address archiving legacy data sets later.

Back to our data model, data is input into the data model using proprietary screens or data interfaces where data can (or should) be validated to check for completeness and correctness before insertion. We ought to be able to rely upon data always being of good quality and fit for purpose. But, for a variety of reasons which I do not go into here, we often find gaps in our data—inconsistencies, duplicates, and missing reference data. So forth, the list goes on, and data correction post-capture is far more expensive than data correction at ingestion. It is important to know that we have lots of data in data models that may or may not be well structured or readily understood, and we need to move the data to Snowflake. And for the curious, types of data models are discussed in Chapter 13.

Having a brief insight into data models and data quality, let's discuss extracting data from our data model and subsequent data interchange.

For live production systems, part of our competitive advantage is derived from having data available in (near) real time in an understandable and easily consumed format. This is where many organizations spend enormous money, time, and effort extracting, transforming, loading then conforming data into usable structures readily understood by consumers. We return to this theme later, but for now, let's focus on the big picture of getting data out from our source data model and into our target data model, further assuming, for simplicity's sake, the source data model and target data model are the same, regardless of source platform, because we have already decided to use Snowflake as our target data warehouse platform.

Seasoned developers know it is relatively easy to transfer data from one system to another. We won't dwell on the mechanics but simply state that we can extract from the source and insert it into the target. Having data does not necessarily convey its meaning, only the presence of it—unless we can infer relationships directly from the data model, which is increasingly rare given vendor predilection for lock-ins.

If you move data without an associated context, all you have done is create another duplicate data set perpetuating the problem we want to solve, arguably one small step closer to breaking down a data silo, but at a high price, as discussed next.

Business Model

You now understand in broad brush strokes what a data model is, how a data model should work, and why we need a data model. Overlaying every data model is the business model, a conceptual, logical model rather than a physical "thing" often used to describe how information flows through a system. This is where functionality interacts with data and gives meaning.

Let's use a simple example. Assume we have a table containing address data. We may implement functionality to validate the postcode/zip code against supplied reference data to ensure each address is valid and there is a house number, the combination of which in the United Kingdom gives the exact place to deliver letters and parcels. We may also have access to the latitude and longitude of each postcode which for the United Kingdom gives geolocation data from which we can use more functionality to derive address density and map our addresses graphically. We can also calculate the distance between a distribution depot and delivery address or calculate travel time between two points for different times of the day using readily available geolocation services.

But there is limited value in knowing the address only. We need to know the context for each address, such as whether residential or commercial, and the name of the company or homeowner. In essence, the relationships from the data model discussed earlier using our tapestry analogy. It is from these relationships we derive information. This business model is a logical overlay to the data model where we can see data ingestion and consumption. I often refer to this simplistically as data on the left, processing, or functionality of some sort in the middle, and data consumption on the right. A business model cannot exist without a data model, and a data model is useless without a business model. In simple terms, the business model gives context and meaning to the data model by providing information from data.

Conventions are important in data modeling because they provide structure and meaning, which often (but not always) translate across domains. As inferred in the previous section, data models can vary wildly in quality and readability. Our objective should always be to apply the KISS Principle: Keep it Simple, Stupid. Alternatively, there is Occam's Razor. Both broadly amount to the same thing.

If vendor lock-ins obscure meaning, the last thing we should consider is replicating, cloning, or reverse-engineering the full business model into our Snowflake copy, because vendors change their data structures, add new features, refactor tables and relationships rendering our copy mechanism invalid with associated dependency management, maintenance, emergency fixes and pain for all consumers.

On a case-by-case basis, we might consider refactoring a small part of the business logic into a suite of reporting objects or views, as this may give sufficient utility to our consumers. This works well, particularly for archived applications discussed later, and may represent an acceptable way forward subject to the risks outlined. Ideally, and in preference, we need some form of abstraction insulating us from a change in both the underlying data model and overlaid business model. It is discussed next.

Application Interface or Reporting Service

You now know issues inherent with understanding and interpreting our data and business models with a partial answer for the business models. Fortunately (with some caveats), we have other options.

Most vendors provide a technical gateway into their business models, referred to as *application programming interfaces* (APIs) or *reporting as a service* (RaaS). Other terminology and differing access paths exist, such as built-in reporting screens, but to keep increasing complexity at bay, we consider both API and RaaS synonymous.

Let's assume we have our access path to extract information from the underlying business model. Downhill from here, right? Wrong. Very wrong because the presence of an API or RaaS access path, assuming we have solved the plumbing as discussed previously, leads us to another set of challenges—turning on the tap.

Naturally, vendors do not make it easy to understand all the nuances and implications of invoking their APIs. While sample code is supplied, it rarely (i.e., never) meets our exact requirements and leaves much information for us to discover by trial and error. Our challenge is finding someone who has the skill, knowledge, and expertise to understand how to invoke the API to derive the required resultant data set. These people are hard to find and harder to engage as they are subject matter experts (SMEs) in high demand and consequently command high fees to engage. Every cost-conscious organization, that is, every single organization in existence, has a small number of SMEs. Each is highly loaded to get the best value from them. We need their managers' engagement to procure SME time, which is harder to attain where the benefits of breaking down data silos have not been explained or understood, and our additional ask is another demand against a precious resource.

Assuming we have access to an SME, our next challenge is explaining what we want from the API, the results of which must be in a mutually acceptable format. We might consider a semi-structured format such as JSON, giving nested records for later processing into flattened relational records. We must also consider the trade-offs with invoking an API, traversing the resultant data set, and recursively calling the same API with slightly different parameters because each round trip takes time, consumes bandwidth, and can fail for a variety of reasons. One answer is to have our SME write a wrapper that resolves all recursive calls and returns a single super-set of data, but this takes time to articulate and understand requirements and then develop an appropriate wrapper, thus compounding our challenge in extracting data.

Depending upon the returned data set format, other considerations arise. Some attributes contain embedded commas, where we have specified a comma-separated values format. Embedded commas in the resultant data set may result in extra columns. For this reason, we prefer pipe-delimited output, assuming the data doesn't contain embedded pipes. You get the idea. Furthermore, APIs and RaaS may not provide 100% coverage of all the information we need from a system.

However, what we should be able to rely upon with APIs and RaaS, is a degree of isolation from changes to both the underlying data model and the business model. As we have discussed in previous sections, vendor changes to both can result in our implementations becoming brittle, causing unexpected failures, emergency changes, and unwanted extra work. A key factor when investigating third-party applications and tooling must be an examination of available SMEs in the marketplace and the desire in the organization to invest in their staff to mitigate execution risk.

We must also consider another vendor lock-in strategy. Rate limiting API and RaaS calls to restrict the number of records returned and the number of times an API or RaaS call can be made in a specific timeframe.

Built-in Reporting Screens

Some vendor products come with built-in reporting tooling with canned reports (those available out of the box) and the capability to develop homegrown screens. Built-in reporting tools can be used to derive data feeds and possibly scheduled too, depending upon the implementation.

Like API and RaaS solutions, built-in reporting screens may provide an attractive option for generating data extracts while also providing a degree of isolation from change to both data model and business model. A further advantage of reporting screens is that they may be developed by power users and do not require SME knowledge. They facilitate closer cooperation from operations staff, providing much-needed contextual information. If you understand why our systems are used, along with typical access paths, you are in a far better position to understand the implications of change.

Custom Views

Database views are stored SQL queries that overlay our data model. They are physical database objects providing an abstraction layer where we may put logic to join tables, pre-filter, aggregate, or summarize data. We often use views to denormalize our data model into more business-friendly terminology and presentation, reducing the technical knowledge required to understand our data.

As with the business model, it may be possible to create a (small) suite of custom reporting views to enable efficient data extract. These require maintenance in the event of underlying data model changes. Likely to be highly dependent upon SME knowledge and at risk of creating more technical debt, custom views are only as good as the underlying SQL used to source the data. Any filters, aggregations, or summarizations should be well described in the documentation.

Application Connectors

Salesforce integration is via Tableau CRM Sync Out Connector, offering out-of-the-box incremental data loads into Snowflake (see Figure 2-5).

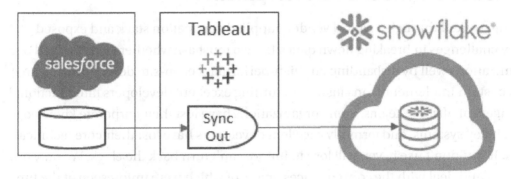

Figure 2-5. *Tableau CRM Sync Out*

Sample scripts are supplied to configure Snowflake, after which Salesforce connection must be configured. Note the type and frequency of data synchronization are determined by Salesforce configuration. This push-only interface from Salesforce into Snowflake managing object creation provides a degree of insulation from Salesforce data model changes. Note that no mention is made of data model delta changes, only object creation. The author has not tested the Tableau CRM Sync Out Connector, but more information is at `www.snowflake.com/blog/integrating-salesforce-data-with-snowflake-using-tableau-crm-sync-out/`.

Third-Party Products

Historically, Snowflake has not seen itself as a data integration tool supplier. Instead, Snowflake typically refers to integration questions to their Partner Connect program accessible via Snowflake User Interface. Many ETL/ELT vendors, such as Boomi and Matillion, provide connectors offering the capability to abstract details away from the real work of implementing API or RaaS calls, in other words, simplifying data interchange.

We observe a clear trend toward consolidating disparate data sets into Snowflake for cross-domain, consistent, strategic reporting, which has not gone unnoticed in the marketplace. Ease of connectivity from disparate data sources and vendor platforms into Snowflake is an area to watch out for developments. Another option is third-party tooling to replicate whole databases and schemas. Partner Connect has several offerings available.

In-House Application Development

Having worked through a typical vendor-supplied application stack and exposed many challenges in breaking down data silos, we must ask whether IT has served our organizations well by disbanding our high-performance, on-site development teams with custom implementations instead of turning excellent developers into functional managers of offshore teams. Some organizations have lost their corporate knowledge of their key systems, and formerly excellent developers have lost their core technical skills. If you don't use it, you will lose it. But we can't turn back the clock. We only move forward and deal with the consequences, many of which were unforeseen at the time decisions were made.

Approaches to Data Capture

There are many approaches to capturing data, and when implementing Snowflake, we should consider a pattern-based approach. But what is a pattern-based approach? How can we develop patterns? And most importantly, why should we bother? I'm glad you asked.

Patterns provide an easily described, documented, repeatable, and consistent integration path to ingesting data sets. The alternative is a custom integration path for each data set, hard to maintain, leverage for reuse, explain to others, and support. Naturally, one size does not fit all. Therefore, we may have several patterns, but not many; ideally a handful or less.

If we want to accelerate our software delivery, reduce our operating risk, and make our systems robust and scalable, we should adopt a pattern-based approach. Furthermore, we may have architectural governance to approve our proposals before allowing changes to occur, and by implementing common patterns, we make their job easier too. Remember the tapestry analogy from Chapter 1? Patterns change our perspective, so we look at the picture, not the reverse side. Patterns and an example are covered in Chapter 12.

Tactical Approach

We often face pressure to deliver as quickly as possible regardless of long-term consequences and technical debt, which is often not addressed. IT is not a perfect discipline, and as you have seen, not going to reduce in complexity any time soon. Quite the opposite. Recognizing the need for business consumers to derive knowledge and wisdom, we are often constrained when delivering quality, robust data sets and relying upon someone cobbling data into a spreadsheet manually. Something is better than nothing, and we can always refactor it later, or so we are told. But in the meantime, we have increased operational risk because the data set is tactical and technical debt, because we must later replace tactical feeds with strategically sourced data.

Adopting a pattern-based approach where a rules-driven engine automagically creates objects and integrates with our continuous delivery pipeline is readily achievable. This approach does not cater to data management principles outlined in Chapter 11 but does provide a starting point for automating our data ingestion capability.

For every book of work item where a tactical approach is the only achievable option in the short term, I strongly recommend tactical feeds are only accepted on the condition funding is made available to remediate at a later stage when strategic data sets become available. Our end-state should be to decommission all tactical feeds and repoint strategic feeds, which is discussed next.

Strategic Approach

What is a strategic data set, and how does it differ from a tactical data set? One definition of a strategic data set is one programmatically sourced from the system which holds the system of record, otherwise referred to as the golden source.

Our preferred approach to data ingestion must be to use strategic data wherever possible and migrate tactical feeds at the earliest opportunity. It takes discipline and determination to reduce technical debt. But we do not serve our organizations well if we allow—or worse, encourage—proliferation and reliance on non-authoritative sourced data sets.

Regardless of our position in our organization, we all have a role to play and should adopt a common approach to reducing our technical debt. Without quality data, we cannot expect accurate information, knowledge may be compromised, and wisdom out of step with reality. Data quality (completeness, volume, validated) and veracity (accurate, timely, reliable) contribute to improved decision-making and better outcomes.

Data Pipeline and Data Content

Previously, you learned why we should separate pipework from the data flow. Where resources are constrained, I advise focusing on the plumbing first, including security impact because this is the biggest challenge, followed by invoking API/RaaS where different challenges await. This approach may also mitigate delivery risk for large amounts of work, since resources are not always available when needed, and we often must wait for availability. We should therefore advance upon a wide front and address those opportunities available to action while waiting for others to open.

Change Data Capture (CDC)

CDC is a technique to capture only those records which have changed, the delta. Depending upon how the originating source handles changed records, CDC may not be available without significant rework, testing, and later deployment; in other words, it's not worth the effort. However, CDC is a very elegant and powerful approach to minimizing data captured for onward processing and should be considered for every interface.

You might also consider implementing Data Vault 2.0, where CDC is required. While non-trivial to deliver a Data Vault 2.0–based design, the outcome is worth the investment if done well. Data Vault 2.0 is covered in Chapter 13.

Bulk Load or Trickle Feed

Some data sets led themselves to a bulk load approach, and other data sets lend themselves to a trickle feed approach, one size does not fit all, and we must determine the appropriate approach for each feed. As a rule of thumb, a monthly feed may be acceptable for infrequently received data; for example, new starters to our organization.

However, inventory changes may need to be intra-day to enable just-in-time shipment to maintain warehouse stock levels. Regardless of the approach, Snowflake has built-in tooling to ease your implementation.

Data Usage

Once we have the data in Snowflake Data Cloud, we must address the questions of ownership and usage. Having identified there may be data from many source systems in an individual line of business or functional domain, we must determine the ground rules for data access.

For some organizations, it is sufficient for a user to belong to a line of business or functional domain to be granted access to all data in the governed data set. A finer-grained approach is preferred for other organizations, where RBAC is created for individual application data sets.

The former approach is something of a blunt instrument and may inadvertently grant access to data to unintended recipients. The latter approach requires more administration (typically via AD group membership) but provides tighter control, and this is the approach we prefer. It is always easier to relax a control than to tighten up controls later.

Row-level security (RLS) controls access to subsets of data in a domain. Typically, RLS is applied to objects created for data consumption. We return to data consumption in Chapter 9.

Migration from Snowflake

Despite being convinced Snowflake Data Cloud is the way forward for all the good reasons explained before, we must consider the possibilities of both Snowflake becoming another data silo. Also, some future products may eclipse Snowflake's core capabilities, to which we may want to re-home our data.

Snowflake as a data silo is an interesting concept. The tooling itself locking up data is anathema to the open market approach adopted by Snowflake, but there is an outside possibility, albeit highly improbable.

Moving data platforms is a very infrequent event in an organization's life. Typically, vendor lock-ins, non-transferable employee skillsets, system interfaces and infrastructure, lack of funding, and appetite impede the bold decision to re-platform. Yet some architectural governance forums prefer to have an exit plan as part of their approval process.

We have already determined SQL skills to be transferrable as the lingua-franca of databases; therefore, our only available data platforms to migrate away from Snowflake are those already in common everyday use, precisely those platforms for data warehousing we are moving from. I'm sure I am not the only one to see the irony here. I find it difficult to imagine what new platform will be available in, say, five years; therefore, the subject of exit planning should be deferred until the question becomes relevant, as any current answer is speculative. Regardless, Snowflake is committed to continued access post exit, which may be enshrined in your master service agreement or simply a reversion to the on-demand charging model.

Summary

This chapter began by explaining why every organization has data silos and identified the types of data silos where value is locked up. We then looked at why organizations have data silos and the perceived benefits thereof, systematically breaking down the arguments and building the case for the Snowflake Data Cloud.

Looking toward the future, where unknowable data types and new forms of media await, you have begun to understand how to address future state data demands. We can now begin to think in terms of *yes* being our first response backed up by the knowledge we can develop tooling to access data, rather than immediately saying *no* because the answers lie outside of our skill, knowledge, expertise, and comfort zone.

This discussion has taken you through some of the complexities in accessing data silos, calling out some key dependencies and challenges we will face, not the least of which is our security posture. And having established the right mindset in preparation for looking deeper into our data silos, let's open the door to the next chapter, which discusses Snowflake's architecture.

PART II

Concepts

CHAPTER 3

Architecture

Understanding how Snowflake works helps you develop your strategy for implementing your contribution to the Snowflake Data Cloud. Snowflake is optimized for data warehouses but not OLTP. The distinction is clearly explained in this chapter. For the curious, search for micro-partitions.

Information is presented early in this book to clarify what is happening "under the hood" as you work through the sample code presented in later chapters. This chapter begins our deep dive into technical matters, and we focus on some items not immediately obvious. I also attempt to provide information at a high enough level to satisfy the curious by asking what, how, and why, later delving into the depths of some obscure, not well-understood parts of Snowflake to assuage the thirst of the technophiles reading this book.

It should be said this chapter is not intended to provide comprehensive product coverage. It is written from my experience learning Snowflake and based on prior knowledge of Oracle. Therefore, this chapter should be read in the context of extending your presumed technical knowledge into the new paradigm Snowflake offers and drawing out information that I, in hindsight, would have benefitted from when learning Snowflake.

Understanding how Snowflake is built from the ground up to take full advantage of cloud capabilities helps you later understand how we integrate with the Snowflake Data Cloud.

Snowflake Service

We start our investigation into Snowflake architecture in the same place as almost every other Snowflake introductory presentation—by examining the three discrete layers implemented by Snowflake: cloud services, multi-cluster compute, and centralized storage. Each layer provides different functionalities and capabilities to Snowflake.

© Andrew Carruthers 2022
A. Carruthers, *Building the Snowflake Data Cloud*, https://doi.org/10.1007/978-1-4842-8593-0_3

We examine each separately. Snowflake introduces its key concepts at `https://docs.snowflake.com/en/user-guide/intro-key-concepts.html`.

The separation of storage and compute uniquely differentiate Snowflake from its competitors. As we work through this chapter, you learn how this fundamental architectural difference is a game-changer.

An important takeaway to note is the cloud agnostic layer. Snowflake clients get to choose which of the three supported cloud platforms (AWS, Azure, Google Cloud) on which to host the Snowflake VPCs. Although all cloud platforms provide the same capability, GCP lags a little behind AWS and Azure for both physical region availability and product feature implementation at the time of writing this book. The Snowflake Status site indicates not only Snowflake's operational status but physical region availability (see `https://status.snowflake.com`).

Nor are we restricted to a single cloud service provider (CSP). We could host Snowflake in all three clouds and seamlessly move data between. Later, you see some potential issues arise with moving data between CSPs. We might also prefer to balance service provision across two or all three CSPs to reduce our application concentration risk or data gravity exposure to a single cloud provider.

The bulk of this book is based upon AWS since this is where my primary expertise lies. Almost all the written materials are transportable. Substitute Azure Blob Storage or Google Cloud Storage for Amazon Simple Storage Service (Amazon S3). There are syntax changes required later when we write some code, but these should be easily understood and resolved.

Snowflake architecture is a hybrid of shared-disk and shared-nothing architectures, with a centralized storage repository for persisted data accessible from all multi-cluster compute nodes in the account, represented by the intersection between centralized storage and multi-cluster compute, as illustrated in Figure 3-1. This means your data is always available for processing in a consistent state while each compute node operates independently.

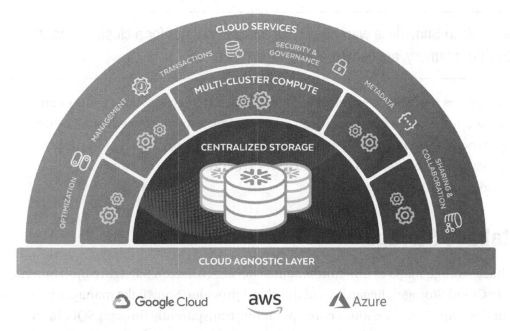

Figure 3-1. *Snowflake architecture*

Cloud Services Layer

Cloud services (`https://docs.snowflake.com/en/user-guide/intro-key-concepts.html#cloud-services`) provide coordination services across all Snowflake activities, including access control, authentication (who can log in, and what can they do), infrastructure management, metadata management, query compilation, query planning, and query optimization.

Multi-Cluster Compute

Multi-cluster compute, also known as *virtual warehouses* (`https://docs.snowflake.com/en/user-guide/intro-key-concepts.html#query-processing`), lets us choose the number of processors and the amount of memory to allocate to processing our queries. The Snowflake approach starkly contrasts the plethora of well-understood databases in our organizations, such as Oracle and SQL Server, where fixed numbers of processors and memory are available. Highly scalable, Snowflake provides tee-shirt-sized warehouse templates to select from also clustering. Snowflake delivers the performance required when we need it, on demand and subject to the parameters set for the warehouse, and scales automatically as our workloads increase.

Note   Each Snowflake warehouse is a named wrapper for a cluster of servers with CPU, memory, and temporary storage (SSD).

All queries require the warehouse to be declared. Queries whose results can be satisfied from cache or metadata do not consume Snowflake credits. All other query invocations consume Snowflake credits. More information is at `https://docs.snowflake.com/en/user-guide/what-are-credits.html#what-are-snowflake-credits`.

Database Storage

Database storage maps to CSP fundamental storage (Amazon S3, Azure Blob Storage, Google Cloud Storage). Regardless of the cloud provider, Snowflake manages all storage interactions for data warehouse core operations transparently through SQL. More information is at `https://docs.snowflake.com/en/user-guide/intro-key-concepts.html#database-storage`.

Snowflake pass-through storage charges from the cloud provider vary according to region and cloud provider, nominally for AWS, and depending upon the region, approximately $23/terabyte. Note additional storage changes are incurred by the Time Travel and Fail-safe features.

Later, you learn how to view the underlying storage for database objects, but for now, it is enough to know Snowflake uses our chosen cloud provider storage to hold our data in a highly secure manner. An infinite amount of storage is available from the moment we log in. There is no need to ask Snowflake for more; storage is always available.

Snowflake Provisioning

Snowflake first establishes its technical footprint across the three supported clouds in disparate physical locations. For more information, please see the following.

- AWS (`https://aws.amazon.com/about-aws/global-infrastructure/regions_az/`)

- Azure (`https://azure.microsoft.com/en-gb/global-infrastructure/geographies/#overview`)

- GCP (`https://cloud.google.com/compute/docs/regions-zones`)

For AWS and Azure, the Snowflake account provisioned is functionally equivalent with minor syntax changes for certain commands. At the time of writing, GCP Snowflake account features lag those for AWS and Azure. However, all Snowflake accounts are provisioned according to a common pattern, as shown in Figure 3-2.

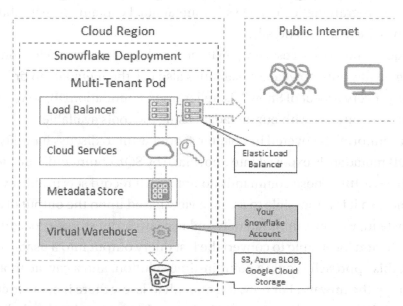

Figure 3-2. *Snowflake provisioning*

Some clues in Figure 3-2 later aid our understanding. First, storage is outside the multi-tenant pod but still in the Snowflake deployment, meaning storage management is devolved from a single Snowflake account, making storage available across multiple Snowflake accounts. This approach underpins shares. Second, the metadata store is stored in a key-pair repository supplying the global services layer, discussed next. Lastly, the diagram calls out the virtual warehouse (your Snowflake account), addressed next. It is the highest-level container available in our Snowflake provision.

Global Services Layer

Underpinning Snowflake and implementing the global services layer is FoundationDB, an open-source key-value data store well suited to OLTP style data management for high-speed read/write and data manipulation. FoundationDB holds our Snowflake account metadata, that is, information about every object, relationship, and security feature, the catalog which documents and articulates our account. Regardless of the cloud provider chosen, the global services layer implementation is consistent across all clouds, ensuring Snowflake interoperability.

From time to time, we need to access the global services layer. Some application development features, for example, used in security monitoring, can only be metadata driven if we can programmatically access our metadata. I explain more in Chapter 6, which dives into the technical details.

Not only do we need to access the global services layer for our application programming needs, but Snowflake capabilities also use the metadata to drive built-in functionality. Every aspect of Snowflake relies upon FoundationDB, so we must understand how to interact with our metadata using the tools available.

Security monitoring is covered in Chapter 4, but I want to discuss it briefly now. FoundationDB metadata is exposed into Snowflake by SQL commands such as LIST, SHOW, and DESCRIBE. These commands do not return record sets usable by other SQL commands, so it is impossible to create a view based upon the output of a SHOW command. Instead, we must think laterally and use JavaScript stored procedures, RESULT_SCAN, and local table to convert the last query output into a usable record set. Note that this approach is only a point-in-time solution, and a caveat applies to waiting to trip up the unwary. For now, it is enough to know there is some friction when accessing the global services layer from SQL. You see how to resolve the challenges later.

Snowhouse

Snowhouse is the in-house tool that Snowflake uses to monitor and manage itself and provide internal metrics on usage telemetry.

Note   At no point can Snowflake staff circumvents account security using Snowsight; therefore, your data always remains secure.

Snowhouse merges every organization's account metadata with data from other Snowflake internal systems ranging from sales and marketing to infrastructure monitoring tooling, providing a holistic view of each account utilization. As an internal management utility, Snowhouse is used for marketing and sales opportunities by sales support engineers to identify consumption spikes, investigate underlying SQL performance issues, and cloud provider virtual machine configuration and performance.

Snowflake is the biggest consumer of Snowflake, confidence building for everyone considering using Snowflake as their Data Cloud, so if contacted by your Snowflake support team, there is sound, evidence-based reasoning for the information imparted. Compare the proactive nature of Snowflake support with other vendors and draw your own conclusions.

Naturally, as end-consumers of Snowflake service provision, we do not have access to Snowhouse, so this section is included for information only.

Snowgrid

Snowgrid is a term you are sure to hear more about soon. This Snowflake in-house technology underpins transparent data sharing regardless of the cloud provider and physical location.

High Availability (HA)

In each region (or geographical location), each CSP implements its core capability across three geographically separate availability zones, segregated but interconnected. Snowflake inherits the underlying CSP architecture as a product deployed in a region, as shown in Figure 3-3. Behind the scenes, Snowflake replicates changes in (near) real time, transparent to end users and seamless operation. Every environment provides resilience and is built for synchronous replication. We cannot see which availability zone our account is hosted on. Simply accept the service as provisioned.

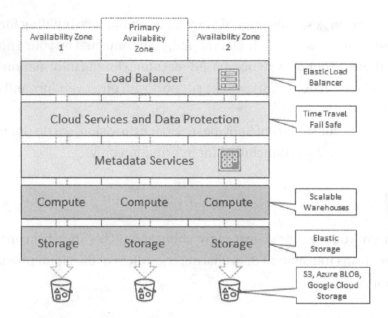

Figure 3-3. *Snowflake high availability*

We won't get into the marketing hype or analysis of around 99.999999999% durability (an act of war has a higher chance of affecting your data) or 99.99% availability (less than an hour of downtime per year), but bring these metrics to your attention both for what cloud providers quote and for further investigation if you so desire.

With three availability zones, and an immutable micro-partition approach (discussed later), Snowflake protects many scenarios, including the following.

- *Customer error* includes accidental deletion of data and objects, loading one or more data files with bad data, or loading data out of sequence.

- *Virtual machine failure* includes triple redundancy for critical cloud services, automatic retries for failed parts of a query.

- *Single availability zone failure* includes multiple availability zones on AWS and multiple availability sets on Azure. Snowflake's service may be degraded, as load balancing is across one or two availability zones depending upon the nature of the failure.

- *Region failure* includes loss of service in a region, cross-region database replication, and failover.

- *Multi-region failure* includes loss of service in a region, cross-region database replication, and failover.

The root cause of region failure would most likely be a Snowflake core service provision outage rather than the whole CSP infrastructure. The real-time Snowflake service status is at `https://status.snowflake.com`.

Patching and Maintenance

Every Snowflake presence is automatically patched and updated with zero downtime, transparent to accounts via a staged release process through which Snowflake updates are first applied to early access accounts 24 hours in advance of Standard accounts and finally Enterprise and upward accounts.

Occasionally Snowflake makes feature changes. These are pre-notified and generally not of great significance. However, for those fortunate to work in large corporates with dedicated Snowflake support staff or support contracts expect to see periodic emails pre-notifying of upcoming changes.

Further information on Snowflake releases is at `https://docs.snowflake.com/en/user-guide/intro-releases.html#snowflake-releases`. Chapter 4 walks through how to enable and disable behavior change releases to allow testing before deployment.

Create Your Trial Account

Most value is obtained by physically doing something. Practice makes perfect, and the repetitive act of typing commands, while tedious, reinforces memory retention. But to type commands, you must have a Snowflake account to practice on. Go to `www.snowflake.com`, click the Start For Free button, and enter a few details to start a 30-day free trial account.

Note   Be sure to select the Business Critical Edition because it is likely the version used by your organization.

We rely upon certain features in this book; therefore, the Business Critical Edition is essential for successfully working through sample code. I do not cover every feature (I must leave something for you to discover yourself!), but some features are essential to our understanding, particularly those related to security features. Figure 3-4 illustrates features available in each Snowflake edition.

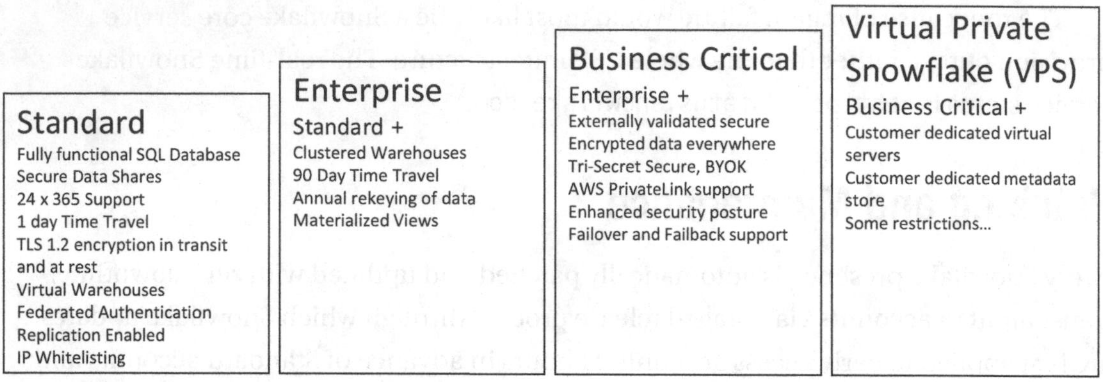

Standard	Enterprise	Business Critical	Virtual Private Snowflake (VPS)
Fully functional SQL Database	Standard +	Enterprise +	Business Critical +
Secure Data Shares	Clustered Warehouses	Externally validated secure	Customer dedicated virtual servers
24 x 365 Support	90 Day Time Travel	Encrypted data everywhere	Customer dedicated metadata store
1 day Time Travel	Annual rekeying of data	Tri-Secret Secure, BYOK	Some restrictions...
TLS 1.2 encryption in transit and at rest	Materialized Views	AWS PrivateLink support	
Virtual Warehouses		Enhanced security posture	
Federated Authentication		Failover and Failback support	
Replication Enabled			
IP Whitelisting			

Figure 3-4. Snowflake editions

All examples in this book have been tested using a trial account on AWS; therefore, I recommend AWS as the chosen platform. Your mileage will vary otherwise.

Organizations

Until recently, every Snowflake account was created in coordination with a Snowflake technical representative. A new feature called Organizations provides a high degree of self-service in managing across all Snowflake accounts for your company. Chapter 4 discusses the Organizations feature, but for now, it is sufficient to note all your accounts can be managed centrally regardless of location and cloud provider(s) using the ORGADMIN role.

Accounts

Snowflake accounts were discussed earlier. Figure 3-2 shows that a virtual warehouse as synonymous with your Snowflake account. Let's unpack this by explaining what an account physically contains and the functionality each layer provides. We think of an account as the highest level of the container and the outermost boundary of our Snowflake provisioned environment. The virtual warehouse is called out in Figure 3-2 as the Snowflake account.

Snowflake provision accounts in one of two billing formats. Snowflake On Demand is the pay-as-you-go option. Every credit is a fixed price. In contrast, a Capacity contract is a bulk credit purchase up-front over a fixed term. At the time of writing, each credit costs $5 with bulk volume discounts. A fuller discussion of this topic is out of scope for this book. Please discuss with your Snowflake sales representative for further details. Snowflake credits are also explained at `https://docs.snowflake.com/en/user-guide/what-are-credits.html#what-are-snowflake-credits`.

Cloud Services

The cloud services layer is a functional suite of components coordinating activities across Snowflake to process user requests, from login to query dispatch. The cloud services layer runs on compute instances provisioned by Snowflake from the cloud provider. Metadata operations do not use compute, and cloud services are charged only if consumption exceeds 10% of daily compute resource consumption. Several services are provisioned in the cloud services layer. Let's discuss them in turn.

Authentication and Access Control

Authentication is the process of verifying the identity of an actor attempting to access Snowflake. An actor may be a named individual user or generic service user for system-to-system interconnectivity.

Snowflake supports multi-factor authentication via Duo, an opt-in feature available to all users.

Access control is enforced by network policies, which can be thought of as a ring fence around your Snowflake account, allowing access to your environment from various end-points and configured at the account level. I discuss setting network policies and how to effectively monitor those set in Chapter 4.

Infrastructure Management

Infrastructure management includes the dynamic (elastic) allocation and deallocation of resources at any time resulting in the exact number of resources required to fulfill the demand for all users and workloads, leaving developers and users free to concentrate on delivering both functionality and business value rather than being distracted by runtime environment issues.

Metadata Management

The metadata repository uses FoundationDB to hold table definitions, references for all micro-partition files for each object, tracking for all versions of the table data in the data retention window, and various statistics. Metadata management services provide query optimization, compilation, planning, and security.

Query Optimization

Snowflake query optimizer performs static pruning in the metadata cache by eliminating micro-partitions based upon query predicates. Filters then identify the exact data to be processed using metadata.

When the result set can satisfy the cache, the records return from the result set cache. For the curious, the metadata cache holds several sets of statistics.

- Tables: Row count, table size in bytes, file reference, and table version

- Micro-partitions: Max/min value range, the number of distinct values, NULL count

- Clustering: Total# micro-partitions, micro-partition overlap values, micro-partition depth

This is used in performance tuning, a topic covered in Chapter 11. Dynamic pruning occurs in a warehouse cache, covered later in this chapter.

Result Cache

The result cache stores the results from queries executed by all warehouses with a retention period of 24 hours. Whenever an exact match query runs, the result set expiry timer is reset, extending retention for 24 hours unless the underlying data has changed. The maximum duration result sets persist for 31 days before being automatically purged, requiring subsequent query execution to reinstate cached data.

Virtual Warehouses

Virtual warehouses are also called multi-cluster compute. Each is a named wrapper for a group (or cluster) of servers with multicore/hyperthreading CPU(s), memory, and temporary storage (SSD). Warehouses are self-tuning *massively parallel processing (MPP)* engines.

Warehouse runtime is where Snowflake makes its money. Therefore, careful attention must be paid to right-sizing and parameter setting, particularly auto-suspend. Not all queries require a running warehouse, as some resolve their results directly from metadata or query cache, but for those which do require a running warehouse, if not already done so, the warehouse instantiates; that is, the cloud services layer provisions the correct number of servers (CPU, memory, and SSD cache) to fulfill the warehouse configuration and begin executing the query.

Snowflake maintains several warehouses in each availability zone, reducing spin-up time on demand. Through Snowhouse statistical analysis, Snowflake accurately predicts usage, but for new regions where usage patterns are not yet established, predictions are less accurate; therefore, provisioning delays of more than 10 seconds may be experienced. Naturally, as usage patterns evolve, predictions become more accurate.

Warehouse minimum runtime is 1 minute, then per second after that. Warehouses can be manually created and managed at the command line or via the user interface.

Note   Check that all warehouses initially have auto-suspend set to 1 minute.

Occasionally warehouses fail to provision during startup. Snowflake detects failures and automatically attempts to repair failed resources. SQL statements start processing when 50% or more of the requested resources have been provisioned.

There is no single answer to identifying optimal warehouse configurations without proper testing. While warehouse scaling provides many answers, other performance options, particularly for large data sets to consider, are outlined later in this chapter. We can say that we require differing warehouse configurations according to the type of activity performed. Data loads and queries have different workload characteristics and must be sized accordingly.

In summary, these broad principles apply to sizing warehouses.

- Scale up increases T-shirt size to improve query performance, process more data, run more complex queries

- Scale out adds clusters to run more queries concurrently, support more users

- Scale across declares more warehouses to isolate workloads on different warehouse sizes to remove resource contention

Single Warehouse Concurrency

> **Note**   Caveat. This section contains (reasonably safe) assumptions regarding underlying cloud provider CPU hardware (AWS EC2, Azure VM). Your mileage may vary according to real-world experience.

An X-small warehouse provisions a single server that contains CPU, memory, and SSD storage. There is no means of identifying the actual hardware—nor should we when operating in a cloud environment. But there is one aspect where the underlying hardware is material to configuring our warehouses: CPU configuration.

CPUs are no longer a single monolithic silicon block capable of processing a single task. Instead, a CPU has multiple cores, and each can handle two or more concurrent threads. For our discussion, let's assume a single CPU has four cores. Each core has two threads, resulting in 16 available concurrent threads, as Figure 3-5 illustrates.

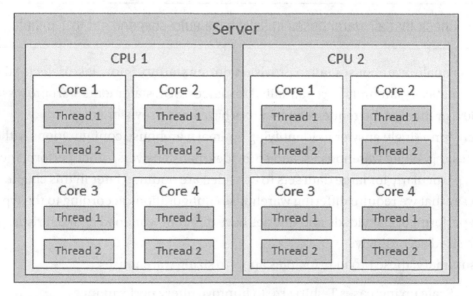

Figure 3-5. *Suggested server configuration*

While each server may have 16 threads available in total, it is also clear each server must retain the capacity to execute other services, including its own operating system. Therefore, we further assume only four of the suggested eight cores are available to Snowflake at any time.

Snowflake Advanced SQL course indicates each server in a cluster has eight concurrent threads available to service queries. When demand for the warehouse exceeds eight concurrent execution requests and clustering is enabled, another warehouse cluster of the same size is instantiated, providing a further eight concurrent threads. Up to ten clusters can be instantiated concurrently, resulting in a maximum of 80 threads for a single X-small warehouse declaration at any time. Active clusters are scaled down when the last working thread has ceased executing its SQL statement subject to auto-suspend setting and not running in maximized mode.

Let's look at an alternate view. There are eight cores in an X-small warehouse, of which we assume four are usable by Snowflake. When a query is submitted to this warehouse, Snowflake tries to split the query into four parts. Each of the cores tries to scan one-fourth of the table. Note that this suggestion is ideal; the query must be splittable, and the warehouse cores must be available for parallelization.

For those seeking a deeper understanding of concurrency, I suggest investigating the MAX_CONCURRENCY_LEVEL parameter at `https://docs.snowflake.com/en/sql-reference/parameters` .html#max-concurrency-level and the blog by Subhrajit Bandyopadhyay at `www.linkedin.com/pulse/snowflake-concurrency-parallel-processing-subhrajit-bandyopadhyay/`.

Scaling

Warehouses are defined from built-in templates in T-shirt sizes ranging from X-small with a single server, and for a 6XL having 512 servers, this is called scaling up and down. The number of servers instantiated doubles for each increase in T-shirt size. Warehouses can be resized dynamically; the T-shirt size clusters can be changed at the command line as desired. Warehouses can also be manually suspended in a session.

A further option (my preferred option) is to have multiple warehouses declared and use the one most appropriate for the workload at hand. This is called *scaling across and back*. You may be asking why the author prefers this approach. First, there is the possibility that a process having resized a warehouse could fail, with the warehouse retaining the resized declaration for the next invocation, which could be inappropriate and costly. Second, for workload segregation, repeatable performance tuning, determining credit consumption for billing purposes, and preserving a consistent system configuration perspective, static warehouse declarations are a better, more appropriate fit. Third, both instances of a re-declared warehouse run briefly while the old warehouse is quiesced, which could be costly for larger warehouses. Fourth, our data warehouses are complex and ever-increasing environments. Multiple warehouse declarations provide workload segregation.

Warehouses can also be clustered. Multiple instances of the same T-shirt size can be declared in the warehouse configuration and dynamically instantiated by Snowflake as load increases. It automatically shuts down when the load decreases. This is called scaling out and in. Equally, a warehouse can be set to run in maximized mode by setting the minimum and the maximum number of clusters to the same value greater than 1, in which case all clusters are automatically instantiated when the warehouse starts.

Warehouse clusters are always of the same cluster type. It is impossible to select different T-shirt sizes in a single warehouse declaration. Figure 3-6 illustrates the number of clusters instantiated and credit consumption per hour to inform our warehouse configuration. I discuss how to monitor credit consumption later.

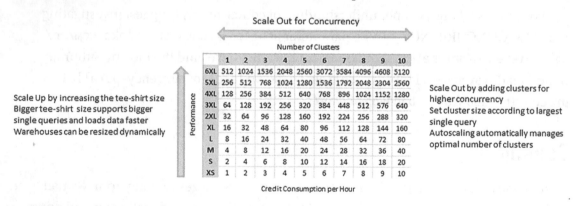

Figure 3-6. *Warehouse scaling options*

For a single large query loading one billion rows, the effect of scaling up is illustrated in Figure 3-7. Increasing the warehouse size improves response time but ultimately at increased cost if the warehouse size increases too much. Warehouse size selection and associated costs must be tested with a representative workload profile recognizing multiple warehouses can (and should) be declared as most appropriately selected for the target workload.

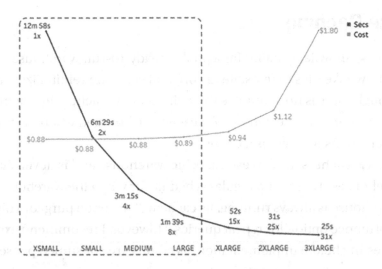

Figure 3-7. *Warehouse scaling up cost/benefit*

The important takeaway from Figure 3-7 is that scalability is linear until the query is no longer resource bound.

The scaling policy is the final warehouse parameter affecting multi-cluster warehouse behavior, which may be either Standard (default) or Economy. However, a deprecated scaling policy called Legacy also exists; its behavior has been changed to Standard and should not be used. A scaling policy affects when clusters are instantiated or shut down. For more information, please refer to the documentation at `https://docs.snowflake.com/en/user-guide/warehouses-multicluster.html#setting-the-scaling-policy-for-a-multi-cluster-warehouse`.

Query Processing

Snowflake query processing engine uses native SQL. A Snowflake session can only have one warehouse operating at a time, which may be either changed or resized, noting earlier comments on the implications of resizing warehouses.

Previous sections discussed some aspects of query processing. Where possible, queries are split according to the nature of the query and available processing cores for parallel processing.

Warehouse Caching

Behind the scenes, Snowflake maintains a pool of ready-to-run warehouses. This explains why Snowflake tries to reinstate the original cache on reinitializing a suspended warehouse, though there is no guarantee this will occur. As cache reinstatement is not guaranteed, the minimum expectation is for cache to be purged when the warehouse is suspended. Cache reinstatement is a bonus.

The warehouse cache is local to each specific warehouse and is never shared with other warehouses. There is no regular schedule for when the warehouse cache is purged. If a warehouse is always running, its cache will never be purged. This may be an attractive performance option for repeat queries. However, I recommend experimenting and thorough testing before implementation, as credits are consumed per second after the first minute of warehouse operation.

Resizing warehouses results in new server provision. Therefore, the cache is purged for the resized warehouse. This behavior is inferred from previously explained behavior where resizing warehouses may result in additional credit consumption while the new servers are instantiated—another reason not to resize warehouses.

Caches are reused if the same (i.e., case sensitive) SQL text/query is reissued unless the Snowflake cache is invalidated by underlying data changes, ensuring the query always returns the latest data.

Query Profiling

Query profiling can be performed on both queries which have completed execution and queries undergoing execution. A profile can only be viewed in the user interface, and current information cannot be extracted programmatically for external tool analysis.

Detailed investigation and explanation of query profiling are beyond the scope of this book. However, I discuss the basic principles in Chapter 11. Further information is at `https://docs.snowflake.com/en/user-guide/ui-snowsight-activity.html`, noting the use of Snowsight.

Monitoring

The classic user interface provides a screen to monitor warehouse load, as shown in Figure 3-8.

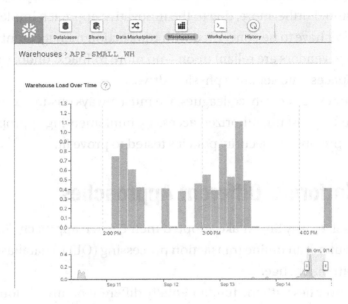

Status	Warehouse Name	Size	Clusters	Scaling Poli...	Runn...	Que...	Auto Suspe...	Auto Resume	Created On ▼	Resumed On	Owner	Comment
Suspended	APP_MED_WH	Medium	min: 1, max: 4	Standard	0	0	1 minute	Yes	9/6/2021, 6:01:10 PM	9/6/2021, 6:01:10 PM	SYSADMIN	
Suspended	APP_SMALL_WH	Small	min: 1, max: 4	Standard	0	0	1 minute	Yes	9/6/2021, 6:01:10 PM	9/6/2021, 6:01:10 PM	SYSADMIN	

Figure 3-8. *Warehouse monitor*

Selecting an individual warehouse drills into utilization for the past 14 days, as shown in Figure 3-9.

Figure 3-9. *Recent warehouse utilization*

Later, I explain how to extract warehouse utilization information and credit consumption from the Snowflake Account Usage store, as it relates to identifying credit consumption. But, for now, it is sufficient to know Snowflake provides monitoring tooling.

Storage

Underpinning every Snowflake implementation, regardless of platform, is storage, the physical place where data lands and is persisted. Our discussion is more AWS-focused. Therefore, substitute Azure Blob Storage or Google Cloud Storage for Amazon S3.

Unlike previous generations of databases, Snowflake takes away all the headache of manually configuring storage on installation, no more configuring a storage area network (SAN) or network-attached storage (NAS), no more databases hanging or transactions failing due to running out of disk space, the elastic nature of cloud storage provision ensures disk is always available, as much as we need, always available at the point of demand. Furthermore, due to the innovative approach implemented by Snowflake, we don't have to be concerned with transaction management in the same way other database vendors are reliant upon—no more rollback, undo, row chaining, or separating tablespaces onto separate physical drives.

As a nod to our cybersecurity colleagues, we must always ensure external storage is properly secured against unauthorized access by implementing appropriate security policies and then proving the security policies tested to prove fit.

Different Platforms, Different Approaches

We must ask ourselves why Snowflake adopted their chosen approach. Snowflake is a data warehouse but not an online transaction processing (OLTP) database; therefore, storage considerations change.

Part of the answer lies with the fundamentally different nature of cloud service storage provision. We do not have to segment our data containers into different logical groups. Snowflake could have imposed similar containers, but the concept no longer applies. Furthermore, data warehousing has been an extension to most historical database vendor products, not an original architectural consideration or initial core capability. Data warehousing may not have been designed into the core product. And the market has matured, and data volumes, velocity, and variety have exploded. Data warehousing is now essential for serious data analysis; therefore, a different approach is warranted. Snowflake positions itself as a data warehouse provider, thus enabling a different storage management paradigm, which is discussed next.

Micro-partitions and Cluster Keys

Snowflake implements storage using micro-partitions, the underlying contiguous units of storage that comprise logical tables. Each micro-partition contains up to 16 MB of data compressed using proprietary compression algorithms. Uncompressed, each micro-partition holds between 50 MB and 500 MB of data organized in hybrid columnar format, optimizing compression according to the column's data type.

Object metadata containing rowcount, table size in bytes, physical file reference, table version, and zone map information is retained for all written micro-partitions according to the Snowflake edition and retention period specified. (Data protection is covered later in this chapter.) For each micro-partition, metadata for the maximum and minimum value range, number of distinct values, and NULL count is retained. Where multiple micro-partitions are written for a single object, metadata includes the total number of micro-partitions, overlap depth, and overlap values. I briefly discuss how metadata is used later, but don't expect a deep dive into Snowflake performance tuning because it is a significant subject and beyond the scope of this book.

Traditional database vendor disk management policy is mutable object storage; the object's contents change according to data manipulation language (DML) operations (insert, update and delete), and the object resizes dynamically, leading to mutable storage containers. Data changes in objects and indexes in separate disk structures lead to proactive monitoring and human/machine intervention for disk management. The Snowflake approach is different; once a disk structure (a *micro-partition*) is written, it is immutable. DML operations result in new micro-partitions being written. Micro-partitions can no longer be physically managed, apart from changing clustering keys, nor can their behavior be influenced.

Understanding the difference between mutable and immutable storage management is key to understanding why OLTP is not considered a good use case for Snowflake. Micro-partition management for individual record operations leads to queueing. Transaction failure as the underlying storage struggles to keep up in high volume DML scenarios. However, careful use of temporary tables to collate OLTP data with a subsequent bulk load into permanent tables may provide an acceptable pattern assuming error handling in the event of failure allows re-run capabilities. This is just a suggestion.

Note   Use Snowflake for data warehousing, not OLTP.

Figure 3-10 illustrates a table with three micro-partitions and a DML operation to the United Kingdom data set, resulting in new micro-partition creation. We build upon this theme throughout this chapter.

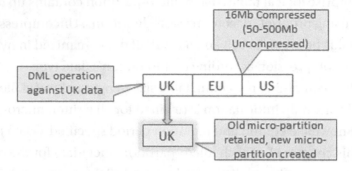

Figure 3-10. *Immutable micro-partitions*

The implications for the Snowflake micro-partition approach are huge. On initial data load, Snowflake automatically clusters the data in micro-partitions without making any assumptions about the data. No histogram or other algorithm is applied, each micro-partition is written and metadata captured. The default approach works well for most scenarios, and most often, particularly for data loads of less than 1 TB, query performance is fine.

Over time, where data has been subject to DML or data volumes exceed 1 TB, performance degradation or initial poor performance may be observed. Under these circumstances, the data may not be optimally clustered in the table according to the query predicates most often used to access the micro-partitions. We might choose to add a clustering key, a set of columns or expressions, on a table explicitly designed to match the most frequently used query predicates. Recognizing one size does not fit all, and alternative query predicate data access paths may be disadvantaged, Snowflake offers materialized views where different clustering strategies can be applied. Your mileage will vary according to your particular scenarios.

Snowflake provides automatic clustering, which monitors the state of clustered tables and, where appropriate, reclusters behind the scenes, but only for tables where a cluster key has been declared. Manual reclustering has been deprecated for all accounts; therefore, it should not be considered. Adding a cluster key is not recommended for all scenarios, and performance testing should be conducted to prove the benefits of any change before release into production environments.

Micro-partitions enable horizontal and vertical partition pruning with static pruning performed in the metadata cache to only include micro-partitions matching query predicates. Dynamic pruning occurs at runtime during query execution based upon join conditions and other constructs. This approach explains the importance of cluster keys and data storage.

Although I discuss how our data in cloud storage is protected later, it is important to note all Snowflake customer data is encrypted by default using the latest security standards and best practices. Snowflake uses strong AES 256-bit encryption with a hierarchical key model rooted in a hardware security module. The Snowflake service rotates the keys regularly, and data can be automatically re-encrypted (rekeyed) regularly. Data encryption and key management are entirely transparent and require no configuration or management. See `https://docs.snowflake.com/en/user-guide/security-encryption.html#encryption-key-management` for further details.

Indexes and Constraints

Snowflake does not support the physical creation of separate indexes as objects stored in addition to the primary table. While unique, primary key and foreign key constraints can be declared, they are not enforced by Snowflake, and no resultant objects are created. The declarations facilitate data self-discovery tools allowing entity relationship diagrams to be automatically created. The only constraint enforced is NOT NULL.

Coming from a traditional RDBMS background with significant data warehousing experience, the absence of enforced constraints at first seems both counter-intuitive and a retrograde step.

In developing Snowflake applications, we face a mindset challenge because we expect our data to be clean at the source in a data warehousing environment. In contrast, in an OLTP environment, we expect our data to be validated before being allowed to enter our database. If we accept this principle, our data warehouse must not allow data updates locally, and all data changes must be made at the source. The absence of enforced constraints no longer matters. It is the declaration themselves that matters. Furthermore, suppose our data warehousing ingestion strategy includes mechanisms to identify data quality issues but not exclude any data regardless of quality. In that case, we have an automated mechanism to both feedback data gaps to the originating system, closing the feedback cycle, and measuring data quality over time.

Materialized Views

As with other RDBMS platforms, Snowflake supports materialized views for Enterprise Edition and higher. We may choose to implement materialized views for a variety of reasons. This section focuses on micro-partitions and not the wider performance tuning considerations recognizing micro-partition pruning is a significant factor in query performance.

Materialized views are maintained using the serverless compute resources discussed later in this chapter. For now, it is enough to know micro-partition maintenance is taken care of by Snowflake and billed accordingly. All we have to do is declare the materialized view with the new cluster key definition. Snowflake takes care of the maintenance.

Note   Materialized views do not have the Time Travel feature.

There are several limitations to Snowflake materialized views when compared to other RDBMS. The most notable is a reference to a single table only. Materialized views cannot reference another materialized view. All limitations are explained at `https://docs.snowflake.com/en/user-guide/views-materialized.html#limitations-on-creating-materialized-views`.

Addressing the full reasons why we would want to implement materialized views is beyond the scope of this book, as the answers lie in the realms of performance tuning and query optimization. The key point is the option to redefine our tables by reclustering, thus supporting different optimal access paths.

Stages

Storage is also referred to as a *stage*. Later, I illustrate mapping S3 storage in our AWS account using storage integration, which along with associated AWS account configuration to establish the trust relationship, provides the first of four storage types, called an *external stage*.

We also referred to S3 storage mapped in our Snowflake account where no AWS configuration was required, the first of which is called an *internal stage*, a *named stage*, or an *internal named stage*. All three labels are used interchangeably and are generally inaccessible outside the Snowflake boundary. Note that future capability is being developed to make internal stages accessible. When data is staged to an internal stage using the PUT command, the data is encrypted on the local machine before being transmitted to the internal stage to ensure security throughout the data transit lifecycle.

There are object types such as tables and materialized views that require storage. Each consumes S3 storage inside our Snowflake account. They are called *table stages*, uniquely named one per object.

Finally, user stages underpin the user interface. They should not be referenced directly. They also consume storage from our accounts.

Note   For external stages, apply and test your security policy before releasing it into production.

Figure 3-11 illustrates the physical location of stages, which is covered later in this chapter.

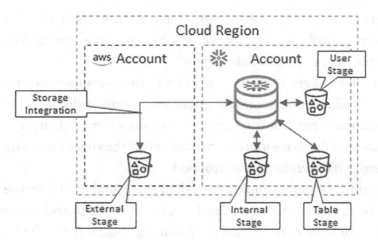

Figure 3-11. *Snowflake stages*

Shares

Chapter 1 discussed Secure Direct Data Share. Now that you understand the immutable nature of micro-partitions and how Snowflake persists objects using cloud storage, let's discuss the first integration pathway to the Snowflake Data Cloud built on Snowgrid, Snowflake's underlying proprietary data interchange platform. Most accounts are enabled by default for sharing. If you encounter an issue, contact Snowflake support.

Note Shares are available for account data copy to any local account in the same cloud provider and region.

The party who owns the data and grants entitlement to those who can use the data is the provider. The party who uses the data is the consumer. Providers always control who can access their data. Snowflake's staff cannot view, override, or take control of a provider's data. Objects in a consumed share cannot be modified or added to. The data can only be queried, not modified. A share is a read-only object for a consumer, and no other action can be performed except reading data from the share by the consumer. Consumers cannot share data with other accounts when provided via share technology, but they can use it and join it to their data sets. Consumers can insert shared data into their tables by selecting from the share and sharing their tables. This is the only way to cascade data sharing.

Secure Direct Data Sharing enables sharing selected objects in a database in your account with other Snowflake accounts, but only for those providers and consumers in the same cloud provider and Snowflake region.

For Snowflake customers, compute is paid for by the consuming account. The subject of non-Snowflake consumers using a reader account is discussed briefly in Chapter 14. the provider pays for consumption. Consumers may create a single database per share. An inbound database must be created before the share can be queried, and created databases can be renamed and dropped.

Most importantly, no data is copied. Metadata alone enables the secure sharing of data in underlying storage. Since no data is copied from the provider account, the consumer Snowflake account is not charged for storage. Just think of the implications: zero-copy, real-time transactional data sharing with clients under your absolute control, screaming data democratization, and enabling citizen scientists. Figure 3-12 illustrates how micro-partitions are shared.

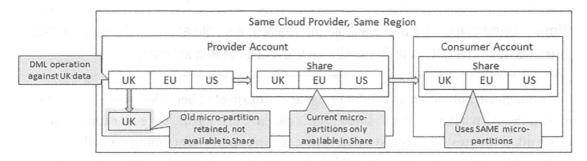

Figure 3-12. *Shared data conceptual model*

Snowflake is constantly working on improving monitoring and exposing useful metrics to its customers. An upcoming enhancement allows providers to gather basic metrics on how their shared data has been consumed.

There are some restrictions. For example, only the most recent micro-partitions are available in the share, and not all schema object types can be shared. But I hope you agree that shares are an incredibly powerful, easy-to-manage, out-of-the-box solution to seamlessly sharing and monetizing data in your organization. Chapter 14 explains how to physically implement shares.

Cloning

Since you now understand the immutable nature of micro-partitions, you can see how Snowflake external stages, tables, schemas, and databases can be cloned using metadata operations alone, resulting in zero-copy cloning. A cloned object is a new, separate object against which all types of permissible operations for the parent object can be performed. The original underlying micro-partitions remain the same until the data is changed in a cloned object. Only when a DML operation occurs against the primary or cloned object are the affected micro-partitions instantiated, and at this point, storage costs are incurred. Cloning is almost instantaneous, a few minutes at most, and there are no limits to the size of the object that can be cloned.

Some limitations apply. Internal named stages cannot be cloned; but tables and their table stages can be cloned. Cloned external stage contents (files) are not copied. Temporary tables and transient tables cannot be cloned as permanent tables. Materialized Views cannot be directly cloned, but if a cloned database or schema contains materialized views, they are cloned. Also, check the documentation for information on how object tags are treated for cloned objects.

Note   Database clones are available internally in a single account. Schema clones are available in a database. Table clones are available in a schema.

Other cloned objects may exhibit unusual or unexpected behavior; more information is at https://docs.snowflake.com/en/user-guide/object-clone.html. Above all, test, retest, then retest again before relying upon your cloning procedures to ensure all edge cases have been considered when running automated test cases and on User Acceptance Testing environments.

Clones may be created for the current timestamp, at or before a timestamp in history, or at or before a SQL statement identifier. For historical object cloning, the point in time chosen must be both in the Time Travel retention period and exist at the point in time chosen. Note your current role must have the appropriate entitlement to the source object. Check the documentation for details.

Cloned databases may be used to quickly spin up like-for-like environments to rapidly test new code or as a starting point for producing sample data sets and all manner of previously unthinkable scenarios due to the inherent time taken by legacy databases to produce replica environments. When coupled with the Time Travel feature, the huge advantages of the Snowflake micro-partition strategy become apparent. Cloned databases usually form the basis of replication, discussed next.

Replication

Replication must be enabled for your account using either an ORGADMIN or an ACCOUNTADMIN role. In larger organizations, the ORGADMIN feature may have been enabled; therefore, replication is probably enabled from a central function. Alternatively, replication can be enabled using the ACCOUNTADMIN role. Account security is addressed in Chapter 4. It goes without saying—but I do for clarity's sake—that a second Snowflake account is required for replication. As stated, accounts may be on any supported cloud provider.

It is fair to say the Snowflake replication story is still a work in progress. The core components to protect data are in place, albeit currently limited to database replication available across cloud providers enabling instant failover and recovery. Account replication is expected in Public Preview H1 2022, with object-level replication planned. Database replication is massively important. The immutable nature of micro-partitions acts as an enabler for data sharing, along with metadata synchronization. However, there are other considerations, such as repointing infrastructure to use the failed over the site, then seamlessly fail back when service is restored. This is where client redirect helps.

Database replication is distinctly different from shares in one fundamental way. Data with a shared object is available in real-time in the consuming account. Data in a database replicated to another account must be refreshed and may take more than 20 seconds to appear in the consuming account as the serverless compute operations completely. Your mileage will vary, and testing must be conducted to ensure replication completes in an acceptable timeframe.

Note   Replicated databases are available for account data copy to any local or remote account, cloud provider, and region.

Some limitations apply. Replication operations fail when object tagging is deployed, and the target account version is lower than the Enterprise Edition. Also, referenced object tags must be in the database to be replicated. Databases with external tables cannot be replicated. The externally referenced objects are not copied, so the replication operation fails; the same for databases created from imported shares. Several object types are not replicated, including users, warehouses, roles, resource monitors, and network policies. The message is clear. Test, test again, and then repeat once more.

Note   Replicated databases do not carry forward roles which must be recreated

---

Having identified several limitations with replicated databases, it is clear there are precursor activities to configure our target account for receiving replicated databases, and several post-replication database import activities must be performed. However, the Snowflake replication story is improving, with less coding expected with future releases.

Regardless of the cloud provider and region chosen, replication between accounts is possible, subject to restrictions outlined in the documentation. Note all secondary replicas are available as read-only until one is made master. From this point, onward is read-write with all other replicas, including the original reverting to read-only.

Same Cloud Provider and Region

This scenario is limited in data protection as both provider and consumer accounts rely on the same underlying HA infrastructure. However, valid scenarios exist where organizations may want to adopt this approach. Where a merger or acquisition provides two distinct accounts for consolidation under a single organization, the security postures differ, and/or the accounts support disparate business use cases. In these scenarios, it can be seen consolidating some data into a single account can be highly beneficial while retaining both security postures independently. Shares may also be used to seamlessly exchange data sets between both accounts at zero cost as the underlying storage is reused and only metadata is copied.

Same Cloud Provider and Different Region

Providing total failover protection in the event a single primary region fails, a replicated secondary may be activated as primary, thus restoring service. In the unlikely event that the CSP or Snowflake suffers from two (or more) availability zone outages, failover may not be possible if both accounts are affected. Having made this statement, the probability of such an event is exceedingly small.

For the same cloud provider in a different region, storage costs are incurred for both accounts as the underlying storage cannot be shared via metadata. Both accounts must instantiate micro-partitions.

Different Cloud Providers and Different Regions

Providing total failover protection in the event a single primary region fails, a replicated secondary may be activated as primary, thus restoring service. Noting two or all of the CSPs would have to suffer concurrent failures to render Snowflake unrecoverable, this highly improbable scenario is thought so extreme as to not be possible.

For different CSPs in different regions, storage costs are incurred for both accounts as the underlying storage cannot be shared via metadata. Both accounts must instantiate micro-partitions. Also, egress costs are incurred when data is moved between cloud providers, which soon add up when refreshes occur.

Storage Summary

Unfortunately, the Snowflake documentation is less than clear when describing shares, cloning, replication, and available options. This section attempts to summarize the options available. Table 3-1 provides a quick summary.

Table 3-1. *Storage Options*

Storage Option	Approach	Storage Cost	Resilience
Share	Same CSP, same region	None, Metadata copy only	Same CSP, HA, single region, real time
Share	Same CSP, different region	Storage cost per account, per region	Same CSP, HA, multi-region
Share	Different CSP	Storage cost per account, per region	Different CSP, HA, multi-region
Clone Database	Same account	None, Metadata copy only initially	Same CSP, HA, single region
Clone Database	Same CSP, different region	Not Available	Not Available
Clone Database	Different CSP	Not Available	Not Available
Replicated Database	Same CSP, same region	None, Metadata copy only	Same CSP, HA, single region, refreshed
Replicated Database	Same CSP, different region	Storage cost per account, per region	Same CSP, HA, multi-region, refreshed
Replicated Database	Different CSP	Storage cost per account, per region	Different CSP, HA, multi-region, refreshed

Serverless Compute

Some Snowflake capabilities do not require virtual warehouses but operate using Snowflake supplied and managed compute resources. Naturally, nothing is for free, and these "behind the scenes" capabilities are still costed and charged to your account.

But which services use serverless compute? The list is growing but includes Snowpipe, automatic clustering, search optimization service, external table metadata refresh, materialized view maintenance, database replication, failover and failback, and most recently, tasks that no longer require a dedicated warehouse to be configured.

We dive into some of these features later. I have already mentioned automatic maintenance for other features. This section serves as a reminder of line items that appear separately on your billing. For a fuller explanation, please see the documentation at `https://docs.snowflake.com/en/user-guide/admin-serverless-billing.html`.

Data Protection

This section discusses the Snowflake features available to protect our data, focusing on physical data recovery. Recall how CSP HA underpins confidence in stated claims of 99.999999999% durability and 99.99% availability. Snowflake relies upon published CSP HA capability to support its product features.

Time Travel

From the discussion on how micro-partitions are managed, you know their immutable nature leads to several extraordinary built-in Snowflake capabilities. The Time Travel feature allows you to revert objects, conduct queries, and clone to a point in time. Note that it is restricted to a 1-day maximum for Standard Edition and up to 90 days for all higher editions and cannot be disabled at the account level but can be disabled for database, schemas, and tables.

Note   It is recommended you set DATA_RETENTION_TIME_IN_DAYS to 90 for your account using the ACCOUNTADMIN role.

Assuming an appropriate retention period is set, a Snowflake administrator can query data at any point in the retained period that has since been updated or deleted; also create clones of databases, schemas, and tables using a specific timestamp, relative timestamp offset, or SQL query ID.

Storage overheads of typically 10% to 15% are incurred when implementing Time Travel. But storage is very cheap, and the benefits far outweigh the cost overhead. If anyone has waited hours for backups to restore a database, this single feature is the one for you. Here you see why the immutable nature of micro-partitions is such a powerful enabler because recovery or cloning is typically sub-minute regardless of the volume of data affected by reverting to an earlier timestamp. We have all experienced mistakes—on production systems while performing releases, inadvertently dropping an object, truncating a table, or incorrectly updating data. The Time Travel feature is an immediate recovery option at our fingertips. There is no need to call a database administrator and request a backup. We have the means to self-serve and move forward, with an immutable 1-year query history to identify the point in time or SQL statement to revert to.

There are some caveats. Referenced objects must be in the available retention period. Restoring tables and schemas is only supported in the current schema or database as set by the in-scope role, even if a fully qualified object name is specified. Also, the user executing the UNDROP command must have ownership privilege on the table granted to the role in use. Finally, the user must have CREATE privilege on the target schema. Note both transient and temporary tables have a maximum Time Travel retention period of 1 day regardless of Snowflake edition. Time Travel is not supported for materialized views or external tables for the (obvious) reason their micro-partitions for materialized views are built from existing table data. For external tables, the storage is not Snowflake managed, only referenced by Snowflake.

Changing the retention period for your account or individual objects changes the value for all lower-level objects where a retention period is not explicitly set. If the retention period at the account level is changed, all databases, schemas, and tables that do not have an explicit retention period automatically inherit the new retention period. The retention period defaults to 1 day and is enabled for all accounts on initial provisioning.

Note   I highly recommend the Time Travel feature be set to 90 days at database creation time.

When a table that has Time Travel enabled is dropped, and if a new table with the same name is created, it fails when the UNDROP table command is executed. You must rename the existing object to enable restoration of the previous version of the object.

Assuming the execution context is set correctly, you can observe a table's version history by executing the following.

```
SHOW TABLES HISTORY LIKE 'table_name' IN database.schema;
```

The Time Travel feature should be used to quickly restore service in the event of a bad data load with affected objects or database being reverted to the point in time immediately before the bad data was loaded, then rolled forward carefully before resuming normal service. No more waiting for database administrators to restore backups or hours spent rolling back transactions.

Fail-safe

The Fail-safe is a data recovery service that provides an additional, non-configurable, 7-day retention period exclusively managed by Snowflake. It represents the final stage of storage lifecycle retention. When objects age out of Time Travel, the underlying micro-partitions are not immediately deallocated and returned to the cloud provider storage pool. Instead, Snowflake retains the micro-partitions—and by inference, associated metadata—and can restore objects, though this may take several hours to complete.

If the Time Travel retention period is reduced, any objects falling outside the reduced period are moved to the Fail-safe service and are no longer directly accessible. Objects moved to Fail-safe are not accessible in the owning account. Objects are only accessible via raising a support ticket to Snowflake. According to Snowflake documentation, "Fail-safe is not provided as a means for accessing historical data after the Time Travel retention period has ended. It is for use only by Snowflake to recover data that may have been lost or damaged due to extreme operational failures."

Caveats apply. Transient, temporary, and external tables have no Fail-safe service period. Fail-safe also incurs storage costs for the storage consumed by objects retained for the 7-day period.

Backup Strategy

Given the available options in Snowflake for data protection, we might be asking ourselves why a backup strategy may still be relevant. There is no single clear-cut answer to this question as several themes, including the right to be forgotten, come into play. What if, outside of the 90-day Time Travel period and the 7-day Fail-safe period, a mistake was made, and the wrong person's details were removed? Should we consider taking a hard backup of all affected objects into offline storage before removing an individual? What does your organization's policy state, and how do we translate policy into the process and operational procedures? Unfortunately, more questions than answers, shaped by your organization's approach, understanding, and maturity of Snowflake, posed to provoke thought.

Disaster Recovery (DR)

Having discussed data protection, let's turn our thoughts to disaster recovery, which we also discussed earlier in our storage summary section. Recognizing our organizations may be subject to external regulatory requirements, and some business colleagues may not fully appreciate the inherent robust nature of Snowflake, we may find ourselves required to implement a full DR policy with complementary processes and procedures.

Our first proactive action should be to inform and educate our colleagues on the fundamentally different nature of CSP-provisioned Snowflake and the inherent protections and tooling baked into each. Having conducted our initial "hearts and minds" campaign, regularly refreshed content, and held repeat sessions, we may find many DR requirements are satisfied by Snowflake's built-in capabilities, reducing our work. However, I recognize this takes time and offer the upcoming information as a starting point for developing your DR strategy.

We must clearly articulate each environment by maintaining a network topology diagram, operations manual, and DR runbook. Each should be periodically reviewed and approved, also stored in a readily accessible form. Our organizations may also have a centralized help desk and ticketing system with procedures to follow during a service outage. This is where our service level agreement documentation dictates both response level and timeline. We must also maintain our technical documentation and ensure our support staff can access the latest detailed information on each aspect of our system. None of this should come as a surprise to seasoned developers.

Business Continuity

Our data is safely stored in Snowflake and, when configured appropriately, immediately accessible at any point in the preceding 90-day period via Time Travel. Depending upon our data replication strategy, we can guarantee our data availability. However, diving deeper into the specifics of shares, clones, and replication, we find some limitations on how specific object types are treated. These must be investigated on a case-by-case basis with Snowflake documentation as the first point of reference recognizing replication features are constantly evolving.

Snowflake cannot consider every possible scenario, and each organization's business continuity requirements and strategy involve a degree of manual activity. It should be said failover and failback in Snowflake are a good deal easier than most other platforms, but still, some work is required.

You may find existing or legacy documentation to assist in identifying essential infrastructure to protect and ensure continuity of service. Organizations often have central risk registers where known weaknesses are stored. These should correlate to applications, interfaces, services, and tooling. Likewise, your organization may have a system catalog containing contact details of application owners and support groups. And for the most forward-looking organizations and those subject to external regulatory requirements, end-to-end system mappings from data ingestion to data consumption may be in place. These sources point to organizational maturity and inputs into delivering robust business continuity plans.

Note   Test your failover and failback scenario regularly by cloning production, replicating, and then failing over the clone to another account. Use automated testing wherever possible.

Snowflake Service Status

The first check should be whether your CSP and location are up and running; see `https://status.snowflake.com`. If your service is affected, failover is a manually invoked customer-controlled activity achieved by metadata update; therefore, it is very quick.

Failover and Failback

Depending upon the type of objects contained in your replicated databases, you may need to run additional cleanup scripts. However, it is hoped all cleanup activity is well known and documented by the support team, so no nasty surprises arise.

We also assume the security model implemented via roles has been applied to the DR replicated databases and all entitlements are consistent with the primary site. Surprisingly, it is common for assumptions to be made that entitlement follow replicated objects when changes are made, which is sadly neglected in many release runbooks.

Naturally, client tooling needs repointing, which is later covered separately. Batch loads, Snowpipe, and other custom loading mechanisms also need consideration to ensure seamless post-failover operation where some feeds may need replaying and others prevented from running twice. Note it may be certain CSP storage infrastructure allows independent replication as AWS S3 does, which may simplify failover and failback.

Note   Just as failover needs careful consideration and planning, so does failback.

Our failover and failback must also consider CSP-specific requirements outside of Snowflake; for example, these may include storage and API integrations, Lambda, and S3 and SQS configurations.

Client Redirect

Client redirect uses Snowflake organization enhancements to URLs for accessing your Snowflake account. A Connection object is created for each account, with only one pointing to the primary account. When failing over, the Connection object is altered to become the primary, causing a behind-the-scenes update to the URL. After the CNAME record (a type of DNS record) is updated, typically in 30 to 60 seconds, all new access via the URL is to the new primary site. Note failover and client redirect are two separate operations.

Extending Snowflake Capability

This section describes some interactions between Snowflake and cloud services essential for implementing the Snowflake Data Cloud. More features are explored in later chapters. For AWS-based Snowflake accounts, S3 storage underpins all our database objects; therefore, we must have a means of accessing S3 buckets from Snowflake. We also need a way to interact with cloud storage outside Snowflake because our data feeds are often supplied as flat files in batch format. Every organization, at some point, resolves its data transfer to the lowest common denominator of flat file batch transfer with the expectation of refactoring later. We start with flat files being the minimum standard for data transfer when moving data from our silos into Snowflake.

Keeping the technical details at bay for now, I must introduce a new component into our discussion, the CSP account, as the services we rely upon are available here, in our case, an AWS account which can be created at `https://aws.amazon.com/free`.

Why should we create an AWS account? Simply put, to access the wealth of capabilities developed and supported by your chosen cloud vendor, which lie outside of Snowflake's core capabilities. While Snowflake delivers huge capabilities with data, and the capabilities are ever expanding, some features that should never become fully integrated, such as Secrets Manager, and others that are a natural fit for Snowflake, such as Lambdas, document parsers, and S3 buckets.

Figure 3-13 shows S3 storage referenced by Snowflake in two locations. Remember, our Snowflake account is a Virtual Private Client (VPC), and our AWS account is also a VPC, two different containers which co-exist in the same (in this case, but could also be in a different) cloud region. But note that Snowflake can reference both S3 buckets. But how?

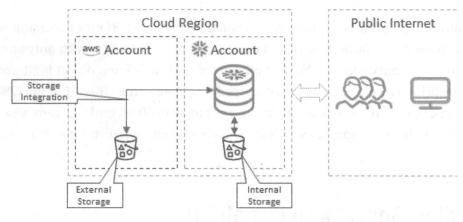

Figure 3-13. *AWS storage integration*

The Snowflake account boundary allows access to unlimited storage. We just have to declare, then reference, our elastic cloud provisions as much as we need. This internal storage can only be referenced in the Snowflake boundary when writing this book. However, it should be noted that upcoming new features can make this internal storage accessible elsewhere.

In our AWS account, we can also create S3 storage into which, from our desktop machines, we can upload files or, using supplied AWS functionality, automagically copy files across from other S3 buckets. We can also upload files programmatically using a variety of tools.

Note   Unprotected files in our AWS account S3 buckets can be seen from the public Internet, so please ensure an appropriate security policy is applied and tested before adding files.

Snowflake has commands to make external storage visible and accessible, called *storage integration* (`https://docs.snowflake.com/en/sql-reference/sql/create-storage-integration.html`). The commands require work in both Snowflake and AWS Management Console to establish the trust relationship. Then, we can drop files into S3 and access them using SQL. This theme is returned to when covering unstructured document support in Chapter 10.

Not only can we access files in external storage from Snowflake, but we can also automatically ingest data when a file lands in our external storage by using Snowpipe, a built-in Snowflake feature; see `https://docs.snowflake.com/en/user-guide/data-load-snowpipe-auto-s3.html`. Figure 3-14 has been extended to incorporate a new AWS service, Simple Queue Service (SQS) (`https://aws.amazon.com/sqs/`), which notifies Snowpipe when files are ready to consume. Snowpipe is discussed in Chapter 8.

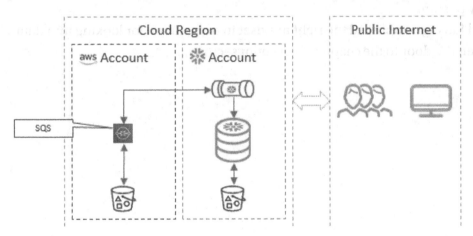

Figure 3-14. *Snowpipe automated ingestion*

Summary

This chapter explained how Snowflake software as a service (SaaS) is deployed centrally in a CSP region and how Snowflake accounts are provisioned. The discussion also gave insight into the Snowflake editions, calling out capabilities we investigate later and answering the question of why we should integrate with an AWS account.

The chapter looked at cloud storage and investigated the different types of stages, understanding their differences and uses, and how Snowflake implements storage using micro-partitioning as these underpin much of the Snowflake Data Cloud core capabilities.

Looking toward the future where unknowable data types and new forms of media await, we understand how to address future state data demands. You can now begin to think in terms of "yes" being your first response backed up by the knowledge we can develop tooling to access data, rather than immediately saying "no" because the answers lie outside of your skill, knowledge, expertise, and comfort zone.

This discussion has taken you through some of the complexities in accessing data silos, calling out some key dependencies and challenges, not the least of which is security posture.

And having established the right mindset in preparation for looking into data silos, let's open the door to the chapter on account security.

CHAPTER 4

Account Security

As provisioned and by default, a Snowflake account is configured for public Internet access. To properly secure our Snowflake account, we should define our security posture, understand where security controls are defined, select those appropriate for our purposes, match and set corresponding Snowflake configuration options, and above all, implement appropriate monitoring to ensure the options set remain until we make an informed change.

It is possible to use Snowflake without touching any security controls. If this book accelerates how to use Snowflake, this section may not appear relevant. As security is central to everything in Snowflake, I argue it is essential to fully understand the why, how, and what of security because without a full understanding of Snowflake security, an incomplete security posture is probable, and mistakes are more likely and the worst of outcomes more certain.

Cybersecurity has always been relevant and should be top on our agenda. Preventing attacks is becoming more and more important. One recent estimate is that 80% of all organizations experience some form of cyberattack each year, and media headlines frequently expose companies and government departments subjected to their data being scrambled.

Note Cybersecurity is everyone's problem—all day, every day.

You should also be aware that Snowflake is constantly being enhanced; new features introduce new security options, and the devil is always in the details. This chapter dives into code. Finally! At last, I hear you say. Something practical for the technophiles to enjoy. I hope!

We take a holistic approach, looking at how to define security controls and sources of information to seed our security posture, from which we can build our monitoring.

© Andrew Carruthers 2022
A. Carruthers, *Building the Snowflake Data Cloud*, https://doi.org/10.1007/978-1-4842-8593-0_4

After implementing selected controls, we must monitor to ensure settings are not inadvertently changed or changed by bad actors.

Sections in this chapter point you to the documentation for features outside Snowflake. I aim to give pointers and information for you to investigate further and provide a starting point recognizing each organization has different tooling and approaches.

Finally, given this chapter's limited space, it is impossible to cover every feature or recent change to behavior (if any).

Security Features

By design, Snowflake has security at its heart. Every aspect of Snowflake has been built with security controls embedded. RBAC is discussed in this chapter, noting fundamental differences in how the Snowflake RBAC model works compared to other RDBMS vendors.

Snowflake-specific security features provide a layered defense but only if we implement controls and monitoring effectively. We can implement the best security posture possible, but if we don't monitor to ensure our controls remain set and raise alerts when we detect a breach, then we might be ok, but we can't be sure. Try explaining to your organization's CIO, regulator, or criminal investigators.

The distinction is very important. Some legacy databases use the words *account* and *user* interchangeably. In Snowflake parlance, an *account* is the outermost container provisioned by Snowflake. The user is either an end user or service user who may be entitled to use objects in Snowflake, but only via a role. There is no capability in Snowflake to grant entitlement directly to a user.

RBAC is covered in the next chapter, which is worth a deep read for two reasons. First, you are most likely migrating from a legacy database vendor where security has not been baked in from the ground up. Second, misconceptions are harder to change than establishing a new concept. I thought I understood RBAC, but it wasn't until I thought about the subject that I realized I had missed some vital information.

System Access

To access any system, you must know the endpoint—the location where the service is provisioned. We are all familiar with Uniform Resource Locators (URLs) and use them daily without thought. Underpinning each URL is an Internet Protocol (IP) address, and each IP address may implement many ports, the most common of which are port 80 (Hypertext Transfer Protocol (HTTP)) and port 443 (Hypertext Transfer Protocol Secure (HTTPS)). Look in your browser and note the prefix. An example of HTTPS is at `https://docs.snowflake.com/en/user-guide/admin-security-fed-auth.html`. Most sites now use HTTPS by default. Classless Inter-Domain Routing (CIDR) is touched upon later. Be aware of the /0 network mask for Snowflake network policies discussed later. Networking is complicated and a subject in its own right. I have introduced some terms of which HTTPS and CIDR are most relevant.

When accessing systems, we must approve each user. In most organizations, this is usually a formal process requiring the user to submit an access request with their manager's approval, which is then processed via another team who provisions the user on either the system or centralized provisioning tool for automated creation with notification back to the user when their request is complete.

Once a user has been notified of their system access, we must consider how they physically connect to the system. This process is called *authentication*. You know that usernames and passwords are inherently insecure. We fool ourselves to think everyone remembers a unique generated password for each system they interact with daily. We must look at authenticating users by other means.

Some of the tools we have available are to implement HTTPS. Imagine a secure pipe connecting your desktop to a known endpoint through which data flows. HTTPS is reliant upon a security certificate. We don't need to know the certificate's contents as it is all handled "behind the scenes" for us, but now you know. Authentication can also be implemented automatically by single sign-on (SSO), discussed later, where we no longer need to log in via username and password. Instead, we rely upon membership of Active Directory (AD) or Okta to determine whether we should be allowed to log in or not.

Entitlement is all about "who can do what." Our organizations have many hundreds of systems. One large European bank has about 10,000 different systems. Their HR department has approximately 147 different systems, and their equities trading department has about 90 different systems.

We need a single place to record entitlement on a role profile basis. In other words, if I perform job function X within my organization, I need access to systems A, B, and

C's entitlement to perform actions J, K, and L for each system. As can be seen, not only is networking complex but both authentication and entitlement are complicated too.

Fortunately, we can also use Microsoft Active Directory Federated Services (ADFS), Okta, and a few others to manage group memberships that map to entitlements in our systems. If only every system could integrate with ADFS or Okta, and user roles were static in our HR systems, we would have a chance of automating user provisioning and entitlement at the point of joining, moving positions internally, and leaving our organizations.

Security Posture

No system can ever be said to meet its objectives without proper requirements for validation. Snowflake security is no exception.

Organizations often have a well-defined, formally articulated security posture, actively maintained as threats evolve, and new features with a security profile are released. Our security posture must include these aims. Preventing data breaches and allowing only appropriately entitled data access. Organizations hold lots of sensitive, confidential, and highly restricted data of immense value to competitors, bad actors, and the curious. Financial penalties incurred by data breaches are limited only by the governing jurisdiction and not restricted to just money; reputation can be more important than a financial penalty, and trust relationships formed over many years are too easily eroded.

Data breaches are not just leaks from internal systems to external locations. Our business colleagues rely upon the integrity of the data contained in our systems, using the information and outcomes to make business critical decisions, so we must ensure our data is not tampered with by bad actors or made accessible via a "back door" into our networks. We may occasionally identify data set usage with inappropriate controls, unauthorized usage, or insecure storage. In this case, we must raise the breach to our cybersecurity colleagues for recording and remediation.

We must also ensure our systems are available to all correctly authorized, entitled customers and system interfaces wherever possible utilizing sophisticated authentication mechanisms proven to be less accessible to attack. Username and password combinations are our least-favored approach to authentication. How many readers use the same password on multiple systems?

We also have legal and regulatory requirements, which adapt to meeting known threats and preparing defenses against perceived threats.

Cybersecurity is an ever-changing and increasingly more complex game of cat and mouse. But working long hours, running remediation calls with senior management attention, and conducting support calls every hour is no fun. Been there, done that, and for those unfamiliar with the scenario, please remain unfamiliar by staying on top of your security posture.

Attack Vectors

Attack vectors occur with staff mobility, and here we pose some common questions to address to reduce the opportunity for cybersecurity attacks.

- How often, when moving to a new position in an organization, has entitlement not been removed, with new entitlement granted for the new position?

- Do managers always recertify their staff on time?

- Are all employee position entitlements always provisioned by Active Directory group (or equivalent) membership?

- Have we provisioned a separate Snowflake environment for production? See the "Change Bundles" section in this chapter for more on this issue of separate environments.

- At what frequency do we scan our infrastructure to ensure our security posture remains intact?

- When did we last conduct a penetration test?

You may be wondering how relevant these questions are to ensuring our Snowflake environment is secure. You find out later in this chapter.

Prevention Not Cure

As the adage says, "Prevention is better than cure." If things go wrong, at least we immediately have some tools available to recover under certain scenarios. Snowflake has an immutable one-year query history of assisting investigations. The Time Travel feature covers any point in history for up to 90 days; if it is not enabled, please do so immediately for your production environments. Finally, Snowflake provisions a fail-safe for an additional seven days of data recovery, noting the need to engage Snowflake support for Fail-safe assistance.

But we are into prevention, not cure, so let's look at defining our security posture, then we have a baseline to work from. Figure 4-1, referenced from `www.nist.gov/cyberframework`, illustrates the constituent parts of our security framework along with dependencies to be resolved.

 If we cannot identify our assets, we cannot protect them

Without identification and protection, we cannot detect potential breaches

Without breach detection we cannot respond to potential breaches

Our ability to recover will be compromised without a robust Cyber Security framework

Figure 4-1. *Security framework*

Our first step is to identify the assets to protect. Some organizations have a central catalog of all systems, technologies, vendors, and products. But many don't, and for those who do, is the product catalog up to date?

Recently, the Log4j zero-day exploit has focused on the mind. More sophisticated organizations also record software versions and have a robust patching and testing cycle to ensure their estate remains current, and patches don't break their systems. With regard to Snowflake, no vulnerabilities for the Log4j zero-day exploit were found, and immediate communications were issued to reassure all customers.

Essential maintenance is the practice of updating code by applying patches. Preventative in nature, and, fortunately for us, it is all taken care of by Snowflake's weekly patch updates, but see the section on change bundles later in this chapter.

Protection involves physical and logical prevention of unauthorized access to hardware, networks, offices, infrastructure, and so on. Protection also establishes guardrails and software specifically designed to prevent unauthorized access, such as firewalls, anti-virus, network policies, multi-factor authentication, break-glass for privileged account use, regular environment scanning, and a host of other preventative measures, of which a subset are applicable for Snowflake.

Alerting relates to monitoring the protections applied and raising an alert when a threshold is reached, or a security posture is breached to inform the monitoring team as soon as possible and enable rapid response. The faster we detect and respond to an alert, the quicker we can recover from the situation.

We have identified Snowflake as the asset to protect as the subject of this book. While our AWS, Azure, and GCP accounts are equally important, they are largely outside this book's scope, but you will find that the same principles apply.

Returning to our focus on Snowflake, where can we find an agreed global suite of standards for reducing cyber risks to Snowflake?

Welcome to the National Institute of Standards and Technology (NIST).

National Institute of Standards and Technology (NIST)

NIST is a U.S. Department of Commerce organization that publishes standards for cybersecurity `www.nist.gov/cyberframework`.

The following quote is from the Snowflake security addendum at `www.snowflake.com/legal/security-addendum/`.

> *Snowflake maintains a comprehensive documented security program based on NIST 800-53 (or industry recognized successor framework), under which Snowflake implements and maintains physical, administrative, and technical safeguards designed to protect the confidentiality, integrity, availability, and security of the Service and Customer Data (the "Security Program").*

NIST is a large site; finding what you need isn't trivial. A great starting point for defining Snowflake security controls and is directly referenced in the Snowflake security addendum is at `https://csrc.nist.gov/Projects/risk-management/sp800-53-controls/release-search#!/controls?version=5.1`

Select Security Controls → All Controls to view the complete list, which at the time of writing runs to 322 separate controls. Naturally, only a subset is appropriate for securing our Snowflake accounts. Each control must be considered in the context of the capabilities Snowflake deliver.

Now that we have identified a trustworthy security control source and reviewed the content, our next step is to identify corresponding Snowflake controls. We then protect our Snowflake account by implementing identified changes, then apply effective monitoring to detect breaches with alerting to inform our support team, who will respond, fix, and remediate, leading to effective service recovery. After which, we can conduct our post-mortem and management reporting.

Our First Control

We may determine the following with an understanding of what our first control is to achieve.

- Our control is to limit permitted activities to prescribed situations and circumstances.

- Our control incorporates Snowflake's best-practice recommendations.

- Our control is to be documented, with implementation and monitoring implied to be delivered by an independent development team.

- There is to be a periodic review of the control policy.

From what you know of Snowflake and the preceding interpretation, you might say the first control (of many) relates to the use of Snowflake-supplied administrative roles and, therefore, could be used to put guardrails around the use of the most highly privileged Snowflake-supplied ACCOUNTADMIN role. If our organization has enabled ORGADMIN role, we might consider extending this control to cover both ACCOUNTADMIN and ORGADMIN roles. Alternatively, we might create a second control with different criteria.

Our organization may require additional factors to be considered when developing our security posture. Some examples include the following.

- Does this control affect Personally Identifiable Information (PII)?

- Does this control affect commercially sensitive information?

In defining control, we must also implement effective monitoring and alerting. This is covered in Chapter 6, which proposes a pattern-based suite of tools.

Our monitoring should encompass the following.

- Detect when the control is breached with an alert being raised, recognizing there are legitimate use cases when this occurs.

- Review alerts raised and determine appropriate action, whether to accept as legitimate, otherwise investigate, escalate, remediate, and repair.

- Record each alert along with the response and periodically report to management.

Having identified the control and its lifecycle, we can now define the specific details, which may look like the following.

SNOWFLAKE_NIST_AC1

Scope and Permitted Usage:

Snowflake supplied role ACCOUNTADMIN must not be used for general administrative use on a day to day basis but is reserved for those necessary operations where no other role can be used. Use of ACCOUNTADMIN role is expected to be pre-planned for system maintenance activities or system monitoring activities only, and is to be pre-notified to Operational Support team before use.

Snowflake Best Practice:

Snowflake recommends ACCOUNTADMIN usage to be assigned to at least two named individuals to prevent loss of role availability to the organization. We suggest a third generic user secured under break-glass conditions is assigned ACCOUNTADMIN role.

Implementation:

It is not possible to prevent the usage of ACCOUNTADMIN role by entitled users.

Review Period:

This policy is to be reviewed annually.

Sensitive Data:

This control does not relate to sensitive data.

Monitoring:

Any use of ACCOUNTADMIN role is to be detected within 5 minutes and notified by email to your_email_group@your_organization.com

Action:

Operational Support team identify whether ACCOUNTADMIN use is warranted, this might be for essential maintenance or software release both under change control. For all other ACCOUNTADMIN uses, identify user and usage, terminate session, escalate to line manager, conduct investigation, remediate and repair.

The exact wording differs according to the requirements, but this is an outline of a typical action.

Note in our initial requirements the implicit assumption of the control being defined by one group, with implementation being devolved to a second. This is good practice and in accordance with the segregation of roles and responsibilities.

Wash, rinse, and repeat for every required control. In conjunction with Snowflake subject matter experts (SMEs), your cybersecurity professionals typically define the appropriate controls.

Snowflake Security Options

Once our controls have been defined, we need to find ways to implement and later monitor, alert, and respond. This section addresses typical controls while explaining "why" we would want to implement each one. The list is not exhaustive. As Snowflake matures, new controls become evident, and as you see later, there is a gap in Snowflake monitoring recommendations.

Network Policies

Our first control, in my opinion, is the most important one to implement. Network policies restrict access to Snowflake from specific IP ranges. Depending upon our security posture and considering how we use our Snowflake accounts, we may take differing views on whether to implement network policies or not. A network policy is mandatory for highly regulated organizations utilizing Snowflake for internal business use only; for other organizations allowing third-party logins from the public Internet, probably not.

Whether required or not, knowing about network policies is wise. They provide defense-in-depth access control via an IP blacklist (applied first) and an IP whitelist (applied second), mitigating inappropriate access and data loss risks.

For this discussion, let's assume we require two network policies; the first ring-fences our Snowflake account allowing access to known IP ranges only, and the second enables access for a specific service user from a known IP range. The corresponding control might be expressed as "All Snowflake Accounts must prevent unauthorized access from locations outside of <your organization> internal network boundary except those from a cybersecurity approved location."

Note   Snowflake interprets CIDR ranges with a /0 mask as 0.0.0.0/0, effectively allowing open public Internet access.

Our network policies must, therefore, not have any CIDR ranges with /0.

When creating a network policy, it is impossible to declare a network policy that blocks the IP range from the currently connected session and attempt to implement the network policy. Also, when using network policies, the optional BLOCKED_IP_LIST is applied first for any connecting session, after which the ALLOWED_IP_LIST is applied.

We can now proceed with confidence in creating our network policies, knowing we cannot lock ourselves out. Our first task is to identify valid IP ranges to allow. These may be from a variety of tools and sources. Your cybersecurity team should know the list and approve your network policy implementation. Naturally, cybersecurity may wish to satisfy themselves that the network policy has correctly been applied after the first implementation and ensure effective monitoring after that. With our approved IP ranges available, we may only need to define the ALLOWED_IP_LIST.

You must issue the following commands to address our Snowflake account network policy.

```
USE ROLE securityadmin;

CREATE NETWORK POLICY <our_policy_name>
ALLOWED_IP_LIST = ( '<ip_address_1>', '<ip_address_2>', ... )
COMMENT = '<Organization Account Network Policy>';

ALTER ACCOUNT SET NETWORK POLICY = <our_policy_name>;
```

While we may have many account network policies declared, we can only have one account network policy in force at a given time.

To enable access from a known, approved Internet location, we require a second network policy, this time for a specific connection. We can declare as many network policies as we wish, each with a specific focus in addition to the single active account network policy. An example may be connecting Power BI from Azure to Snowflake on AWS. The east-west connectivity is from a known IP range, which your O365 administrators will know.

To address the Power BI network policy, we issue the following commands.

```
USE ROLE securityadmin;

CREATE NETWORK POLICY <powerbi_policy_name>
ALLOWED_IP_LIST = ( '<ip_address_1>', '<ip_address_2>', ... )
COMMENT = '<PowerBI Network Policy>';
```

We can assign our Power BI network policy using the following commands.

```
USE ROLE useradmin;

ALTER USER <powerbi_service_user>
SET NETWORK POLICY = <powerbi_policy_name>;
```

When the Power BI service user attempts to log in, their IP is checked against the ALLOWED_IP_LIST range for the assigned network policy.

Naturally, we want to be able to view the declared network policies.

```
USE ROLE securityadmin;

SHOW NETWORK POLICIES IN ACCOUNT;
SHOW NETWORK POLICIES;

DESCRIBE NETWORK POLICY <powerbi_policy_name>;
```

Finally, we may wish to remove a network policy.

```
USE ROLE securityadmin;

ALTER ACCOUNT UNSET network_policy;

ALTER USER <powerbi_service_user> UNSET network_policy;

DROP NETWORK POLICY IF EXISTS <our_policy_name>;
DROP NETWORK POLICY IF EXISTS <powerbi_policy_name>;
```

Using these commands, we can manage our network policies. Monitoring and alerting are addressed later. There are a few hoops to jump through in common with implementing other monitoring patterns.

For monitoring in Chapter 6, this setting is referred to as SNOWFLAKE_NIST_NP1.

Preventing Unauthorized Data Unload

Our next control might be to prevent data from being unloaded to user-specified Internet locations. Of course, your security posture and use cases may need to allow data to be unloaded, in which case this control should be ignored. User-specified Internet locations can be any supported endpoint. Effectively, your user determines where they wish to unload data; for most organizations, it could be a primary source of data leaks.

We can implement this control using the following commands.

```
USE ROLE accountadmin;

ALTER ACCOUNT SET prevent_unload_to_inline_url = TRUE;
```

Equally, the following removes this control.

```
ALTER ACCOUNT UNSET prevent_unload_to_inline_url;
```

For monitoring in Chapter 6, this setting is SNOWFLAKE_NIST_AC2.

Restricting Data Unload to Specified Locations

Our next control might restrict data unloads to specified, system-mapped Internet locations. Of course, your security posture and use cases may need to allow data to be unloaded to any user-defined location, in which case this control should be ignored. System mapped Internet locations can be any supported endpoint mapped via storage integration only, thus restricting data egress to known locations.

The advantages of implementing this control should be obvious. The rigor associated with software development ensures the locations are reviewed and approved before storage integrations are implemented.

We can implement this control using the following commands.

```
USE ROLE accountadmin;

ALTER ACCOUNT SET require_storage_integration_for_stage_creation  = TRUE;
ALTER ACCOUNT SET require_storage_integration_for_stage_operation = TRUE;
```

Equally, the following removes this control.

```
ALTER ACCOUNT UNSET require_storage_integration_for_stage_creation;
ALTER ACCOUNT UNSET require_storage_integration_for_stage_operation;
```

For monitoring in Chapter 6, this setting is SNOWFLAKE_NIST_AC3.

Single Sign-On (SSO)

When our Snowflake account is provisioned, all users are provisioned with a username and password. You know that usernames and passwords are vulnerable to bad actors acquiring our credentials, potentially leading to data loss, reputational impact, and financial penalties. Our cybersecurity colleagues rightly insist we protect our credentials. SSO is one of the tools we can use where we no longer rely upon username and password but instead authenticate via centralized tooling.

Snowflake SSO documentation is at https://docs.snowflake.com/en/user-guide/admin-security-fed-auth-overview.html. Implementing SSO relies upon having federated authentication. This section explains the steps required to integrate ADFS and some troubleshooting information. Naturally, SSO integration varies according to available tooling, so apologies in advance for those using alternative SSO providers, space (and time) do not permit a wider examination of all available options.

Note   Snowflake supports SSO over either Public or PrivateLink but not both at the same time; see the following for more information on PrivateLink.

Security Assertion Markup Language (SAML) is an open standard that allows identity providers (IdP) to pass authorization credentials to service providers (SPs). In our case, Snowflake is the service provider, and Microsoft ADFS is the identity provider. Due to the segregation of roles and responsibilities in organizations, setting up SSO requires both SME knowledge and administrative access to ADFS.

Step 1. Configure Identity Provider in ADFS

The first step is to generate an IdP certificate for your organization. For that, you might need to work with a subject matter expert with the right experience, and that expert might benefit from the guidance provided by Snowflake at `https://docs.snowflake.com/en/user-guide/admin-security-fed-auth-configure-idp.html`.

The IdP certificate you generate in this step is then used in step 3.

Step 2. Configure Snowflake Users

Snowflake users must match the IdP declaration. We assume email addresses are the currency unit identifying Snowflake users and ADFS users. When creating a new Snowflake user, you must set LOGIN_NAME to match the ADFS user, the password to NULL, and MUST_CHANGE_PASSWORD to FALSE. Note the DISPLAY_NAME can be different and substitute your own information.

```
USE ROLE useradmin;

CREATE USER IF NOT EXISTS abc
PASSWORD              = NULL
DISPLAY_NAME          = 'Abc Def'
LOGIN_NAME            = 'abc.def@your_adfs_domain.com'
MUST_CHANGE_PASSWORD = FALSE
DEFAULT_ROLE          = dummy_role;
```

Alternatively, the following is for an existing user.

```
USE ROLE useradmin;

ALTER USER abc SET
LOGIN_NAME            = 'abc.def@your_adfs_domain.com',
PASSWORD              = NULL
MUST_CHANGE_PASSWORD = FALSE;
```

Step 3. Specify IdP Information

Using the IdP certificate, you can now set the SAML_IDENTITY_PROVIDER value: https://docs.snowflake.com/en/sql-reference/parameters.html#saml-identity-provider. Note the certificate is a very long alphanumeric string.

```
USE ROLE accountadmin;

ALTER ACCOUNT SET saml_identity_provider =
{
  "certificate": "<your IdP certificate from Step 1 here>",
  "ssoUrl": "https://<your ADFS URL here>",
  "type"  : "ADFS",
  "label" : "SSO"
};
```

Note the label can only contain letters and numbers. The label cannot contain spaces and underscores.

Like all global service layer parameters, the current value can be accessed using the SHOW command.

```
USE ROLE accountadmin;

SHOW PARAMETERS LIKE 'SAML_IDENTITY_PROVIDER' IN ACCOUNT;
```

Step 4. Enable SSO

Once the preceding steps have been completed, SSO can be enabled.

```
USE ROLE accountadmin;

ALTER ACCOUNT SET sso_login_page = TRUE;
```

The SSO sign-in box appears when logging into Snowflake, as shown in Figure 4-2.

Figure 4-2. Snowflake SSO enabled login

Future attempts to log in to Snowflake should produce the login dialog shown in Figure 4-2, where you should now be able to click the "Sign in using SSO" button. A dialog box appears, prompting for username and password, as shown in Figure 4-3.

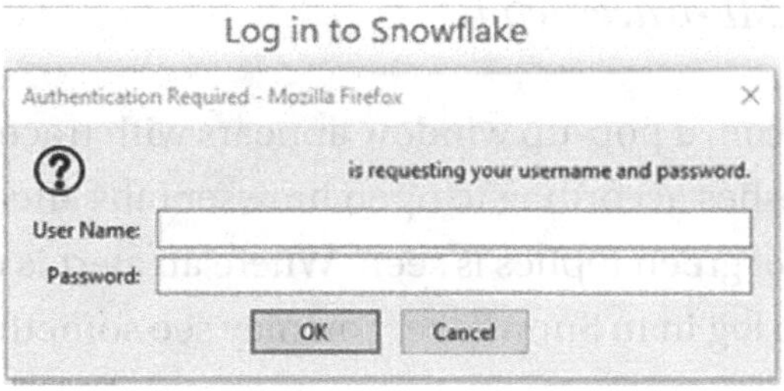

Figure 4-3. Snowflake login dialog box

Enter credentials for your organization domain login, which should authenticate against your IdP and allow access to Snowflake.

Troubleshooting

Occasionally, things go wrong with configuring SSO. This section provides troubleshooting information. I cannot cover every scenario but provide information on tools to help diagnose the root cause.

Using Firefox, search for "saml tracer firefox". You should see results similar to those in Figure 4-4.

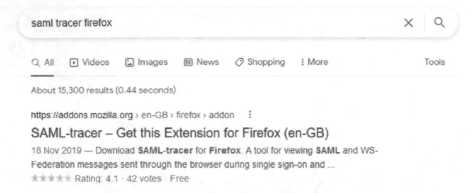

Figure 4-4. Firefox SAML-tracer

Download and install the browser SAML-tracer add-in, and when complete, in the top right-hand corner of your Firefox browser, look for the icon highlighted in Figure 4-5.

Figure 4-5. Firefox SAML-tracer icon

When you click the icon, a pop-up window appears with trace information. Note the trace automatically refreshes according to open browser tabs and pages displayed. For successful traces, a wall of green replies is seen. Where an alert is raised, the response is amber or red. Attempt to log in to Snowflake. You may see something like what's shown in Figure 4-6.

Figure 4-6. SAML-tracer messages

Click each message and navigate to the SAML tab at the bottom of the tracer, where you see something similar to Figure 4-7.

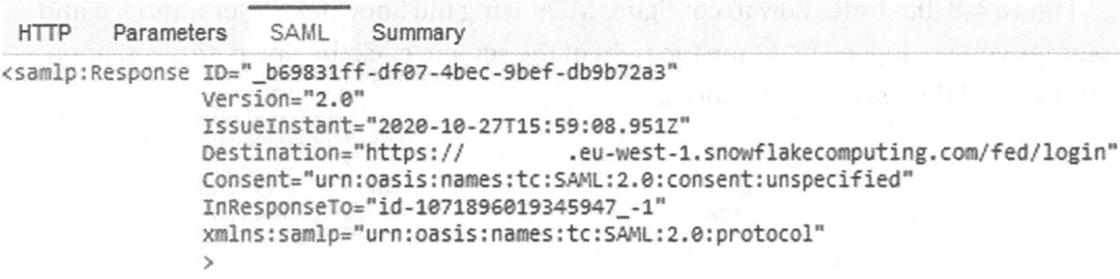

| HTTP | Parameters | SAML | Summary |

```
<samlp:Response ID="_b69831ff-df07-4bec-9bef-db9b72a3"
                Version="2.0"
                IssueInstant="2020-10-27T15:59:08.951Z"
                Destination="https://        .eu-west-1.snowflakecomputing.com/fed/login"
                Consent="urn:oasis:names:tc:SAML:2.0:consent:unspecified"
                InResponseTo="id-1071896019345947_-1"
                xmlns:samlp="urn:oasis:names:tc:SAML:2.0:protocol"
                >
```

Figure 4-7. *SAML-tracer message content*

You may need to work with our ADFS administrator to resolve the issue. Snowflake may also provide some clues though you need to enable the Snowflake Account Usage store, as described in Chapter 6, first. The following example code identifies login attempts along with the error code.

```
USE ROLE accountadmin;

SELECT *
FROM    snowflake.account_usage.login_history
WHERE   error_code IS NOT NULL
ORDER BY event_timestamp DESC;
```

Identify the error_code from the information presented at https://docs.snowflake.com/en/user-guide/errors-saml.html.

Further information is at https://docs.snowflake.com/en/user-guide/admin-security-fed-auth-use.html#using-sso-with-aws-privatelink-or-azure-private-link.

Multi-Factor Authentication (MFA)

MFA is not currently enabled by default. Each user must configure MFA manually. Snowflake documentation covering all aspects of MFA is at https://docs.snowflake.com/en/user-guide/security-mfa.html.

Note   Snowflake strongly recommends users with an ACCOUNTADMIN role be required to use MFA.

Figure 4-8 illustrates how to configure MFA using the Snowflake User Interface and assumes you are logged in. At the top right of the screen, click the down arrow to open Options, and then select Preferences.

Figure 4-8. *Snowflake user interface options*

After confirming your email address, click the "Enroll in MFA" link, as shown in Figure 4-9.

Multi-factor Authentication

Enroll in MFA, edit the phone number associated with your MFA account.

Status   Not Enrolled Enroll in MFA

Phone   -

Figure 4-9. *Snowflake user interface preferences*

Download and install DUO Mobile onto your phone. A QR code is sent. Scan the QR code. Note you may be prompted to upgrade your phone operating system. Once enrolled, refresh the Preferences option. You see that your phone number is displayed.

System for Cross-domain Identity Management (SCIM)

Without SCIM integration, you cannot automate the creation and removal of Snowflake users and roles maintained in Active Directory. Supported IdPs are Microsoft Azure AD and Okta. SCIM integration automates the exchange of identity information between two endpoints.

Note   I use the term *endpoint* to describe access points to any network that malicious actors can exploit.

In this example, we integrate Snowflake and Azure AD to automatically provision and de-provision users and roles, thus automating user and RBAC provisioning via Azure AD. Snowflake documentation is at `https://docs.snowflake.com/en/user-guide/scim-azure.html`. Figure 4-10 illustrates interactions between the various components required.

Figure 4-10. *SCIM schematic*

The first step in Snowflake is to provision a role in the following code called aad_provisioner, specifically created for Azure AD provisioning, and grant entitlement for the new role to create users and create roles.

```
USE ROLE accountadmin;

CREATE OR REPLACE ROLE aad_provisioner;

GRANT CREATE USER ON ACCOUNT TO ROLE aad_provisioner;
GRANT CREATE ROLE ON ACCOUNT TO ROLE aad_provisioner;
```

We also assign our new role to ACCOUNTADMIN to follow Snowflake's best practices.

```
GRANT ROLE aad_provisioner TO ROLE accountadmin;
```

Our next step is to establish security integration, a Snowflake object providing an interface between Snowflake and, in our example, Azure AD.

```
CREATE OR REPLACE SECURITY INTEGRATION aad_provisioning
TYPE        = SCIM
SCIM_CLIENT = 'azure'
RUN_AS_ROLE = 'AAD_PROVISIONER';
```

Now generate the SCIM token to update in Azure AD. More information is at `https://docs.snowflake.com/en/sql-reference/functions/system_generate_scim_access_token.html`.

```
SELECT system$generate_scim_access_token ( 'AAD_PROVISIONING' );
```

The returned value should look something like the value shown in Figure 4-11.

Figure 4-11. *SCIM access token*

For the remaining steps, refer to Azure documentation at `https://docs.microsoft.com/en-us/azure/active-directory/saas-apps/snowflake-provisioning-tutorial`, update Azure AD with the token generated in Figure 4-11 then test.

PrivateLink

PrivateLink is an AWS service for creating VPC direct connections with secure endpoints between the AWS account and Snowflake without traversing the public Internet. The corresponding component for direct connections with secure endpoints between AWS accounts and on-prem is AWS Direct Connect, as illustrated in Figure 4-12.

Figure 4-12. *Secure network schematic*

PrivateLink requires Snowflake support assistance and may take up to two working days to provision. Please refer to the documentation at `https://docs.snowflake.com/en/user-guide/admin-security-privatelink.html` for further information. Note that your corporate network configuration may require changes to allow connectivity.

If you experience connectivity issues, you may also need to use SnowCD for diagnosis and open ports on firewalls, specifically 80 and 443. Further information is at `https://docs.snowflake.com/en/user-guide/snowcd.html#snowcd-connectivity-diagnostic-tool`.

Data Encryption

Data encryption is a primary means of ensuring our data remains safe. Snowflake utilizes underlying cloud service provider storage, which for AWS is S3 buckets. We have not yet discussed *how* our data is protected in S3. We assume everyone knows *why*.

Snowflake takes security seriously, very seriously indeed. What is covered in the next few pages cannot do justice in explaining the immense work Snowflake has put into security. This whitepaper requires registration to download and is well worth investing the effort to read `www.snowflake.com/resource/industrial-strength-security-by-default/`.

Tri-Secret Secure

By default, Snowflake encrypts data using four key levels.

- The root key maintained by the Snowflake hardware security module

- The account master key is individually assigned to each Snowflake account

- The table master key is individually assigned to each object storing data

- The file key individually assigned to each S3 file

Figure 4-13 is from Snowflake documentation at `https://docs.snowflake.com/en/user-guide/security-encryption-manage.html#understanding-encryption-key-management-in-snowflake`. It shows the relationship between the keys.

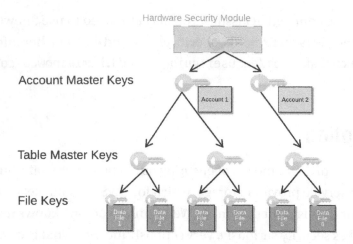

Figure 4-13. *Default key hierarchy*

Periodic Rekeying

One benefit of rekeying is the total duration for which a key is actively used is limited, thus making any external attack far more difficult to perpetrate. Furthermore, periodic rekeying allows Snowflake to increase encryption key sizes and utilize improved encryption algorithms since the previous key generation occurred. Rekeying ensures that all customer data, new and old, is encrypted with the latest security technology.

Periodic rekeying replaces active keys with new keys on a 30-day basis, retiring the old account master key and table master keys automatically, behind the scenes, no fuss, no interaction required, all managed automatically and transparently by Snowflake. Periodic rekeying requires the Enterprise Edition and higher. Further information is at https://docs.snowflake.com/en/user-guide/security-encryption. html#encryption-key-rotation.

Following NIST recommendations, Snowflake ensures all customer data, regardless of when the data was stored, remains encrypted with the latest security technology.

We may find our internal data classifications and protection requirements mandate periodic rekeying as an essential guardrail, especially where data classification information is unavailable. Setting periodic rekeying is a best practice and should be adopted wherever possible.

The following command enables periodic rekeying.

```
USE ROLE accountadmin;
```

```
ALTER ACCOUNT SET periodic_data_rekeying = TRUE;
```

Naturally, we would want to monitor the account setting remains in force, which I discuss in Chapter 6.

Customer Managed Key (CMK)/Bring Your Own Key (BYOK)

A further optional guardrail is available to implement for customers using Business Critical Edition and higher. You have seen how Snowflake protects our data using a hierarchy of keys, and there is one further level of protection offered: the ability for the customer to add their own key, otherwise known as CMK or BYOK. The advantage of implementing CMK is the ability to disable the account where the CMK has been deployed at customer discretion by disabling the key. Note that the key is required to access data within the account, and without which, Snowflake cannot help.

Before getting into the details of implementing CMK, you must consider how the customer manages the key. CMK is generated and must be stored securely. Not all organizations have a team to manage encryption keys, so how can a locally managed key be created and maintained securely?

Using AWS as our platform, these steps provide an outline.

1. Generate the key.

    a. Create a custom policy to restrict the deletion of the key

    b. Create a custom IAM role and attach the policy

    c. Create Key Management Store (KMS) CMK as

        i. Symmetric KMS key

        ii. Name and Description

        iii. Labels

        iv. Choose the IAM role to be attached

        v. Set usage permissions

        vi. Review and create CMK

2. Share the KMS CMK ARN with Snowflake.

    a. Raise a support ticket

3. Snowflake provides a key policy code to be added to the key policy

4.   Snowflake confirms account rekeying is complete

Naturally, this guide cannot be prescriptive and does not inform controls around AWS account security.

S3 Security Profile and Scanning

Strictly speaking, AWS S3 bucket security is not a Snowflake-specific issue, but in the context of external stages is mentioned to provide a holistic view of application security. Referenced throughout this book, S3 is a gateway into Snowflake.

Every organization has a policy or template security setting for S3 buckets with appropriate security scanning tools such as Checkpoint Software Technologies CloudGuard. Naturally, configuration and use of any scanning tools are beyond the scope of this book but are mentioned for completeness because artifacts loaded into S3 must also be protected from unauthorized access.

Penetration Test (Pen Test)

The subject of penetration testing occasionally reappears in organizations where new team members, management, and oversight look to reaffirm we have all our controls in place and pen testing is up to date. From a delivery perspective, pen testing is paradoxically out of our hands. It is not enough to make this statement without explaining why, and this section provides context and reasoning.

Understanding the objectives of pen testing provides part of the answer, with this definition from www.ncsc.gov.uk/guidance/penetration-testing.

Note   A method for gaining assurance in the security of an IT system by attempting to breach some or all that system's security, using the same tools and techniques as an adversary might.

Stress testing the Snowflake environment is naturally in Snowflake Inc.'s best interest, and much continual effort is expended to ensure Snowflake remains secure. But in so doing, the tools and techniques must remain confidential. Any bad actor would love to know which tools and techniques are deployed, and where gaps in coverage may identify weakness or opportunity. The last thing any product vendor needs is a zero-day exploit. Does anyone remember Log4j?

We must rely upon another means of validating the Snowflake approach as the direct understanding of tooling and approach are inaccessible. Fortunately, Snowflake provides external verification and attestations that we can rely upon, as shown in Figure 4-14.

INFOSEC & COMPLIANCE

All reports, attestations, documentation, and certifications

Third Party Reports & Certifications

- Snowflake SOC 2 Type II Report
- Snowflake SOC 1 Type II Report
- Snowflake PCI-DSS-AOC-Final Report
- HIPAA Report (proving ability to enter BAA)
- Snowflake's ISO 27001 Certificate
- FedRAMP Moderate (on OMB MAX)
- CyberGRX Report
- Penetration Test Results

ISO/IEC 27001

HIPAA

HITRUST in Progress

PCI-DSS

 FedRAMP

FedRAMP
Moderate

(Available from OMB MAX)

Snowflake's Policy Documentation

- Snowflake Security Policy
- https://www.snowflake.com/legal/ for Acceptable Use, Support, and more

Snowflake Internal Controls & Testing

- DRP, BCP, and Information System Contingency Plans
- Security Incident Process
- Staff Training, Onboarding, and Access Policies

Snowflake Self Assessment Reports

- CAIQ
- SIG Lite
- Red Team Pen Tests

SOC 2 Type II
12 Month Coverage Period
SOC 1 Type II
6 Month Coverage Period

Figure 4-14. *Snowflake security attestations*

But this section relates to pen tests. Apart from the generally available proofs, *how* can our organization be assured at a detailed level?

Figure 4-15 provides an overview of the actions Snowflake performs on our behalf.

INFRASTRUCTURE SECURITY & MONITORING

How is the Snowflake Infosec Team monitoring the service?

Snowflake's Internal Critical Security Controls Dashboard - Real-Time Risk Visibility

- Access Control, Security Assessment & Authorization, Configuration Management, Security Awareness, and so on—all represented on a single Dashboard

- Real-time monitoring of data loaded into Snowflake from internal and other relevant data sources

Snowflake uses CIS Benchmark Templates for configuration hardening

- Service configuration information is collected centrally in Snowflake

- Continuously and automatically tracked, unplanned changes cause alerts

- Part of Snowflake Security Compliance Team's Dashboard

Snowflake performs 7–12 third-party penetration tests per year

- Comprehensive Web Application Penetration Test – Annually

- Internal Network Penetration Test – Annually

- Major Functionality Penetration Tests – As major functionality is released as part of the SDLC

Snowflake performs weekly vulnerability scans on infrastructure

- Vulnerabilities are remediated per Security Policy

- Remediation trends tracked using Snowflake

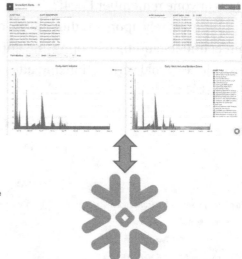

Figure 4-15. *Snowflake security actions*

Contractual negotiations between organizations and Snowflake Inc. include a provision for disclosing details of pen tests conducted to named individuals. The named individuals should be cybersecurity specialists as the information disclosed is highly sensitive. Therefore, the recipient list must be kept short.

The final body of evidence we can rely upon is Snowflake's field security CTOs, specialists available to explain those topics of interest to our cybersecurity colleagues in presentation and document formats.

With this explanation in mind, I trust there is sufficient evidence to satisfy our immediate concerns while providing information on how to dig deeper. Finally, more information can be found at `www.snowflake.com/product/security-and-trust-center/` from which some of the preceding content is sourced.

Time Travel

A brief discussion of the Time Travel account security feature is important. As discussed in Chapter 3, at a minimum, you must ensure your production environments have the Time Travel feature set to 90 days. It is recommended that all other environments have it enabled for the occasional mishaps during development. Yes, we have all been there. For monitoring in Chapter 6, this setting is SNOWFLAKE_NIST_TT1.

Change Bundles

Change bundles make behavior changes to your application code base and are pre-announced to registered users. See https://community.snowflake.com/s/article/Pending-Behavior-Change-Log.

Note   Change bundles are applied at the account level; therefore, a single account holding both production and non-production environments should be given additional consideration before applying change bundles.

There are circumstances where a change bundle may need to be enabled as a precursor for other activities. At the time of writing, for example, to participate in some private previews, a named change bundle must be enabled. To manage change bundles, use change bundle '2021_10' as an example to run the following commands one line at a time.

```
/****************************/
/**** Enable change bundle ****/
/****************************/
USE ROLE accountadmin;

-- Check BEFORE status
SELECT system$behavior_change_bundle_status('2021_10');

-- Apply change bundle
SELECT system$enable_behavior_change_bundle('2021_10');

-- Check AFTER status
SELECT system$behavior_change_bundle_status('2021_10');
```

Naturally, the AFTER status should return ENABLED, after which point testing can begin. Do not forget to remove all test objects and roles created beforehand

To disable the change bundle, run the following commands one line at a time.

```
/****************************/
/**** Disable change bundle ****/
/****************************/
USE ROLE accountadmin;
```

```
-- Check BEFORE status
SELECT system$behavior_change_bundle_status('2021_10');

-- Apply change bundle
SELECT system$disable_behavior_change_bundle('2021_10');

-- Check AFTER status
SELECT system$behavior_change_bundle_status('2021_10');
```

Naturally, the AFTER status should return DISABLED.

Note   Always ensure test cases are removed after testing.

Summary

This chapter began by identifying how and why cybersecurity attacks occur, available resources to identify security requirements, defining corresponding Snowflake cybersecurity controls, and some examples of implementing controls with sample code.

You also looked at several Snowflake guardrails provided to allow us to control our environments, along with a troubleshooting guide for SSO.

The discussion included explanations of underlying storage security, focusing on AWS S3. You also looked at penetration testing, explaining the security context, actions Snowflake conduct behind the scenes on our behalf, and the means available to satisfy ourselves. Snowflake is inherently secure.

Finally, having dipped your toes into a very deep subject, and hopefully, you were given a decent account, let's move to Chapter 5. Stay tuned!

CHAPTER 5

Role-Based Access Control (RBAC)

You now know that security is baked into Snowflake from the ground up. In this chapter, you discover how everything is an object and how roles govern everything—object definition, object ownership, and object interaction.

I do not cover every aspect of *role-based access control* (RBAC) because the subject is complex and subtly different from other legacy *relational database management system* (RDBMS) implementations. In Snowflake, there are no shortcuts. We call out every entitlement explicitly. There are no assumptions or wildcard entitlements such as SELECT ANY OBJECT.

Through many years of practical hands-on experience, I have seen many poor ways to configure RBAC and discovered a few good patterns that work regardless of RDBMS (albeit with some "gotchas" along the way). For those of you learning Snowflake without any prior RDBMS experience, I hope to impart best practices as we walk through practical examples.

Note   The sample code is intended to be a hands-on tutorial. The most benefit is gained by running the code interactively.

Please ensure your account is the Business Critical Edition. The examples here depend on Business Critical Edition features. Otherwise, please re-register using a different email address by selecting the Business Critical option at `https://signup.snowflake.com`.

Where possible, I have provided context to explain why we do things the way we do. A bland walk-through of syntax does not a programmer make, nor will information, knowledge, and wisdom be imparted.

© Andrew Carruthers 2022
A. Carruthers, *Building the Snowflake Data Cloud*, https://doi.org/10.1007/978-1-4842-8593-0_5

Out-of-the-Box RBAC

In Snowflake, everything is an object—even roles you have heard about and perhaps seen in action elsewhere. I assume you have worked through Chapter 3 before landing here, and you are familiar with Snowflake's basic concepts and principles.

Dispelling Myths

Prior experience is usually a good thing to bring to Snowflake, particularly hands-on, where experience can be tested and proven either valid or invalid. Assumptions are not usually good to bring into Snowflake. A more measured approach is to test assumptions. Caveat emptor. Buyer beware. Let's address some misconceptions next.

Grants to Roles

Many readers with the experience of legacy RDBMS may think we have a full understanding of RBAC yet have operated under a hybrid model where entitlement can be granted in two ways: directly to a user or to a role that may be granted to a user. With Snowflake, only grants to a role can be made, and no grants directly to a user can be made, as Figure 5-1 illustrates.

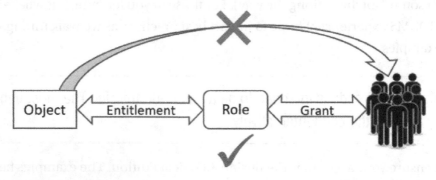

Figure 5-1. *Object entitlement grants*

Please prove the preceding assertion because the distinction is important and underpins the Snowflake RBAC model.

We can therefore say with confidence. Snowflake only implements an RBAC model, not an entitlement-based model.

We might ask ourselves why the direct entitlement and RBAC role model distinction are important. Consider how organizations manage their workforce with many monthly joiners, movers, and leavers. Part of every organization's natural lifecycle is the flow of staff. While joiners and leavers are relatively simple to manage, our most complex use case is where staff change positions internally in our organization.

From a cybersecurity perspective, we want to automate the provisioning of an entitlement according to the business or technical role a user holds. One way we achieve this is by Active Directory group (AD group) membership. Let's assume we have AD groups that automatically provision a corresponding Snowflake user, along with one or more Snowflake roles. You can see how changing positions in our organization results in AD group membership; therefore, corresponding Snowflake role assignments should change automatically.

If Snowflake allowed direct entitlement grants to users, these would occur in the Snowflake account with no reference to AD groups. Anyone changing their position in our organization would retain entitlement to which their new position may not be allowed, therefore presenting a security breach and requiring manual cleanup, not something we ought to be doing.

Super User Entitlements

In legacy RDMBS, it is usual for elevated roles to be able to access data to which their role has not been explicitly granted; for example, some elevated entitlement roles have SELECT ANY TABLE granted by default, and removing this vendor implemented capability is unwise either potentially rendering support contract invalid due to interference with core capability, or limiting vendor support options when troubleshooting. Likewise, removing entitlement without advising those using the elevated privilege role is confusing.

Better to not have an issue to solve in the first place. In Snowflake, assuming the user has been granted an elevated role and is using the elevated role, unless entitlement has been explicitly granted to select from an object, then data cannot be seen. I will explain this more later, but for now, it is enough to know the distinction.

System Migration

Do not assume RBAC models are equivalent when porting applications from legacy RDBMS. They are not, with Snowflake implementing a stricter RBAC model than many other vendors. Do take time to understand RBAC differences and ensure your Snowflake RBAC model reflects your desired behavior. Understanding the differences make your system migration run much smoother than you might otherwise experience and you are better informed to adjust and adapt if required.

Tenancy Model

By default, of the box, Snowflake is provisioned with two or more, but certainly less than a handful, of users with an ACCOUNTADMIN role whose responsibility is to make the new account conform to the organization's agreed controls discussed in Chapter 4.

Recognizing there is an awful lot of work in specifying, designing, and developing controls and effective management, there is one further concept that our organizations may wish to implement: whether to implement separate accounts for each federated line of business (LoB) in an operating division (single tenant), or whether to integrate many federated LoBs in an operating division in a single account (multi-tenant). The decision is not "one size fits all," and both models can co-exist. It comes down to the preferred operating model or the inherited legacy if the tenancy model was not initially available. To fully discuss the tenancy options mode, we must first examine the Snowflake default roles, from which you can understand the fundamental concepts and future issues that may arise.

Snowflake Default Roles

Snowflake supplies several out-of-the-box roles. See the Snowflake documentation at `https://docs.snowflake.com/en/user-guide/security-access-control-overview.html#system-defined-roles`. For many use cases, the Snowflake default roles are sufficient to manage any account (specifically single tenant, however, for a multi-tenant environment, the use cases will differ).

Figure 5-2 illustrates the Snowflake role hierarchy.

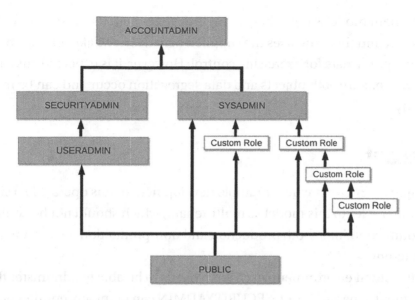

Figure 5-2. *Snowflake supplied roles*

Following are descriptions of the roles shown in Figure 5-2.

- ORGADMIN (not shown): Allows centralized account administration

- ACCOUNTADMIN: Account parameter declaration, guardrails, and so forth

- SYSADMIN: Database, warehouse, and other object administration

- SECURUTYADMIN: Role and entitlement administration

- USERADMIN: User administration

- PUBLIC: Pseudo-role that can own objects and is granted to all users

Custom roles are those we create to manage our applications. We work through a practical example later in this chapter.

SIngle Tenant

A single tenant model is a Snowflake account with only one LoB implementing their application. This is not to be confused with a single account containing both production and non-production environments which is distinctly different and discussed separately later in this chapter.

A single tenant model is one where a single development team controls what goes into the account and therefore uses the supplied default Snowflake roles as these are sufficient and appropriate for exercising control. However, it is expected custom roles are implemented to ensure both objects and data segregation occur and can be maintained independently.

Multi-Tenant

A more complex scenario is where multiple development teams operate in a single account. Again, I refer to this model as multi-tenant, which should not be confused with a single account containing both production and non-production environments that are distinctly different.

In a multi-tenant environment, each team needs to be able to administer their own environment separately. But, SECURITYADMIN can grant any entitlement and role to any other role or user regardless of tenancy. There is no way to devolve granting entitlement to a subset of components in a multi-tenant-based model. SECURITYADMIN cannot be decomposed. The only way forward is for all tenants to agree on how SECURITYADMIN role is used and by whom, which requires more coordination between teams.

Furthermore, when using the ACCOUNTADMIN role, all account level settings apply to all tenants. For example, a network policy may need to be set where each tenant requires different IP ranges opening according to their tooling. In some cases, there may be opposing views on account level settings. For example, one tenant may require data to be unloaded to S3. Another tenant may require the prevention of data unloading to S3. Another example is change bundles (Snowflake feature changes or bug fixes) applied at the account and not at the tenant level. We recently experienced a bug in a change bundle affecting result sets, caught by our operations support team, as change bundles are an "all or nothing" proposition for the account.

Note   SECURITYADMIN can grant any role to any user.

Assuming all competing requirements can be resolved, multi-tenant accounts can work successfully. The RBAC approach is to create subsidiary administrative roles wherever possible, mimicking Snowflake roles but restricted to a subset of entitlement, one set of custom roles for each tenant.

Account Management

The Snowflake model does not enforce one particular view of the world. It is possible to have single tenant, multi-tenant, separate production, or co-located production with non-production accounts. One size does not fit all, and every organization is free to make its own decisions.

From experience with legacy RDBMS implementations, the traditional dev, test, UAT, production, and DR account approach is familiar and easily explained. We have been doing RDBMS implementations this way for decades, and in highly regulated environments, there is a natural tendency to segregate production from non-production environments. But Snowflake allows us to think differently and conflate our segregated environments into a single account, thus providing the same capability but in a higher-level container. This approach reduces the number of accounts we need to manage.

We might also have scenarios where dummy data cannot be created or live data cannot easily be masked, leading to UAT environments being connected to a production environment. Our account strategy must consider all known options, and RBAC caters to new options that arise over time.

Considering our previous comments on both challenges we will face and regulatory requirements, the decision is yours. However, if unsure, we advise separating production from non-production because your role naming convention will be simpler, particularly if fully segregating into dev, test, UAT, and production.

Access Model

Having given due consideration to the potential issues with tenancy models and segregation of production from non-production environments, we move on to addressing how data can be segregated while providing a flexible, integrated approach. The following suggestion does not prescribe one approach over another.

Typically, we have at least three roles interacting with our data at any time, plus an object owner role and possibly a data masking role and row-level security roles. This section articulates how data moves in a typical application and how the different roles facilitate interaction with system objects while maintaining entitlement separation: the concepts become clearer as you work through the examples.

Discretionary Access Control

According to Wikipedia, *discretionary access control* is "a means of restricting access to objects based on the identity of subjects and/or groups to which they belong" (`https://en.wikipedia.org/wiki/Discretionary_access_control`). In other words, we can decide who has access to what, a theme I will unpack later. Each Snowflake object is owned by a role, and entitlement can be granted against an object to one or more roles, see `https://docs.snowflake.com/en/user-guide/security-access-control-overview.html`.

Overview

Figure 5-3 illustrates a simple architecture that hides a degree of technical complexity explained in corresponding sections. Your real-world implementation will certainly be a lot more complex. With three objects in-scope, we load data on the left, process data in the middle and present data for a report on the right. For now, we will ignore both data masking and row-level security roles, returning to address these in a later chapter.

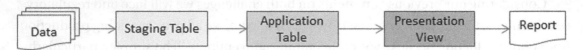

Figure 5-3. *Abstract object layout*

Figure 5-3 only specifies objects, not databases, schemas, or roles, which overlay in Figure 5-4, showing the container hierarchy: Account ➤ Database ➤ Schema ➤ Object (Table or View) and the three associated roles for schema interaction, along with the object owner role. For clarity's sake, we define objects with the object owner role and manipulate the data in the objects using the three data roles.

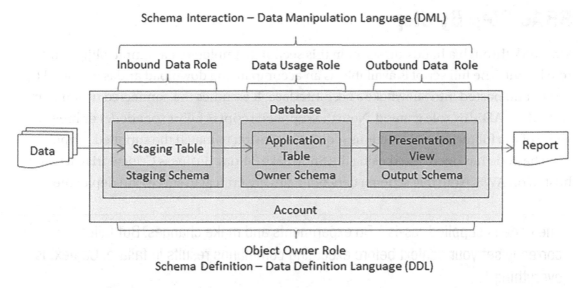

Figure 5-4. *Sample object and role layout*

Figure 5-4 differentiates between object creation and object usage. You must never conflate object creation and object usage roles; otherwise, we lose control of our system configuration management. In other words, the objects must remain immutable (DDL) except for authorized change, whereas the content is mutable through normal system interaction (DML).

This brings us to a very important point. Since Snowflake is a data warehousing product typically implemented for bulk data load, manipulation, and subsequent read, should we allow ad-hoc data correction in Snowflake, or should we defer to only allow data modification in source systems? The answer is clear. If we allow data modification in our data warehouse, then we ultimately become responsible for all such data modifications, and our data warehouse is inconsistent with the source. While we may identify and record data quality issues by programmatic means, we must not exclude data and always faithfully represent all received data regardless of quality. We return to this theme later in this book.

Let's now turn to set up a sample environment and get our hands dirty with some code. The following sample code was developed incrementally: build "something," then test, wash, rinse, repeat. Don't expect "big bang" implementations to work the first time. Instead, bite off small pieces and incrementally build out application capability. Do not forget to integrate your test cases; it is very important. Although omitted in this example, it is crucial for continuous integration.

RBAC Step By Step

You work through a hands-on script in this section to implement a sample object and role layout. The full script is available as an accompanying download and is intended to be cut and pasted into Snowflake's user interface. It assumes the connected user has an ACCOUNTADMIN role granted. Note the supplied script is fully expanded. For brevity, some of the following code sections require expansion to match the supplied script.

The code has been tested using the Snowflake 30-day Business Critical trial account hosted on AWS. Note that schema dependencies are critical to successful deployment.

The code is supplied "as-is." Do experiments and make changes. But failure to correctly set your context before executing commands results in failure. Context is everything.

Preamble

Our script starts with some declarations enabling later automation. Remembering the scope of this book is limited, it is left for you, dear reader, to expand upon what we present. Also, note the script can be modified to become a test harness for your own purposes later with minimal intervention. Just change the SET declarations.

```
SET test_database       = 'TEST';
SET test_staging_schema = 'TEST.staging_owner';
SET test_owner_schema   = 'TEST.test_owner';
SET test_reader_schema  = 'TEST.reader_owner';
SET test_warehouse      = 'test_wh';
SET test_load_role      = 'test_load_role';
SET test_owner_role     = 'test_owner_role';
SET test_reader_role    = 'test_reader_role';
SET test_object_role    = 'test_object_role';
```

The SET command allows us to declare a label that resolves its actual value when the script runs. You see how the labels are interpreted shortly, providing a single point where we can make declarations for later automation.

Database, Warehouse, and Schema

Let's create several Snowflake objects. These are containers for objects whose taxonomy is shown in Figure 5-4, along with a warehouse used for query execution. We start by using the SYSADMIN role, which should be set as this role owns all the objects. A brief digression. In some legacy RDBMS, the connected user owns declared objects. In Snowflake, all objects are owned by the active role.

```
USE ROLE sysadmin;

CREATE OR REPLACE DATABASE IDENTIFIER ( $test_database ) DATA_RETENTION_
TIME_IN_DAYS = 90;

CREATE OR REPLACE WAREHOUSE IDENTIFIER ( $test_warehouse ) WITH
WAREHOUSE_SIZE      = 'X-SMALL'
AUTO_SUSPEND        = 60
AUTO_RESUME         = TRUE
MIN_CLUSTER_COUNT   = 1
MAX_CLUSTER_COUNT   = 4
SCALING_POLICY      = 'STANDARD'
INITIALLY_SUSPENDED = TRUE;

CREATE OR REPLACE SCHEMA IDENTIFIER ( $test_staging_schema );
CREATE OR REPLACE SCHEMA IDENTIFIER ( $test_owner_schema   );
CREATE OR REPLACE SCHEMA IDENTIFIER ( $test_reader_schema  );
```

This command set changes the role to SYSADMIN, creates a database with 90-day Time Travel, creates a single extra-small warehouse, and creates three schemas in the database. Refresh the browser to see the new database and schemas. Note that the use of the IDENTIFIER keyword expands the labels. Using the SYADMIN role ensures all created objects remain owned by SYSADMIN and is a safeguard against unexpected manipulation.

Your worksheet should reflect Figure 5-5 when refreshed.

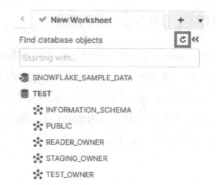

Figure 5-5. *Refreshed database and schema*

You may wish to consider how ELT tools resize warehouses dynamically with objects owned by SYSADMIN and refactor object ownership accordingly. Static warehouse declarations ensure all warehouses remain fixed until explicitly (by design) are changed. Therefore account configuration is always known. If warehouses are redefined, system integrity could be compromised in the event of a failing process not resetting a warehouse to the expected baseline condition resulting in either poor performance or excessive costs.

Roles

Let's switch roles to SECURITYADMIN to create custom roles for later use resolving IDENTIFIER labels.

```
USE ROLE securityadmin;

CREATE ROLE IF NOT EXISTS IDENTIFIER ( $test_load_role   )      COMMENT =
'TEST.test_load Role';
CREATE ROLE IF NOT EXISTS IDENTIFIER ( $test_owner_role  )      COMMENT =
'TEST.test_owner Role';
CREATE ROLE IF NOT EXISTS IDENTIFIER ( $test_reader_role )      COMMENT =
'TEST.test_reader Role';
CREATE ROLE IF NOT EXISTS IDENTIFIER ( $test_object_role )      COMMENT =
'TEST.test_object Role';
```

In line with Snowflake's best practices, we assign our newly created roles to SECURITYADMIN. You may choose to have a custom local admin role as an intermediary. Some homework for you!

```
GRANT ROLE IDENTIFIER ( $test_load_role   ) TO ROLE securityadmin;
GRANT ROLE IDENTIFIER ( $test_owner_role  ) TO ROLE securityadmin;
GRANT ROLE IDENTIFIER ( $test_reader_role ) TO ROLE securityadmin;
GRANT ROLE IDENTIFIER ( $test_object_role ) TO ROLE securityadmin;
```

Role Grants

With a SECURITYADMIN role, let's enable each role to use the database and warehouse. Note that this section must be repeated for each of the four roles that resolve IDENTIFIER labels.

```
GRANT USAGE   ON DATABASE  IDENTIFIER ( $test_database       ) TO ROLE
IDENTIFIER ( $test_load_role   );
GRANT USAGE   ON WAREHOUSE IDENTIFIER ( $test_warehouse      ) TO ROLE
IDENTIFIER ( $test_load_role   );
GRANT OPERATE ON WAREHOUSE IDENTIFIER ( $test_warehouse      ) TO ROLE
IDENTIFIER ( $test_load_role   );
GRANT USAGE   ON SCHEMA    IDENTIFIER ( $test_staging_schema ) TO ROLE
IDENTIFIER ( $test_load_role   );
```

Now we grant specific object entitlements for each of the three schemas to the object owner role, repeat for each schema resolving IDENTIFIER labels,

```
GRANT USAGE                 ON SCHEMA IDENTIFIER ( $test_staging_
schema ) TO ROLE IDENTIFIER ( $test_object_role );
GRANT MONITOR               ON SCHEMA IDENTIFIER ( $test_staging_
schema ) TO ROLE IDENTIFIER ( $test_object_role );
GRANT MODIFY                ON SCHEMA IDENTIFIER ( $tcst_staging_
schema ) TO ROLE IDENTIFIER ( $test_object_role );
GRANT CREATE TABLE          ON SCHEMA IDENTIFIER ( $test_staging_
schema ) TO ROLE IDENTIFIER ( $test_object_role );
GRANT CREATE VIEW           ON SCHEMA IDENTIFIER ( $test_staging_
schema ) TO ROLE IDENTIFIER ( $test_object_role );
```

```
GRANT CREATE SEQUENCE              ON SCHEMA IDENTIFIER ( $test_staging_
schema ) TO ROLE IDENTIFIER ( $test_object_role );
GRANT CREATE FUNCTION              ON SCHEMA IDENTIFIER ( $test_staging_
schema ) TO ROLE IDENTIFIER ( $test_object_role );
GRANT CREATE PROCEDURE             ON SCHEMA IDENTIFIER ( $test_staging_
schema ) TO ROLE IDENTIFIER ( $test_object_role );
GRANT CREATE STREAM                ON SCHEMA IDENTIFIER ( $test_staging_
schema ) TO ROLE IDENTIFIER ( $test_object_role );
GRANT CREATE MATERIALIZED VIEW  ON SCHEMA IDENTIFIER ( $test_staging_
schema ) TO ROLE IDENTIFIER ( $test_object_role );
GRANT CREATE FILE FORMAT           ON SCHEMA IDENTIFIER ( $test_staging_
schema ) TO ROLE IDENTIFIER ( $test_object_role );
```

Assign Roles to a User

When all four roles have been created and all entitlements granted, we must enable one or more users to interact with our empty schemas. We do this by granting each role to nominated users. Note the syntax may vary according to how users have been configured. The following assumes AD group provisioning with an email address or user used at the login page. (It is in the top-right displayed role in the classic console.)

```
GRANT ROLE IDENTIFIER ( $test_load_role   ) TO USER "<your_email_
address >";
GRANT ROLE IDENTIFIER ( $test_owner_role  ) TO USER "<your_email_address>";
GRANT ROLE IDENTIFIER ( $test_reader_role ) TO USER "<your_email_
address >";
GRANT ROLE IDENTIFIER ( $test_object_role ) TO USER "<your_email_
address >";
```

The following is an alternative syntax for a locally created user.

```
GRANT ROLE IDENTIFIER ( $test_load_role   ) TO USER <your_name>;
GRANT ROLE IDENTIFIER ( $test_owner_role  ) TO USER <your_name>;
GRANT ROLE IDENTIFIER ( $test_reader_role ) TO USER <your_name>;
GRANT ROLE IDENTIFIER ( $test_object_role ) TO USER <your_name>;
```

Object Creation

Now that we have our template structures in place, our focus moves to populating schemas with objects, and for this, we must use the correct role, as each role has a specific function and purpose. In this example, test_object_role is the correct role to use.

```
USE ROLE      IDENTIFIER ( $test_object_role    );
USE DATABASE  IDENTIFIER ( $test_database       );
USE SCHEMA    IDENTIFIER ( $test_staging_schema );
USE WAREHOUSE IDENTIFIER ( $test_warehouse      );
```

Note that there are four statements. From experience, when interacting with Snowflake, I prefer to explicitly call out each one. I occasionally refer to these four statements as the "magic 4." The combination always correctly sets my execution context and provides a repeatable baseline. Whenever my colleagues ask for help with fixing their code, the first thing I ask for is their context. the second is the SQL statement, "Give me the magic 4, and the query." Then I can confidently re-run to debug the issue. Without context, all bets are off.

This example declares our role, which allows the creation and maintenance of objects for all three schemas, sets the database and schema with which we interact, and for completeness, sets the warehouse in case we later wish to execute a query.

Let's move on to creating our first object: a staging table.

```
CREATE OR REPLACE TABLE stg_test_load
(
id                                 STRING(255),
description                        STRING(255),
value                              STRING(255),
stg_last_updated                   STRING(255)
);
```

We might not set the database and schema but instead use the full canonical notation to declare the table. This approach creates hard-coding of the path in object declaration scripts, which is inconvenient and inflexible, particularly with large numbers of objects to deploy. For completeness, I show the full canonical notation in the next object declaration.

```
CREATE OR REPLACE TABLE TEST.staging_owner.stg_test_load
```

```
(
id                                      STRING(255),
description                             STRING(255),
value                                   STRING(255),
stg_last_updated                        STRING(255)
);
```

We might ask ourselves why a staging table is required. Surely we just want to load data and "get on with the job." Good question. I'm glad you asked.

In our example table, note the attribute data type is STRING. When loading data, we may not know the data type; for example, our data may be supplied in numeric or dates in various formats. The problem with declaring any table where data enters our system is knowing the attribute data type for the provided data. We always want to load the highest possible number of records without rejections. Therefore, the most appropriate data type to default our staging table attributes to is STRING, from which we can later conform the data into more appropriate data types.

Now that a table is created, we must grant entitlements.

```
GRANT SELECT, INSERT, TRUNCATE ON stg_test_load TO IDENTIFIER ( $test_load_
role );
GRANT SELECT, REFERENCES       ON stg_test_load TO IDENTIFIER ( $test_
owner_role );
```

What did we just do? The first GRANT allows our load role to clear the staging table, create new data, and query what is in the staging table. Note there is no entitlement to change the staging table definition nor to update or delete data, which is, by design, separating the capability to modify objects from the capability to use objects. Our second GRANT allows the data usage role to query and allows other (as yet undefined) objects to reference the staged data table. Remembering our discussion on security, we apply the principle of least entitlement to achieve our objective.

We have segregated our data access as per Figure 5-6 and began to define how data flows through our system.

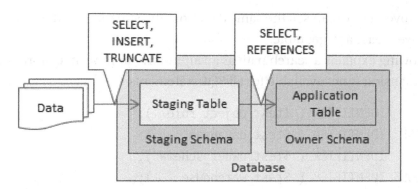

Figure 5-6. Sample security setup

Owner Schema Build

At this point, we might wish to automate our data ingestion by implementing Snowflake components such as a stream (created earlier) triggering a task, which calls a JavaScript stored procedure to ingest data into our owner schema, as Figure 5-7 shows.

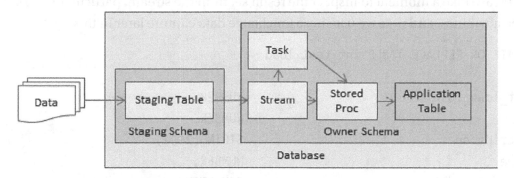

Figure 5-7. Sample automation

Adding a stream, task, and stored procedure exposes unexpected entitlement requirements. The stream is straightforwardly used to detect the presence of data in a table; therefore, stream declaration occurs in the same schema as the source table. The task is akin to a scheduled job; in our case, it is triggered by the presence of data in the stream, which then calls a stored procedure. Our example could be a single SQL statement, but we are building the Snowflake Data Cloud and need to build a more sophisticated demonstration for re-use.

Set the context to a data usage role and create a sequence. Prove that we can select data from the staging table and create a stream noting a fully qualified path to the source table. And prove that we can see the same data from the stream as in the source table, after which we create a target table.

The following explains a search path as an alternative approach. Unfortunately, identifier expansion is not available for all SQL statements.

```
USE ROLE      IDENTIFIER ( $test_object_role  );
USE DATABASE  IDENTIFIER ( $test_database     );
USE SCHEMA    IDENTIFIER ( $test_owner_schema );
USE WAREHOUSE IDENTIFIER ( $test_warehouse    );
CREATE SEQUENCE seq_test_load_id START WITH 100000;

SELECT * FROM TEST.staging_owner.stg_test_load;

CREATE OR REPLACE STREAM strm_stg_test_load ON TABLE TEST.staging_owner.
stg_test_load;

SELECT * FROM strm_stg_test_load;
```

Please take a moment to inspect the result set from the stream, particularly the last three attributes, and note we use these for change data capture later in this book.

```
CREATE OR REPLACE TABLE int_test_load
(
test_load_id                      NUMBER,
id                                NUMBER,
description                       STRING(255),
value                            NUMBER,
stg_last_updated                  TIMESTAMP
);
```

Having proven that we can extract data from our stream, let's look at how to work with stored procedures and later interact with a task. Stored procedures allow you to call multiple SQL statements rather than a single SQL statement, providing significant flexibility and code re-use, which is discussed in a later chapter. However, using JavaScript stored procedures may also prevent data lineage tooling from self-discovery, and some architects prefer not to have any logic in the database.

Our first stored procedure performs a simple INSERT as a SELECT statement and is indirect insofar as the stored procedure body encapsulates the logic; therefore, it is a black box, but an incredibly powerful black box when given a little thought.

```
CREATE OR REPLACE PROCEDURE sp_merge_test_load()
RETURNS string
LANGUAGE javascript
EXECUTE AS CALLER
AS
$$
   var sql_stmt   = "";
   var err_state = "";
   var result     = "";

   sql_stmt += "INSERT INTO TEST.test_owner.int_test_load\n"
   sql_stmt += "SELECT seq_test_load_id.NEXTVAL,            \n"
   sql_stmt += "          id,                               \n"
   sql_stmt += "          description,                      \n"
   sql_stmt += "          value,                            \n"
   sql_stmt += "          stg_last_updated                  \n"
   sql_stmt += "FROM    strm_stg_test_load;"

   stmt = snowflake.createStatement( { sqlText: sql_stmt } );

   try
   {
      stmt.execute();

      result = "Success";
   }
   catch(err)
   {
      err_state += "\nFail Code: "    + err.code;
      err_state += "\nState: "        + err.state;
      err_state += "\nMessage : "     + err.message;
      err_state += "\nStack Trace:\n" + err.StackTraceTxt;
```

```
      result = err_state;
   }
   return result;
$$;
```

Note that stored procedure compilation does not resolve dependencies nor prove logic is correct, so debugging can only occur when invoked. For complex stored procedures, add logging code, which I leave for your further investigation.

Staging Data Load

We must isolate ingesting data from the consumption of data. An inbound data role should not be able to interact with data in upstream schemas nor, by default, consume data from outbound objects. The segregation of roles and responsibilities is a complex subject. I can only give a flavor of the full requirements here.

Switching to our inbound data role and staging schema, let's load data into our staging table. A sample sequence of operations would typically remove all existing data first.

There are several ways to load data. The simplest is to upload a CSV file using the user interface. Another way is via direct SQL INSERT statement, as follows.

```
USE ROLE      IDENTIFIER ( $test_load_role     );
USE DATABASE  IDENTIFIER ( $test_database      );
USE SCHEMA    IDENTIFIER ( $test_staging_schema );
USE WAREHOUSE IDENTIFIER ( $test_warehouse     );

TRUNCATE TABLE stg_test_load;

INSERT INTO stg_test_load
VALUES
( 1000, 'Test Record 1','Stream1','2021-12-08 16:39:11.700' ),
( 1001, 'Test Record 2','Stream2','2021-12-08 16:39:11.700' );
```

You were missing values "Expecting 4 got 3"

Despite our INSERT statement executing successfully, we should check our records have been loaded. Assume nothing, always prove assertions, and reinforce positive thought processes and behavior.

```
SELECT * FROM stg_test_load;
```

We cannot select from the stream declared against the staging table. This is intended for upstream data consumption only. Our inbound data role is intended to land data into staging tables, but not onward process landed data into upstream schema tables.

Test Stored Procedure

Having declared our stored procedure, we now need to test it. Do so by executing the following code.

```
USE ROLE      IDENTIFIER ( $test_object_role  );  -- Object Owner Role
USE DATABASE  IDENTIFIER ( $test_database     );
USE SCHEMA    IDENTIFIER ( $test_owner_schema );  -- Owner Schema
USE WAREHOUSE IDENTIFIER ( $test_warehouse    );

SELECT * FROM strm_stg_test_load;

CALL sp_merge_test_load();
```

The output should be "Success". Beware—there are scenarios where stored procedures appear to have been executed successfully. But, due to incomplete internal error handling, they may have an internal failure hence the need to add logging.

Check the stream contents.

```
SELECT * FROM strm_stg_test_load;
```

What happened? Where did our data go? Streams hold information and metadata about the actual data loaded into the table they are based upon. When the data is consumed from the stream, the contents are flushed, and the stream does not persist content, which makes streams ideal for change data capture.

However, our stored procedure worked as expected. To prove (and you should), query the target table into which our data was loaded and check the results.

```
SELECT * FROM int_test_load;
```

Create a Task

Now we need to invoke the stored procedure from our task. For your reference, the documentation is at `https://docs.snowflake.com/en/user-guide/tasks-intro.html`, noting that serverless tasks are in public preview at the time of writing.

Let's address the creation of a task. We need additional entitlements granted to the schema and role that owns the task.

```
USE ROLE securityadmin;

GRANT CREATE TASK ON SCHEMA IDENTIFIER ( $test_owner_schema    ) TO ROLE
IDENTIFIER ( $test_object_role );
```

Then revert to the context for creating our task, our owner schema.

```
USE ROLE      IDENTIFIER ( $test_object_role  );
USE DATABASE  IDENTIFIER ( $test_database     );
USE SCHEMA    IDENTIFIER ( $test_owner_schema );
USE WAREHOUSE IDENTIFIER ( $test_warehouse    );
```

Back to the familiar mantra of proving our entry conditions. Do we have data available?

```
SELECT * FROM strm_stg_test_load;
```

If not, repeat the "Staging Data Load" section, reset the context (the preceding USE statements), and requery stream.

Two commands are useful in debugging tasks. Run them before proceeding because the output is informative.

```
SHOW streams;

SELECT system$stream_has_data( 'STRM_STG_TEST_LOAD' );
```

Moving on to creating our task to automate data ingestion.

```
CREATE OR REPLACE TASK task_stg_test_load
WAREHOUSE = test_wh
SCHEDULE  = '1 minute'
WHEN system$stream_has_data ( 'STRM_STG_TEST_LOAD' )
AS
CALL sp_merge_test_load();
```

On creation, every task is suspended. Run the SHOW command to check the state attribute.

```
SHOW tasks;
```

Enabling a task requires the state to be updated. We do this by altering the task. However, while the object owner role is entitled to create a task, there is no entitlement to execute the task.

There are several choices. We could use the ACCOUNTADMIN role to enable all tasks or create another role with entitlement to execute tasks and grant the role to the object owner role, or we could grant entitlement to execute tasks directly to the object owner role. Because this is a sample code walk-through, we grant directly to the object owner role, noting the use of the ACCOUNTADMIN role.

```
USE ROLE accountadmin;

GRANT EXECUTE TASK ON ACCOUNT TO ROLE IDENTIFIER ( $test_object_role );
```

Then revert to the context for creating the owner schema task.

```
USE ROLE      IDENTIFIER ( $test_object_role  );
USE DATABASE  IDENTIFIER ( $test_database      );
USE SCHEMA    IDENTIFIER ( $test_owner_schema );
USE WAREHOUSE IDENTIFIER ( $test_warehouse     );
```

We can now set out task to execute.

```
ALTER TASK task_stg_test_load RESUME;
```

We must check the task status to ensure the status has started.

```
SHOW tasks;
```

Having set our task running, we can insert rows into the staging table for periodic insertion into the application table. The stream detects new rows and enables the task to run. Try it. Don't forget to set the correct context for adding rows and checking the task status.

Not forgetting to query our application table.

```
SELECT * FROM int_test_load;
```

The penultimate step is to enable data consumption, making data accessible to consumers. As with every configuration change, we set our context switching to the output schema before defining our secure view.

```
USE ROLE      IDENTIFIER ( $test_object_role   );
USE DATABASE  IDENTIFIER ( $test_database       );
USE SCHEMA    IDENTIFIER ( $test_reader_schema );
USE WAREHOUSE IDENTIFIER ( $test_warehouse      );
```

Create a secure view to prevent the SQL query from being viewed by any other role in Snowflake. This ensures the query remains hidden.

```
CREATE OR REPLACE SECURE VIEW v_secure_test_load COPY GRANTS
AS
SELECT * FROM TEST.test_owner.int_test_load;
```

And check our new view returns the expected data set.

```
SELECT * FROM v_secure_test_load;
```

Then GRANT SELECT and USAGE to the role our end users will use to access the SECURE VIEW.

```
USE ROLE SYSADMIN;
```

```
GRANT USAGE ON SCHEMA reader_owner TO ROLE IDENTIFIER ( $test_
reader_role );
GRANT SELECT ON v_secure_test_load TO ROLE IDENTIFIER ( $test_
reader_role );
```

Figure 5-8 illustrates the end-to-end process we have implemented.

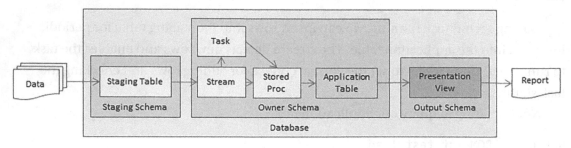

Figure 5-8. *End-to-end process*

End User/Service User

The last step is to create a user. Please use a sensible password and set MUST_CHANGE_PASSWORD = TRUE. Note the DEFAULT_ROLE and DEFAULT_WAREHOUSE settings define object access.

```
USE ROLE useradmin;

CREATE OR REPLACE USER test
PASSWORD            = 'test'
DISPLAY_NAME        = 'Test User'
EMAIL               = 'test@test.xyz'
DEFAULT_ROLE        = 'test_reader_role'
DEFAULT_NAMESPACE   = 'TEST.reader_owner'
DEFAULT_WAREHOUSE   = 'test_wh'
COMMENT             = 'Test user'
MUST_CHANGE_PASSWORD = FALSE;
```

Setting DEFAULT_ROLE for a user does not automatically entitle the user to USE the role.

```
USE ROLE securityadmin;

GRANT ROLE test_reader_role TO USER test;
```

GRANT S

Using a different browser, log in to Snowflake using the test user, set the context, and query the SECURE VIEW.

```
USE ROLE       test_reader_role;
USE DATABASE   TEST;
USE WAREHOUSE  test_wh;
USE SCHEMA     reader_owner;

SELECT * FROM TEST.reader_owner.v_secure_test_load;
```

Troubleshooting

Eventually, you will need to know why something has broken, how to fix it, and what to do. This section contains a few tips and commands to help.

Let's start with the "magic 4" context. I assume you are now familiar with the necessary commands to run this query.

```
SELECT current_user(), current_role();
```

To identify entitlement granted to the current role, run this command.

```
SHOW GRANTS TO ROLE test_reader_role;
```

Re-run the failed SQL statement.

Check the History button.

Re-run the supplied script to baseline the implementation, then roll forward your changes to the point of failure.

Please also see search_path later in this chapter.

Cleanup

Finally, you want to remove all objects.

```
USE ROLE sysadmin;

DROP DATABASE IDENTIFIER ( $test_database );

DROP WAREHOUSE IDENTIFIER ( $test_warehouse );

USE ROLE securityadmin;

DROP ROLE IDENTIFIER ( $test_load_role   );
DROP ROLE IDENTIFIER ( $test_owner_role  );
DROP ROLE IDENTIFIER ( $test_reader_role );
DROP ROLE IDENTIFIER ( $test_object_role );

USE ROLE useradmin;

DROP USER test;
```

Search Path

This chapter used implicit pathing derived from our context (or "magic 4") and absolute pathing to reference objects in different schemas. In your real-world Snowflake implementations, you need to know both pathing options.

Some legacy databases use synonyms to abstract an absolute path to a named object. This is a shortcut from one schema to another constrained locally, or it may be a public declaration available to all users. Snowflake does not use synonyms but instead uses search_path, an environment variable with fixed order of schemas to search to resolve object names. Both approaches have pros and cons, and for Snowflake, if search_path is used, the context must be set before execution.

To examine the search_path setting, run the following query.

```
SHOW PARAMETERS LIKE 'search_path';
```

Further explanation is available at `https://docs.snowflake.com/en/sql-reference/name-resolution.html`.

Summary

This chapter introduced RBAC and dispelled some myths usually derived from legacy RDBMS implementations. It then outlined different ways to implement Snowflake tenancy and account operating models according to business and technical drivers.

We then looked at a practical, hands-on scripted implementation. On one level, explain RBAC through example. On another level, providing a working template of how data is ingested, manipulated, and finally reported on using Snowflake. We also introduced automation via streams and tasks, explaining how to use each component, albeit relatively simplistic. Where possible, "gotchas" were called out and remediated before introducing some tools to troubleshoot implementation issues noting the example given uses multiple schemas and a reasonable degree of complexity. And then, we cleaned up our test case as all good examples should conclude.

Next is a chapter I took great pleasure in writing because I wrote a full suite of monitoring tooling, and I learned an awful lot about Snowflake and how to handle difficult coding challenges using JavaScript, streams, tasks, and a few other techniques.

CHAPTER 6

Account Usage Store

In Chapter 4, you worked through some examples of account controls and learned how to implement Snowflake features to keep the bad guys out and protect your accounts. Chapter 4 is also referenced to implement corresponding monitoring and alerting.

Let's take a step back and ask ourselves what other operational functions we can facilitate by building out monitoring, and see whether the Account Usage store can fulfill our needs. The functional requirements typically include the following.

- Audit: Both internal and external system audits would want to access the Account Usage store and run ad hoc queries. We cannot know the full suite of future requirements, but we can make an educated guess at the likely use cases an audit function would want to satisfy.

- Consumption charging: Every finance department wants to know how many credits have been consumed and by whom and looks at consumption growth trends over time. The procurement team will want to know consumption numbers to project future trends and inform contract renewal negotiations.

- Metadata cataloging: Enterprise data model mapping, object and attribute capture, and data lineage are valid use cases.

- Event monitoring: This chapter's primary purpose is tooling, which frequently monitors activity and raises alerts.

Please ensure your account is the Business Critical Edition. The examples here are dependent upon Business Critical Edition features. Otherwise, please re-register using a different email address by selecting the Business Critical option at `https://signup.snowflake.com`.

© Andrew Carruthers 2022
A. Carruthers, *Building the Snowflake Data Cloud*, https://doi.org/10.1007/978-1-4842-8593-0_6

Snowflake continually enhances the Account Usage views and capability. Wherever possible, we want to build the ability to future-proof usage and enable new monitoring requirements with minimal re-work. I suggest the operational support team conduct monthly reviews of Account Usage views to identify changes proactively. Better to be on top of change and present our cybersecurity colleagues with an opportunity to tighten controls and add value.

Automating each control using streams and tasks brings its own challenges and the obligation to record the status and logging of each task run. For simplicity and to reduce the page count, I have not implemented a status setting or log in each procedure. I leave this for you to investigate but note that you may encounter concurrency issues with multiple high-frequency processes updating the same status and logging table. Likewise, the code presented is a skeleton. You may wish to implement additional attributes such as URLs to confluence pages containing control information or an active_flag for each parameter. Just note the need to extend all dependent objects such as views and MERGE statements.

And add comments omitted due to page constraints. Hopefully, the code is sufficiently self-explanatory.

This chapter could—and probably should—be a book in its own right. The more I dig into the Account Usage store, the more I find of interest. After writing the monitoring suite over a year ago, it is a delight revisiting this subject, and I hope this chapter inspires you to investigate further. Please also see `https://docs.snowflake.com/en/sql-reference/account-usage.html`.

Let's now discuss Snowflake Account Usage, what it is, how to use it, and offer practical hands-on suggestions to make your job easier when building monitoring tooling.

What Is the Account Usage Store?

All RDBMS have a metadata catalog, Oracle has data dictionary views, dynamic performance views (v$), and a plethora of k$ and x$ tables—complex and hard to navigate. SQL Server and MySQL have the information_schema, and Snowflake has followed the same pattern by implementing an information schema for each created database.

Except for a single Snowflake account level holistic source of information, Snowflake supplies an imported database confusingly called SNOWFLAKE, where the Account Usage store resides, maintained by Snowflake, which also contains an information schema. Please note there are some features to be aware of. Account Usage views have a latency of between 45 minutes and 3 hours. Therefore, transactions can take time to appear in the views but typically store information for up to one year. But there is a way to gain access to data in real time by using information_schema, noting data is only stored for 14 days. I demonstrate both approaches. The distinction is important. The Account Usage store is an aggregated view of Snowflake's centrally maintained data presented as a shared database. The information schema is driven from the global services layer accessing the FoundationDB key-value store. We use both throughout this chapter.

Depending upon the role in use, the imported Snowflake database may not be visible, which changes the role. ACCOUNTADMIN is the only role enabled by default to access Account Usage store content.

```
USE ROLE accountadmin;
```

Then refresh as shown in Figure 6-1, after which the Snowflake database is viewable. SNOWFLAKE and SNOWFLAKE_SAMPLE_DATA are both imported databases. An imported database is one supplied by Snowflake as part of the initial account delivery. Like any other object, they can be dropped.

Note   Do not drop the SNOWFLAKE database!

You may drop SNOWFLAKE_SAMPLE_DATA, but note UNDROP is not supported for an imported database that must be re-created along with granting imported entitlement to roles.

Figure 6-1. *Snowflake Account Usage*

By default, only the ACCOUNTADMIN role can access the Account Usage views. Other roles can see the Snowflake database but not access the contents. Immutable history is restricted to one year, which is insufficient to support audit requests for greater than one year. We have a further requirement to periodically capture and retain all history.

Let's dive into the Snowflake Account Usage store, along with an investigation of the global services layer. In line with the hands-on approach, I offer a script and walk-through with a strong recommendation to cut and paste each section as a practical learning experience. The code looks familiar to what is in Chapter 5, but the implementation is quite different.

Best Practices

Snowflake best practice suggests using the one-year immutable history to check events against, but I suggest this does not fulfill all requirements. There are scenarios where account settings have aged out of the history and are not available for comparison or reinstatement back to original values. I show you how to mitigate against this risk in the history retention section. In other words, always expect the unexpected, don't trust anything, prove assertions and verify the recommendations match your requirements. Those words of wisdom have served me well in nearly 30 years of development.

Accessing the Account Usage Store

As discussed in Chapter 5, the only way to grant entitlement is via a role. The first step is to design our approach, as shown in Figure 6-2.

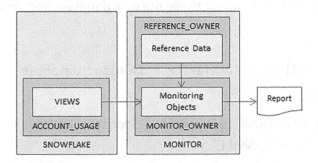

Figure 6-2. *Example Account Usage store access pattern*

Our approach is to create a new database (MONITOR) containing two schemas: one for static reference data and the second to wrap the source Account Usage views to enrich, contextualize, summarize, and pre-filter the source data to make reporting easier.

While we must explicitly enable access to the Account Usage store, each database information schema is immediately accessible.

Preamble

Our script starts with some declarations enabling later automation.

```
SET monitor_database          = 'MONITOR';
SET monitor_reference_schema  = 'MONITOR.reference_owner';
SET monitor_owner_schema      = 'MONITOR.monitor_owner';
SET monitor_warehouse         = 'monitor_wh';
SET monitor_reference_role    = 'monitor_reference_role';
SET monitor_owner_role        = 'monitor_owner_role';
SET monitor_reader_role       = 'monitor_reader_role';
```

The SET command allows us to declare a label that resolves its actual value when the script runs. You see how the labels are interpreted shortly, providing a single point where we can make declarations for later automation.

Database, Warehouse, and Schema

The following is a command set that changes role to SYSADMIN, then creates a database with 90-day time travel. Also created is a single, extra-small warehouse with two schemas in the database. Refresh your browser to see the new database and schemas.

```
USE ROLE sysadmin;

CREATE OR REPLACE DATABASE IDENTIFIER ( $monitor_database ) DATA_RETENTION_
TIME_IN_DAYS = 90;

CREATE OR REPLACE WAREHOUSE IDENTIFIER ( $monitor_warehouse ) WITH
WAREHOUSE_SIZE        = 'X-SMALL'
AUTO_SUSPEND          = 60
AUTO_RESUME           = TRUE
MIN_CLUSTER_COUNT     = 1
MAX_CLUSTER_COUNT     = 4
SCALING_POLICY        = 'STANDARD'
INITIALLY_SUSPENDED = TRUE;

CREATE OR REPLACE SCHEMA IDENTIFIER ( $monitor_reference_schema );
CREATE OR REPLACE SCHEMA IDENTIFIER ( $monitor_owner_schema      );
```

In the preceding example, the labels are expanded by using the IDENTIFIER keyword. Using the SYADMIN role ensures all created objects remain owned by SYSADMIN and is a safeguard against unexpected manipulation.

Your worksheet should reflect Figure 6-3 when refreshed.

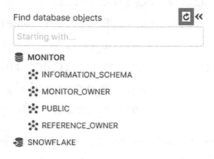

Figure 6-3. *Refreshed database and schema*

Roles

Let's switch roles to SECURITYADMIN to create custom roles for later use resolving IDENTIFIER labels.

```
USE ROLE securityadmin;
CREATE OR REPLACE ROLE IDENTIFIER ( $monitor_reference_role )
COMMENT = 'MONITOR.monitor_reference Role';
CREATE OR REPLACE ROLE IDENTIFIER ( $monitor_owner_role     )
COMMENT = 'MONITOR.monitor_owner Role';
CREATE OR REPLACE ROLE IDENTIFIER ( $monitor_reader_role    )
COMMENT = 'MONITOR.monitor_reader Role';
```

In line with Snowflake's best practices, let's assign our newly created roles to SECURITYADMIN. (Note that we may choose to have a custom local admin role as an intermediary role—some homework for you!)

```
GRANT ROLE IDENTIFIER ( $monitor_reference_role ) TO ROLE securityadmin;
GRANT ROLE IDENTIFIER ( $monitor_owner_role     ) TO ROLE securityadmin;
GRANT ROLE IDENTIFIER ( $monitor_reader_role    ) TO ROLE securityadmin;
```

Role Grants

With the SECURITYADMIN role in use, let's enable each role to use the database and warehouse.

```
GRANT USAGE   ON DATABASE  IDENTIFIER ( $monitor_database        )
TO ROLE IDENTIFIER ( $monitor_reference_role );
GRANT USAGE   ON WAREHOUSE IDENTIFIER ( $monitor_warehouse       )
TO ROLE IDENTIFIER ( $monitor_reference_role );
GRANT OPERATE ON WAREHOUSE IDENTIFIER ( $monitor_warehouse       )
TO ROLE IDENTIFIER ( $monitor_reference_role );
GRANT USAGE   ON SCHEMA    IDENTIFIER ( $monitor_reference_schema )
TO ROLE IDENTIFIER ( $monitor_reference_role );

GRANT USAGE   ON DATABASE  IDENTIFIER ( $monitor_database        )
TO ROLE IDENTIFIER ( $monitor_owner_role );
```

```
GRANT USAGE   ON WAREHOUSE IDENTIFIER ( $monitor_warehouse        )
TO ROLE IDENTIFIER ( $monitor_owner_role );
GRANT OPERATE ON WAREHOUSE IDENTIFIER ( $monitor_warehouse        )
TO ROLE IDENTIFIER ( $monitor_owner_role );
GRANT USAGE   ON SCHEMA    IDENTIFIER ( $monitor_reference_schema )
TO ROLE IDENTIFIER ( $monitor_owner_role  );
GRANT USAGE   ON SCHEMA    IDENTIFIER ( $monitor_owner_schema     )
TO ROLE IDENTIFIER ( $monitor_owner_role  );

GRANT USAGE    ON DATABASE   IDENTIFIER ( $monitor_database        )
TO ROLE IDENTIFIER ( $monitor_reader_role );
GRANT USAGE   ON WAREHOUSE IDENTIFIER ( $monitor_warehouse        )
TO ROLE IDENTIFIER ( $monitor_reader_role );
GRANT OPERATE ON WAREHOUSE IDENTIFIER ( $monitor_warehouse        )
TO ROLE IDENTIFIER ( $monitor_reader_role );
GRANT USAGE    ON SCHEMA     IDENTIFIER ( $monitor_owner_schema     )
TO ROLE IDENTIFIER ( $monitor_reader_role );
```

Now let's grant specific object entitlements for each of the two schemas.

```
GRANT CREATE TABLE              ON SCHEMA IDENTIFIER ( $monitor_reference_
schema ) TO ROLE IDENTIFIER ( $monitor_reference_role );
GRANT CREATE VIEW               ON SCHEMA IDENTIFIER ( $monitor_reference_
schema ) TO ROLE IDENTIFIER ( $monitor_reference_role );
GRANT CREATE SEQUENCE           ON SCHEMA IDENTIFIER ( $monitor_reference_
schema ) TO ROLE IDENTIFIER ( $monitor_reference_role );
GRANT CREATE STREAM             ON SCHEMA IDENTIFIER ( $monitor_reference_
schema ) TO ROLE IDENTIFIER ( $monitor_reference_role );
GRANT CREATE MATERIALIZED VIEW ON SCHEMA IDENTIFIER ( $monitor_reference_
schema ) TO ROLE IDENTIFIER ( $monitor_reference_role );

GRANT CREATE TABLE               ON SCHEMA IDENTIFIER ( $monitor_owner_
schema   ) TO ROLE IDENTIFIER ( $monitor_owner_role );
GRANT CREATE VIEW                ON SCHEMA IDENTIFIER ( $monitor_owner_
schema   ) TO ROLE IDENTIFIER ( $monitor_owner_role );
GRANT CREATE SEQUENCE            ON SCHEMA IDENTIFIER ( $monitor_owner_
schema   ) TO ROLE IDENTIFIER ( $monitor_owner_role );
```

```
GRANT CREATE FUNCTION           ON SCHEMA IDENTIFIER ( $monitor_owner_
schema   ) TO ROLE IDENTIFIER ( $monitor_owner_role );
GRANT CREATE PROCEDURE          ON SCHEMA IDENTIFIER ( $monitor_owner_
schema   ) TO ROLE IDENTIFIER ( $monitor_owner_role );
GRANT CREATE STREAM             ON SCHEMA IDENTIFIER ( $monitor_owner_
schema   ) TO ROLE IDENTIFIER ( $monitor_owner_role );
GRANT CREATE MATERIALIZED VIEW  ON SCHEMA IDENTIFIER ( $monitor_owner_
schema   ) TO ROLE IDENTIFIER ( $monitor_owner_role );
GRANT CREATE FILE FORMAT        ON SCHEMA IDENTIFIER ( $monitor_owner_
schema   ) TO ROLE IDENTIFIER ( $monitor_owner_role );
```

Assigning Roles to Users

When all three roles have been created and all entitlement granted, we must enable one or more users to interact with our empty schemas. We do this by granting each role to nominated users.

```
GRANT ROLE IDENTIFIER ( $monitor_reference_role ) TO USER <your_name>;
GRANT ROLE IDENTIFIER ( $monitor_owner_role     ) TO USER <your_name>;
GRANT ROLE IDENTIFIER ( $monitor_reader_role    ) TO USER <your_name>;
```

Enabling the Account Usage Store

The following command entitles monitor_owner_role to access the Account Usage store.

```
GRANT IMPORTED PRIVILEGES ON DATABASE snowflake TO ROLE IDENTIFIER (
$monitor_owner_role );
```

Switch to monitor_owner_role as this is the only custom role that can view Account Usage store objects and then run a sample query.

```
USE ROLE      IDENTIFIER ( $monitor_owner_role   );
USE DATABASE  IDENTIFIER ( $monitor_database     );
USE SCHEMA    IDENTIFIER ( $monitor_owner_schema );
USE WAREHOUSE IDENTIFIER ( $monitor_warehouse    );
```

We can now query the Account Usage views, in this example, selecting metadata for all active databases in our account.

```
SELECT * FROM SNOWFLAKE.account_usage.databases WHERE deleted IS NULL;
```

Building Reference Data Sets

In the previous sections, we walked through setting up our environment. Now let's focus on the practical implementation of monitoring objects. To do this, we must understand what we are monitoring and why.

Starting with a simple use case, our first objective is to examine the sample controls developed in Chapter 4 and determine whether a pattern-based approach can be derived, enabling future reuse. A design pattern is a code template that can easily be replicated, extended, or, better still, be data driven.

From experience, there are six different patterns to be developed that cover all known scenarios. Let's begin our investigation into these patterns, starting with the simplest and adding notes to build up to the most complex. Note space does not permit a full explanation of each pattern, and this chapter builds upon principles outlined in Chapter 5, albeit in a different format.

With a fully historized reference data set, we achieve two objectives. First, we establish a temporal history for audit purposes because settings age out of the history after a year, and this is a weakness we must correct. Second, in the unlikely event a parameter is incorrectly set, we have a record of what the current value should be and of all historical values.

We must identify a way of ensuring parameter values are recorded for comparison, hence the need for reference data created in the reference_owner schema. We also introduce a way of historizing our reference data, we provide historical data to ensure the data lineage is preserved, and a brief digression is in order. Data warehouses often re-create a report at any point where the data warehouse holds information. But, if we only hold the most recent values for a record, we cannot store the history too. Enter Slowly Changing Dimension (SCD) modeling. A full explanation is beyond the scope of this chapter, but more information is at https://en.wikipedia.org/wiki/Slowly_changing_dimension. For our purposes, we need to know account parameter values at any point in history. SCD2 provides the answer; hence all historization for reference data is SCD2 based.

Finally, to ease building out each control pattern, we need to identify a way of grouping like controls in our reference data. Please take time to understand the companion script, Chapter_6_Account_Usage_Demo.sql, and all will become clear.

Referencing Schema Objects

Our reference schema contains three core tables along with their historized counterparts and a single secure view that denormalizes the data model for ease of use. A full explanation of denormalization is at https://en.wikipedia.org/wiki/ Denormalization. Figure 6-4 shows the three tables and secure view we will create, along with their relationships.

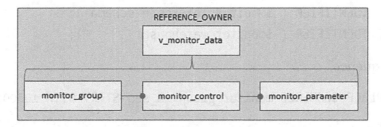

Figure 6-4. *Sample reference owner schema objects*

The general pattern for creating a single table SCD2 implementation is shown in Figure 6-5. Please refer to the accompanying script, Chapter_6_Account_Usage_Demo. sql, for the complete implementation of all three tables and a secure view. Note Chapter_6_Account_Usage_Demo.sql takes a few minutes to run as there are many SQL statements.

Figure 6-5 shows the full suite of SQL statements and data flow required to implement our example.

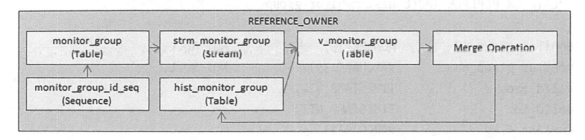

Figure 6-5. *Sample reference owner data flow*

Let's overview Figure 6-5. We create sequence monitor_group_id to generate a surrogate key. Then we create tables monitor_group and hist_monitor_group also stream strm_monitor_group, which captures records as they are inserted into the monitor_group table. We define view v_monitor_group against both stream and history table, which does all the hard work of identifying delta records. With all objects created, we insert data into monitor_group and finally run the merge statement, which loads the SCD2 records into hist_monitor_group. Don't worry. All will become clear. Just know there are dependencies on object creation order, which I now explain.

Change the role to reference_owner role.

```
USE ROLE       IDENTIFIER ( $monitor_reference_role   );
USE DATABASE   IDENTIFIER ( $monitor_database         );
USE SCHEMA     IDENTIFIER ( $monitor_reference_schema );
USE WAREHOUSE  IDENTIFIER ( $monitor_warehouse        );
```

Create a sequence.

```
CREATE OR REPLACE SEQUENCE monitor_group_id_seq START WITH 10000;
```

Create tables.

```
CREATE OR REPLACE TABLE monitor_group
(
monitor_group_id        NUMBER(10) PRIMARY KEY NOT NULL,
monitor_group_name      VARCHAR(255)           NOT NULL,
last_updated            TIMESTAMP_NTZ DEFAULT current_
timestamp()::TIMESTAMP_NTZ NOT NULL,
CONSTRAINT monitor_group_u1 UNIQUE ( monitor_group_name )
);

CREATE OR REPLACE TABLE hist_monitor_group
(
monitor_group_id        NUMBER(10)             NOT NULL,
monitor_group_name      VARCHAR(255)           NOT NULL,
valid_from              TIMESTAMP_NTZ,
valid_to                TIMESTAMP_NTZ,
current_flag            VARCHAR(1)
);
```

Create a stream.

```
CREATE OR REPLACE STREAM strm_monitor_group ON TABLE monitor_group;
```

Create a history view. Note that UNION ALL joins three sections. The first code block identifies records for INSERTs, the second for UPDATEs, and the third for DELETEs. An UPDATE is two actions: a logical DELETE and an INSERT. Hence in v_monitor_group declaration, dml_flag is never set to 'U'.

```
CREATE OR REPLACE VIEW v_monitor_group
AS
SELECT monitor_group_id,
       monitor_group_name,
       valid_from,
       valid_to,
       current_flag,
       'I' AS dml_type
FROM   (
       SELECT monitor_group_id,
              monitor_group_name,
              last_updated          AS valid_from,
              LAG ( last_updated ) OVER ( PARTITION BY monitor_group_id
              ORDER BY last_updated DESC ) AS valid_to_raw,
              CASE
                 WHEN valid_to_raw IS NULL
                    THEN '9999-12-31'::TIMESTAMP_NTZ
                    ELSE valid_to_raw
              END AS valid_to,
              CASE
                 WHEN valid_to_raw IS NULL
                    THEN 'Y'
                    ELSE 'N'
              END AS current_flag,
              'I' AS dml_type
       FROM   (
              SELECT strm.monitor_group_id,
                     strm.monitor_group_name,
```

```
                     strm.last_updated
             FROM    strm_monitor_group     strm
             WHERE   strm.metadata$action    = 'INSERT'
             AND     strm.metadata$isupdate = 'FALSE'
             )
        )
UNION ALL
SELECT monitor_group_id,
       monitor_group_name,
       valid_from,
       valid_to,
       current_flag,
       dml_type
FROM    (
        SELECT monitor_group_id,
               monitor_group_name,
               valid_from,
               LAG ( valid_from ) OVER ( PARTITION BY monitor_group_id ORDER
               BY valid_from DESC ) AS valid_to_raw,
               CASE
                  WHEN valid_to_raw IS NULL
                     THEN '9999-12-31'::TIMESTAMP_NTZ
                     ELSE valid_to_raw
               END AS valid_to,
               CASE
                  WHEN valid_to_raw IS NULL
                     THEN 'Y'
                     ELSE 'N'
               END AS current_flag,
               dml_type
        FROM    (
               SELECT strm.monitor_group_id,
                      strm.monitor_group_name,
                      strm.last_updated AS valid_from,
                      'I' AS dml_type
```

```
          FROM     strm_monitor_group      strm
          WHERE    strm.metadata$action   = 'INSERT'
          AND      strm.metadata$isupdate = 'TRUE'
          UNION ALL
          SELECT tgt.monitor_group_id,
                  tgt.monitor_group_name,
                  tgt.valid_from,
                  'D' AS dml_type
          FROM     hist_monitor_group tgt
          WHERE    tgt.monitor_group_id IN
                  (
                  SELECT DISTINCT strm.monitor_group_id
                  FROM    strm_monitor_group       strm
                  WHERE   strm.metadata$action    = 'INSERT'
                  AND     strm.metadata$isupdate = 'TRUE'
                  )
          AND      tgt.current_flag = 'Y'
                  )
        )
UNION ALL
SELECT strm.monitor_group_id,
       strm.monitor_group_name,
       tgt.valid_from,
       current_timestamp()::TIMESTAMP_NTZ AS valid_to,
       NULL,
       'D' AS dml_type
FROM    hist_monitor_group tgt
INNER JOIN strm_monitor_group strm
   ON  tgt.monitor_group_id   = strm.monitor_group_id
WHERE   strm.metadata$action    = 'DELETE'
AND     strm.metadata$isupdate = 'FALSE'
AND     tgt.current_flag        = 'Y';
```

Determine monitor groups, for example, we have account parameters, network policies, and time travel. there are others to add later when we encounter them, but for now, these three suffice. create INSERT statements for each.

```
INSERT INTO monitor_group ( monitor_group_id, monitor_group_name ) VALUES (
monitor_group_id_seq.NEXTVAL, 'ACCOUNT PARAMETER' );
INSERT INTO monitor_group ( monitor_group_id, monitor_group_name ) VALUES (
monitor_group_id_seq.NEXTVAL, 'NETWORK POLICY' );
INSERT INTO monitor_group ( monitor_group_id, monitor_group_name ) VALUES (
monitor_group_id_seq.NEXTVAL, 'TIME TRAVEL' );
```

Run the MERGE statement to historize the data, as dml_flag is never set to 'U', we only need to consider INSERT and (logical) DELETE operations. Note the number of rows inserted and number of rows updated in the returned result set.

```
MERGE INTO hist_monitor_group tgt
USING v_monitor_group strm
ON    tgt.monitor_group_id   = strm.monitor_group_id
AND   tgt.valid_from         = strm.valid_from
AND   tgt.monitor_group_name = strm.monitor_group_name
WHEN MATCHED AND strm.dml_type = 'D' THEN
UPDATE
SET   tgt.valid_to     = strm.valid_to,
      tgt.current_flag = 'N'
WHEN NOT MATCHED AND strm.dml_type = 'I' THEN
INSERT
(
tgt.monitor_group_id,
tgt.monitor_group_name,
tgt.valid_from,
tgt.valid_to,
tgt.current_flag
)
VALUES
(
strm.monitor_group_id,
strm.monitor_group_name,
strm.valid_from,
strm.valid_to,
strm.current_flag
);
```

Now query the history table to ensure the MERGE created the three records noting the valid_to date is set to '9999-12-31' and current_flag = 'Y'.

```
SELECT * FROM hist_monitor_group;
```

Testing

Please refer to the corresponding script, Chapter_6_Account_Usage_Demo.sql, where additional SQL commands are found to INSERT, UPDATE, and DELETE test parameters, always remembering to run the MERGE statement which historizes the data. Additional steps for automation (and possibly overkill as maintenance is a very low volume occasional activity) would be to create a task; see Chapter 5 for an example.

Completing the Reference Schema Data Flow

Due to space constraints, I have only provided a single table walk-through. Two more tables conforming to the same pattern, including foreign key references, are required to be built contained in the Chapter_6_Account_Usage_Demo.sql corresponding script.

Note   If you add attributes to the sample code, ensure the MERGE statement is extended to capture the new attributes.

You should begin to see that we build our code according to patterns. Adopting this approach leads us to consider writing code to generate the patterns ensuring repeatability and a consistent rule-based approach to development. We explore this concept later in this book as the implication for building the Snowflake data cloud are profound.

When complete, we can build the secure view on top to abstract our data model into a single usable object.

Creating a Secure View

Assuming Chapter_6_Account_Usage_Demo.sql has been run and all objects created, let's now build a secure view to denormalize our reference data which delivers a single object for lookups. We use a secure view to prevent the underlying SQL from being visible (a security feature). Views provide abstraction meaning less code to write when referencing the underlying model.

```
USE ROLE       IDENTIFIER ( $monitor_reference_role );
USE DATABASE   IDENTIFIER ( $monitor_database        );
USE SCHEMA     IDENTIFIER ( $monitor_owner_schema   );
USE WAREHOUSE IDENTIFIER ( $monitor_warehouse       );
CREATE OR REPLACE SECURE VIEW v_monitor_data COPY GRANTS
AS
SELECT mg.monitor_group_id,
       mg.monitor_group_name,
       mc.monitor_control_id,
       mc.monitor_control_name,
       mp.monitor_parameter_id,
       mp.monitor_parameter_name,
       mp.monitor_parameter_value
FROM   monitor_group           mg,
       monitor_control         mc,
       monitor_parameter       mp
WHERE  mg.monitor_group_id     = mc.monitor_group_id
AND    mc.monitor_control_id   = mp.monitor_control_id;
```

We should check v_monitor_data returns expected values.

```
SELECT * FROM v_monitor_data;
```

And finally, grant entitlement to monitor_owner_role.

```
GRANT SELECT, REFERENCES ON v_monitor_data TO ROLE IDENTIFIER ( $monitor_
owner_role );
```

Storing Monitored Output

This section describes how to store result sets and maintain a full audit history of monitoring activity and extension when our monitoring requirements change. Figure 6-6 illustrates the components required to identify changes to our baseline security configuration held in reference data, then compared to the actual values set on our account, with any differences reported for consumption by monitoring teams.

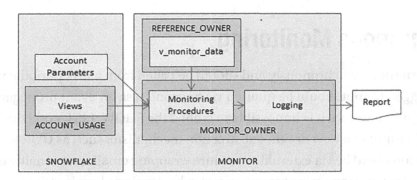

Figure 6-6. *Logging monitored output*

We begin (as usual) by setting our context.

```
USE ROLE      IDENTIFIER ( $monitor_owner_role   );
USE DATABASE  IDENTIFIER ( $monitor_database     );
USE SCHEMA    IDENTIFIER ( $monitor_owner_schema );
USE WAREHOUSE IDENTIFIER ( $monitor_warehouse    );
```

We must determine what is to be logged. The reference data, actual values, and context, plus anything useful in assisting future investigation such as record creation timestamp. From these requirements, our surrogate key sequence and primary logging table look like the following.

```
CREATE OR REPLACE SEQUENCE monitor_log_id_seq START WITH 10000;

CREATE OR REPLACE TABLE monitor_log
(
monitor_log_id          NUMBER     PRIMARY KEY NOT NULL,
event_description        STRING     NOT NULL,
event_result             STRING     NOT NULL,
monitor_control_name     STRING     NOT NULL,
```

```
monitor_parameter_name  STRING     NOT NULL,
monitor_parameter_value STRING     NOT NULL,
last_updated            TIMESTAMP_NTZ DEFAULT current_
timestamp()::TIMESTAMP_NTZ NOT NULL
);
```

There is a second logging table for one complex use case described later.

Asynchronous Monitoring

These pattern runs asynchronously and should be called from a task periodically. Then the logged output should be queried via a second task for event propagation. We adopt this approach since it is impossible to embed the SHOW, LIST, and DESCRIBE commands' output in a view or other abstraction mechanism, such as UDF.

Publication could be via external procedure wrapping email functionality or writing to an external stage. Monitoring screens can also be developed against the monitor_log table for use by your operational support team.

Account Parameters

The following pattern is for simple account parameter monitoring, TRUE/FALSE flag setting for a named parameter.

You may recall in Chapter 4 that we developed controls to prevent data from being unloaded to user-specified Internet locations and restrict data unloads to specified, system-mapped Internet locations.

```
USE ROLE accountadmin;

ALTER ACCOUNT SET prevent_unload_to_inline_url                      = TRUE;
ALTER ACCOUNT SET require_storage_integration_for_stage_creation  = TRUE;
ALTER ACCOUNT SET require_storage_integration_for_stage_operation = TRUE;
```

Two separate controls with a common implementation pattern. we are setting account level parameters.

Note our sample script Chapter_6_Account_Usage_Demo.sql initially creates only three monitor parameters in support of account parameter monitoring, providing our first set of controls to monitor.

```
USE ROLE       IDENTIFIER ( $monitor_owner_role   );
USE DATABASE   IDENTIFIER ( $monitor_database     );
USE SCHEMA     IDENTIFIER ( $monitor_owner_schema );
USE WAREHOUSE IDENTIFIER ( $monitor_warehouse     );

SELECT *
FROM MONITOR.reference_owner.v_monitor_data
WHERE monitor_parameter_value = 'TRUE';
```

Note fully qualified object reference. We should see three rows returned as per Figure 6-7.

Row	MONITOR_GROUP_ID	MONITOR_GROUP_NAME	MONITOR_CONTROL_ID	MONITOR_CONTROL_NAME	MONITOR_PARAMETER_ID	MONITOR_PARAMETER_NAME	MONITOR_PARAMETER_VALUE
1	10000	ACCOUNT PARAMETER	10001	SNOWFLAKE_NIST_AC2	10000	PREVENT_UNLOAD_TO_INLINE_URL	TRUE
2	10000	ACCOUNT PARAMETER	10002	SNOWFLAKE_NIST_AC3	10001	REQUIRE_STORAGE_INTEGRATION_FOR_STAGE_CREATION	TRUE
3	10000	ACCOUNT PARAMETER	10002	SNOWFLAKE_NIST_AC3	10002	REQUIRE_STORAGE_INTEGRATION_FOR_STAGE_OPERATION	TRUE

Figure 6-7. *Expected v_monitor_data result set*

But how do we compare the reference data to the actual values?

First, we must run the SHOW command to access the Snowflake global services layer based on FoundationDB, the key-pair store that underpins all Snowflake metadata operations. We return to this theme later in this chapter.

```
SHOW PARAMETERS LIKE 'prevent_unload_to_inline_url' IN ACCOUNT;
```

However, there is a problem. SHOW is not a true SQL command since we cannot use it in a SQL statement. For example, we cannot create a view based on it. This does not work. Create or replace view v_show_parameters as SHOW parameters. However, we can use RESULT_SCAN to convert SHOW output to a result set, but we must understand the interaction. When RESULT_SCAN is used with last_query_id, we cannot run commands in between as the query_id changes.

```
SELECT "value" FROM TABLE ( RESULT_SCAN ( last_query_id()));
```

Instead, we must use a stored procedure to programmatically extract the value for a named parameter as the two statements run consecutively, noting we extend this code pattern later for network parameter monitoring.

```
CREATE OR REPLACE PROCEDURE sp_get_parameter_value ( P_PARAMETER STRING )
RETURNS STRING
LANGUAGE javascript
```

```
EXECUTE AS CALLER
AS
$$
    var sql_stmt  = "SHOW PARAMETERS LIKE '" + P_PARAMETER + "'IN ACCOUNT";
    var show_stmt = snowflake.createStatement ({ sqlText:sql_stmt });

    show_res = show_stmt.execute();
    show_op  = show_res.next();

    var sql_stmt  = `SELECT "value" FROM TABLE ( RESULT_SCAN ( last_query_
    id()));`;
    var exec_stmt = snowflake.createStatement ({ sqlText:sql_stmt });

    rec_set = exec_stmt.execute();
    rec_op  = rec_set.next();

    return rec_set.getColumnValue(1);
$$;
```

The following tests the stored procedure.

```
CALL sp_get_parameter_value ( 'prevent_unload_to_inline_url' );
```

The stored procedure should return "true".

You may be wondering whether the stored procedure can be converted into a function and embedded in a SQL statement. In short, the answer at the time of writing, using a function, is not possible (but would have been so much simpler).

Having explained and implemented all dependent objects on which account parameter monitoring relies, Figure 6-8 illustrates how the components relate with sp_check_parameters being the next deliverable for development.

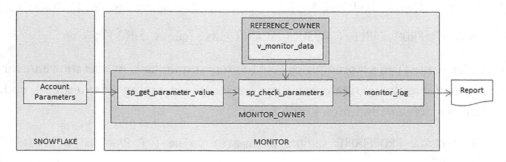

Figure 6-8. *Account parameter monitoring*

Flowchart to explain the logic contained in sp_check_parameters in Figure 6-9.

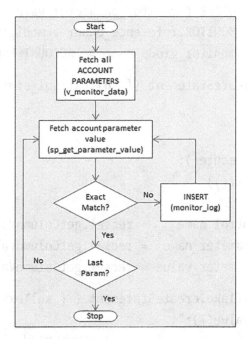

Figure 6-9. *sp_check_parameters logic*

Let's implement the sp_check_parameters stored procedure using JavaScript.

```
CREATE OR REPLACE PROCEDURE sp_check_parameters() RETURNS STRING
LANGUAGE javascript
EXECUTE AS CALLER
AS
$$
    var sql_stmt  = "";
    var stmt      = "";
    var recset    = "";
    var result    = "";

    var monitor_control_name    = "";
    var monitor_parameter_name  = "";
    var monitor_parameter_value = "";
    var parameter_value         = "";

    sql_stmt += "SELECT monitor_control_name,\n"
```

```
sql_stmt += "          UPPER ( monitor_parameter_name  ),\n"
sql_stmt += "          UPPER ( monitor_parameter_value )\n"
sql_stmt += "FROM   MONITOR.reference_owner.v_monitor_data\n"
sql_stmt += "WHERE  monitor_group_name = 'ACCOUNT PARAMETER';";

stmt = snowflake.createStatement ({ sqlText:sql_stmt });

try
{
    recset = stmt.execute();
    while(recset.next())
    {
        monitor_control_name    = recset.getColumnValue(1);
        monitor_parameter_name  = recset.getColumnValue(2);
        monitor_parameter_value = recset.getColumnValue(3);

        stmt = snowflake.createStatement ( { sqlText: "CALL sp_get_
        parameter_value(?);",

                                          binds:[monitor_parameter_
                                          name] } );

        res = stmt.execute();
        res.next();

        parameter_value = res.getColumnValue(1);

        sql_stmt  = "INSERT INTO monitor_log\n"
        sql_stmt += "SELECT monitor_log_id_seq.NEXTVAL,\n"
        sql_stmt += "        '" + monitor_parameter_name  + "',\n"
        sql_stmt += "        '" + parameter_value         + "',\n"
        sql_stmt += "        '" + monitor_control_name    + "',\n"
        sql_stmt += "        '" + monitor_parameter_name  + "',\n"
        sql_stmt += "        '" + monitor_parameter_value + "',\n"
        sql_stmt += "        current_timestamp()::TIMESTAMP_NTZ\n"
        sql_stmt += "FROM   dual\n"
        sql_stmt += "WHERE  UPPER ( '" + parameter_value + "' ) <> UPPER
        ( '" + monitor_parameter_value + "' );";
```

```
        stmt = snowflake.createStatement ({ sqlText:sql_stmt });

        try
        {
            stmt.execute();
            result = "Success";
        }
        catch { result = sql_stmt; }
    }
    result = "Success";
}
catch { result = sql_stmt; }
return result;
$$;
```

Call the stored procedure to test. This should not insert any records in monitor_log.

```
CALL sp_check_parameters();

SELECT * FROM monitor_log;
```

Finally, mis-set a parameter then retest.

```
USE ROLE accountadmin;
ALTER ACCOUNT SET prevent_unload_to_inline_url = FALSE;

USE ROLE      IDENTIFIER ( $monitor_owner_role   );
USE DATABASE  IDENTIFIER ( $monitor_database     );
USE SCHEMA    IDENTIFIER ( $monitor_owner_schema );
USE WAREHOUSE IDENTIFIER ( $monitor_warehouse    );

CALL sp_check_parameters();

SELECT * FROM monitor_log;
```

You should see a single record, as shown in Figure 6-10.

MONITOR_LOG_ID	EVENT_DESCRIPTION	EVENT_RESULT	MONITOR_CONTROL_NAME	MONITOR_PARAMETER_NAME	MONITOR_PARAMETER_VALUE	LAST_UPDATED
10000	PREVENT_UNLOAD_TO_INLINE_URL	false	SNOWFLAKE_NIST_AC2	PREVENT_UNLOAD_TO_INLINE_URL	TRUE	2021-12-27 06:37:31.440

Figure 6-10. *Parameter mismatch test*

Network Parameters

Network policy monitoring is a variation on account parameter monitoring with two extra steps. Just as sp_get_parameter_value uses the SHOW command with consecutive RESULT_SCAN query, interrogating network policies requires the same constructs followed by DESCRIBE and RESULT_SCAN query.

Add reference data changing the IP range to suit your local environment, and noting your Snowflake account network policy should also be set.

```
USE ROLE      IDENTIFIER ( $monitor_reference_role   );
USE DATABASE  IDENTIFIER ( $monitor_database          );
USE SCHEMA    IDENTIFIER ( $monitor_reference_schema );
USE WAREHOUSE IDENTIFIER ( $monitor_warehouse         );

INSERT INTO monitor_parameter ( monitor_parameter_id, monitor_control_id,
monitor_parameter_name, monitor_parameter_value ) SELECT monitor_parameter_
id_seq.NEXTVAL, monitor_control_id, 'MY_NETWORK_POLICY', '192.168.0.0' FROM
monitor_control WHERE monitor_control_name = 'SNOWFLAKE_NIST_NP1';
```

The queries required (in order) are as follows.

```
USE ROLE accountadmin;

CREATE OR REPLACE NETWORK POLICY my_network_policy
ALLOWED_IP_LIST=( '192.168.0.0' );

SHOW NETWORK POLICIES IN ACCOUNT;

SELECT "name" FROM TABLE ( RESULT_SCAN ( last_query_id()));

DESCRIBE NETWORK POLICY my_network_policy;

SELECT "name", "value" FROM TABLE ( RESULT_SCAN ( last_query_id()));
```

Build the dynamic INSERT INTO monitor_log using same pattern from sp_check_parameters, call and test your stored procedure.

System for Cross-Domain Identity Management (SCIM) Events

Monitoring SCIM events is distinctly different from all other monitoring insofar as accessing rest_event_history is only allowable using ACCOUNTADMIN role. We can achieve the same outcome with some lateral thinking, a task (which runs as ACCOUNTADMIN), and a separate logging table. Please refer to https://docs. snowflake.com/en/sql-reference/functions/rest_event_history.html for more details.

For this example, we use SNOWFLAKE_NIST_RE1 as the control reference.

Create a sequence and table to capture rest_event_history output.

```
USE ROLE      IDENTIFIER ( $monitor_owner_role   );
USE DATABASE  IDENTIFIER ( $monitor_database     );
USE SCHEMA    IDENTIFIER ( $monitor_owner_schema );
USE WAREHOUSE IDENTIFIER ( $monitor_warehouse    );

-- Create Sequence
CREATE OR REPLACE SEQUENCE hist_rest_event_history_id_seq START WITH 10000;

CREATE OR REPLACE TABLE hist_rest_event_history
(
hist_rest_event_history     NUMBER,
event_timestamp             TIMESTAMP_LTZ,
event_id                    NUMBER,
event_type                  TEXT,
endpoint                    TEXT,
method                      TEXT,
status                      TEXT,
error_code                  TEXT,
details                     TEXT,
client_ip                   TEXT,
actor_name                  TEXT,
actor_domain                TEXT,
resource_name               TEXT,
resource_domain             TEXT
);
```

Recognizing the need to report events in the past hour, using the ACCOUNTADMIN role, create a view joining SNOWFLAKE.information_schema.rest_event_history to our reference data. Note the use of the TABLE function.

```
USE ROLE accountadmin;

CREATE OR REPLACE SECURE VIEW v_rest_event_history COPY GRANTS
AS
SELECT  reh.event_timestamp,
        reh.event_id,
        reh.event_type,
        reh.endpoint,
        reh.method,
        reh.status,
        reh.error_code,
        reh.details,
        reh.client_ip,
        reh.actor_name,
        reh.actor_domain,
        reh.resource_name,
        reh.resource_domain,
        current_timestamp()::TIMESTAMP_NTZ   current_timestamp
FROM    TABLE ( snowflake.information_schema.rest_event_history (
                        'scim',
                        DATEADD ( 'minutes', -60, current_timestamp()),
                        current_timestamp(),
                        1000 )
            ) reh;
```

An alternative way to derive the time since the last run is to use your local monitor status table if implemented. The INSERT statement should reference the view using a stored procedure and TASK to execute periodically.

Synchronous Monitoring

The remainder of our monitoring configuration is much more straightforward and relies upon accessing the Account Usage views. We still rely upon reference data to determine our baseline criteria reflected in the following sections.

Time Travel

Database time-travel monitoring does not require SHOW/DESCRIBE processing. Instead, it is a straightforward join between v_monitor_data and Account Usage store (assuming corresponding reference data has been created first and the monitoring databases have 90-day time travel enabled).

```
USE ROLE       IDENTIFIER ( $monitor_reference_role   );
USE DATABASE   IDENTIFIER ( $monitor_database         );
USE SCHEMA     IDENTIFIER ( $monitor_reference_schema );
USE WAREHOUSE  IDENTIFIER ( $monitor_warehouse        );

INSERT INTO monitor_parameter ( monitor_parameter_id, monitor_control_id,
monitor_parameter_name, monitor_parameter_value ) SELECT monitor_parameter_
id_seq.NEXTVAL, monitor_control_id, 'MONITOR', '90' FROM monitor_control
WHERE monitor_control_name = 'SNOWFLAKE_NIST_TT1';
INSERT INTO monitor_parameter ( monitor_parameter_id, monitor_control_id,
monitor_parameter_name, monitor_parameter_value ) SELECT monitor_parameter_
id_seq.NEXTVAL, monitor_control_id, 'TEST', '90' FROM monitor_control WHERE
monitor_control_name = 'SNOWFLAKE_NIST_TT1';
```

Use the following query to test, noting the Account Usage views can take up to 3 hours to refresh.

```
USE ROLE       IDENTIFIER ( $monitor_owner_role   );
USE DATABASE   IDENTIFIER ( $monitor_database     );
USE SCHEMA     IDENTIFIER ( $monitor_owner_schema );
USE WAREHOUSE  IDENTIFIER ( $monitor_warehouse    );

INSERT INTO monitor_log
SELECT monitor_log_id_seq.NEXTVAL,
       d.database_name,
```

```
           d.retention_time,
           v.monitor_control_name,
           v.monitor_parameter_name,
           v.monitor_parameter_value,
           current_timestamp()::TIMESTAMP_NTZ
FROM       MONITOR.reference_owner.v_monitor_data v,
           snowflake.account_usage.databases      d
WHERE      v.monitor_parameter_name                = d.database_name
AND        v.monitor_parameter_value              != d.retention_time
AND        v.monitor_group_name                    = 'TIME TRAVEL'
AND        d.deleted IS NULL;
```

Testing can also be accomplished by setting != to = in this query.

Events

Many monitoring requirements can be achieved by directly querying account_usage_store or information_schema and joining the reference data. Examples of both approaches are offered next. Remember the key differences are Account Usage views have a latency of between 45 minutes and 3 hours and typically store information for up to one year. In contrast, the information schema has zero latency, but data is only stored for 14 days.

Reference Chapter 4 and our first control, SNOWFLAKE_NIST_AC1.

```
USE ROLE       IDENTIFIER ( $monitor_owner_role   );
USE DATABASE   IDENTIFIER ( $monitor_database      );
USE SCHEMA     IDENTIFIER ( $monitor_owner_schema );
USE WAREHOUSE  IDENTIFIER ( $monitor_warehouse     );
```

Create secure view v_snowflake_nist_ac1.

```
CREATE OR REPLACE SECURE VIEW v_snowflake_nist_ac1 COPY GRANTS
COMMENT = 'Reference: Chapter 4 Account Security'
AS
SELECT 'SNOWFLAKE_NIST_AC1'                          AS control_name,
       'MONITOR.monitor_owner.v_snowflake_nist_ac1'  AS control_object,
       start_time,
```

```
        role_name,
        database_name,
        schema_name,
        user_name,
        query_text,
        query_id
FROM    snowflake.account_usage.query_history
WHERE   role_name = 'ACCOUNTADMIN';
```

Note the absence of timeband filters. We should not limit the lowest Level Account Usage views, but we may later choose to apply timeband filters. This is a building block for later use.

Check view v_snowflake_nist_ac1 has content:

```
SELECT * FROM v_snowflake_nist_ac1;
```

The equivalent real-time information schema view looks like this.

```
CREATE OR REPLACE SECURE VIEW v_rt_snowflake_nist_ac1 COPY GRANTS
COMMENT = 'Reference: Chapter 4 Account Security and our first control'
AS
SELECT 'SNOWFLAKE_NIST_AC1'                          AS control_name,
        'MONITOR.monitor_owner.v_rt_snowflake_nist_ac1'  AS control_object,
        start_time,
        role_name,
        database_name,
        schema_name,
        user_name,
        query_text,
        query_id
FROM    TABLE ( snowflake.information_schema.query_history ( DATEADD (
'days', -1, current timestamp()), current_timestamp()))
WHERE   role_name = 'ACCOUNTADMIN';
```

Note the use of the TABLE operator converting function output to tabular content. More information is at https://docs.snowflake.com/en/sql-reference/functions-table.html and parameters to derive only the last day activity.

Naturally, we would develop one wrapper view for each control and consolidate it into a single master view, joining our reference data against which we may wish to apply filters.

```
CREATE OR REPLACE SECURE VIEW v_rt_snowflake_controls COPY GRANTS
AS
SELECT ac1.control_name,
       ac1.control_object,
       ac1.start_time,
       ac1.role_name,
       ac1.database_name,
       ac1.schema_name,
       ac1.user_name,
       ac1.query_text,
       ac1.query_id,
       rd.monitor_group_name,
       rd.monitor_control_name,
       rd.monitor_parameter_name,
       rd.monitor_parameter_value
FROM   MONITOR.monitor_owner.v_rt_snowflake_nist_ac1    ac1,
       MONITOR.reference_owner.v_monitor_data              rd
WHERE  ac1.control_name          = rd.monitor_control_name
AND    rd.monitor_control_name   = 'SNOWFLAKE_NIST_AC1'
AND    rd.monitor_parameter_name = 'ACCOUNTADMIN'
AND    ac1.start_time            >= DATEADD ( 'minutes', rd.monitor_
parameter_value, current_timestamp()::TIMESTAMP_NTZ );
```

Only one source view is shown. All others should be added via UNION ALL. Also, note the filter on start_time, restricting results to the past 60 minutes set by reference data.

Check view v_rt_snowflake_controls has content:

```
SELECT * FROM v_rt_snowflake_controls;
```

External Monitoring Tools

Some third-party tools require direct access to Snowflake Account Usage views to function. We should not allow direct access to any Snowflake objects but instead, create wrapper views and custom roles to limit data access to minimal objects required to achieve each objective. While this approach may be uncomfortable for vendors, we must not compromise our account security and must insist on vendors conforming to our security policies. Some vendors make provision for customization through a configuration file, as with Datadog.

An example view for specific monitoring purposes would include all source view attributes and some custom attributes, which may look like the following.

```
CREATE OR REPLACE SECURE VIEW login_history COPY GRANTS
AS
SELECT lh.*,
       current_account()||'.'||user_name          AS object_name,
       'Login History Wrapper View'                AS source_name,
       'snowflake.account_usage.login_history'  AS source_path
FROM   snowflake.account_usage.login_history lh;
```

Check view login_history has content:

```
SELECT * FROM login_history;
```

Note the removal of the v_ prefix because we want to present our view to the consuming user in (almost) native format to make integration easy.

We must grant entitlement on our views to a role.

```
GRANT SELECT ON login_history TO ROLE IDENTIFIER ( $monitor_reader_role );
```

Finally, grant the role to a user, which I leave you to complete.

Cleanup

Finally, you want to remove all objects.

```
USE ROLE sysadmin;

DROP DATABASE IDENTIFIER ( $monitor_database );

DROP WAREHOUSE IDENTIFIER ( $monitor_warehouse );

USE ROLE securityadmin;

DROP ROLE IDENTIFIER ( $monitor_reference_role  );
DROP ROLE IDENTIFIER ( $monitor_reader_role );
DROP ROLE IDENTIFIER ( $monitor_owner_role );
```

History Retention

We may need to hold the full history of every action in our account from inception to the current date. While Snowflake retains a one-year immutable history, we may need to periodically copy data to local tables before information ages out.

The following list of snowflake.account_usage views should be reviewed for completeness in scope and attribute changes before invoking data copy. I suggest that data be copied in March and September to avoid year-end change freeze blackouts.

- task_history

- access_history

- automatic_clustering_history

- copy_history

- database_storage_usage_history

- data_transfer_history

- load_history

- login_history

- materialized_view_refresh_history

- metering_daily_history

- metering_history

- pipe_usage_history

- query_history

- replication_usage_history

- search_optimization_history

- serverless_task_history

- sessions

- stage_storage_usage_history

- storage_usage

- warehouse_events_history

- warehouse_load_history

- warehouse_metering_history

Summary

This chapter introduced the Account Usage store and built a new database and associated schemas for later use. You then enabled access to the Account Usage store.

Next, you developed a reference schema with metadata storing our account level settings and denormalizing it for later use, demonstrating a sample but extensible data model.

A full monitoring solution is not possible with the limited space available. The page budget has been significantly exceeded to make things clear. It is well worth the effort. I hope you agree and have plenty to pique your interest. I have included additional code in the accompanying file to assist your understanding.

Let's turn our attention to ingesting data, where you learn more techniques on your journey to building the Snowflake Data Cloud.

PART III

Tools

CHAPTER 7

Ingesting Data

This chapter explains some common configuration, tooling, and ingestion patterns and later expands our scripts to deliver sample code generators for immediate and future use. The previous chapters looked at how to secure our account and monitor our security settings with historized data, providing robust mechanisms to allow full recovery to any point in time. Using scripts, you have become familiar with some implementation patterns, which we will now build upon. But familiarity breeds contempt, and from personal experience, I have found it far too easy to make mistakes. While writing the previous chapter, I discovered a (minor) bug in my code that went undiscovered for too long.

Please ensure your account was created for AWS with the Business Critical Edition. The examples here are dependent upon Business Critical Edition features. Otherwise, please re-register using a different email address and selecting the Business Critical option. You may register at `https://signup.snowflake.com`.

This chapter follows what was covered in Chapter 3 and assumes our cloud service provider (CSP) is AWS. Signing up for a new AWS account requires a mobile phone number and debit card information. The final step is where your account type and billing option are set. Simply select the desired option. I used the AWS Free Tier. Confirmation is via email, which takes a few minutes while the AWS account is being provisioned. After confirmation is received, log in to your AWS account as root, where you should be presented with the AWS Management Console.

We also use SnowSQL. More information is at `https://docs.snowflake.com/en/user-guide/snowsql-install-config.html`.

Working through the examples provided in this chapter is not trivial, as several new concepts and AWS Management Console are introduced. Once more, space does not permit a full scripted walk-through and screenshot of each step. Instead, reference should be made to the Snowflake documentation contextually linked for the latest information. Note that Snowflake support may also assist in the event of issues arising.

© Andrew Carruthers 2022
A. Carruthers, *Building the Snowflake Data Cloud*, https://doi.org/10.1007/978-1-4842-8593-0_7

There is another reason why we do not conduct a full walk-through. Your organization may have default policies for key management and S3 bucket security, and the implementation steps may vary from those articulated in Snowflake documentation.

Note   At the time of writing, the AWS Management Console screenshots in Snowflake documentation do not match live AWS Management Console.

Secure Connectivity

Whenever a connection is made into Snowflake, and regardless of the type of connection made, Snowflake converts the connection to HTTPS SSL TLS1.2, thus rendering the connection secure and encrypted. However, data transiting through the connection is not always guaranteed to be encrypted. There are use cases such as when using the user interface to load files into an internal stage where Snowflake also encrypted the files before transit.

For the avoidance of doubt, file compression is not encryption. Later in this chapter, you see references to GZIP, a file compression utility.

Snowflake and CSP Interconnect

Secure connectivity between an AWS account and Snowflake is not essential but called out as best practice to preserve our security posture. You may wish to develop security requirements in line with those established in Chapter 4 to mandate secure connectivity between a CSP account and Snowflake.

Note   Features and capabilities are regularly enhanced; please check the documentation.

Multiple connections can be made for all forms of interconnectivity between Snowflake and CSP. For example, from a single Snowflake account, there may be PrivateLink connections to several AWS accounts.

AWS PrivateLink

Chapter 4 also discussed AWS PrivateLink. More information is at `https://docs.snowflake.com/en/user-guide/admin-security-privatelink.html`. Configuring PrivateLink can be problematic. In the first instance, please refer to the troubleshooting guide at `https://community.snowflake.com/s/article/AWS-PrivateLink-and-Snowflake-detailed-troubleshooting-Guide`.

Figure 7-1 illustrates how Direct Connect and PrivateLink enable connectivity between on-prem and our Snowflake account. The connections are to our AWS account's virtual private cloud (VPC).

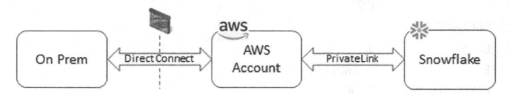

Figure 7-1. *Secure network schematic*

Azure Private Link

Secure communications between both Snowflake and Azure are via Private Link at `https://docs.snowflake.com/en/user-guide/privatelink-azure.html`. Configuring Private Link can be problematic. In the first instance, please refer to the troubleshooting guide at `https://community.snowflake.com/s/article/Azure-Private-Link-and-Snowflake-detailed-troubleshooting-Guide`.

GCP Private Service Connect

Secure communications between Snowflake and GCP are via GCP Private Secure Connect at `https://docs.snowflake.com/en/user-guide/private-service-connect-google.html#label-gcp-psc-considerations`.

Handling Data Files

Without diving too deep into this subject right now, when we load data, we may want to perform certain operations such as automatically removing a header record, setting the record delimiter to a specific character, or defining default behavior where attributes are missing. I now introduce the Snowflake file format object. Documentation is at `https://docs.snowflake.com/en/sql-reference/sql/create-file-format.html`.

Declaring a file format allows us to specify behavior and set default behavior when we automate file loading. More on this later, but for now, let's look at an example.

```
USE ROLE      sysadmin;
USE DATABASE TEST;
USE SCHEMA    public;

CREATE OR REPLACE FILE FORMAT TEST.public.test_pipe_format
TYPE                = CSV
FIELD_DELIMITER     = '|'
SKIP_HEADER         = 1
NULL_IF             = ( 'NULL', 'null' )
EMPTY_FIELD_AS_NULL = TRUE
SKIP_BLANK_LINES    = TRUE;
```

What did we just do? The specified file format is almost the same as at `https://docs.snowflake.com/en/sql-reference/sql/create-file-format.html#examples` but with an additional parameter to ignore and not load any empty lines, and the compression option was removed. With our file format declared, we may now use it in our STAGE declaration, which you see later in this chapter.

We may want to review our file format to ensure behavior is as expected.

```
SHOW FILE FORMATS LIKE 'test_pipe_format';
```

Figure 7-2 shows the output.

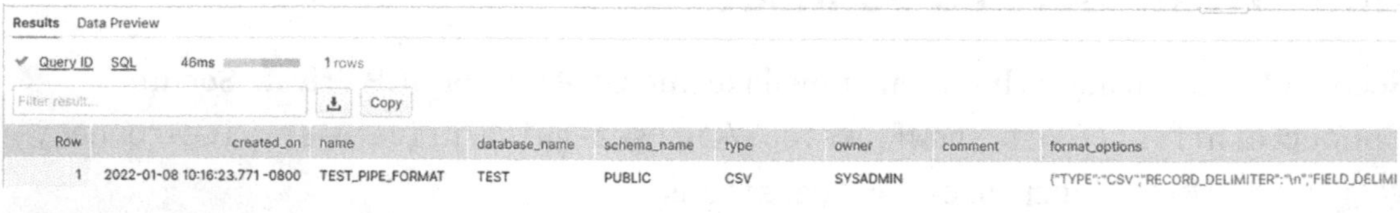

Figure 7-2. *File format output*

Clicking format_options reveals the details. Figure 7-3 shows that compression is set to AUTO.

Details

```
1  {"TYPE":"CSV","RECORD_DELIMITER":"\n","FIELD_DELIMITER":"|","FILE_EXTE
   NSION":null,"SKIP_HEADER":1,"DATE_FORMAT":"AUTO","TIME_FORMAT":"AUTO",
   "TIMESTAMP_FORMAT":"AUTO","BINARY_FORMAT":"HEX","ESCAPE":"NONE","ESCAP
   E_UNENCLOSED_FIELD":"
   \\","TRIM_SPACE":false,"FIELD_OPTIONALLY_ENCLOSED_BY":"NONE","NULL_IF"
   :
   ["NULL","null"],"COMPRESSION":"AUTO","ERROR_ON_COLUMN_COUNT_MISMATCH":
   true,"VALIDATE_UTF8":true,"SKIP_BLANK_LINES":true,"REPLACE_INVALID_CHA
   RACTERS":false,"EMPTY_FIELD_AS_NULL":true,"SKIP_BYTE_ORDER_MARK":true,
   "ENCODING":"UTF8"}
```

Figure 7-3. *File format details*

We may choose to alter the default setting, and as you will see later, using compression saves storage costs, but any unloaded files require uncompressing before use.

External Storage

Several ways exist to land data in Snowflake, some of which are discussed later. This section examines how Snowflake interacts with external storage. The (simplified) diagram in Figure 7-4 illustrates the key points.

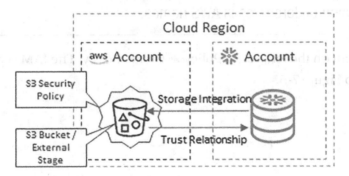

Figure 7-4. *AWS storage integration*

You may be wondering why we need to use an external stage. Can't we use an internal stage? What is the purpose of S3 buckets?

With cloud applications, you have to think differently. To use an internal stage, applications, processes, and tools must connect directly to Snowflake. This means opening a network policy and then establishing connectivity—introducing risk. Using an external stage linked to S3 storage hosted on our AWS account provides additional security and flexibility. For example, using AWS-supplied components to interact with the S3 bucket, automatic data replication can be achieved or generate an event when a file lands.

Connectivity is not limited to a single AWS account integrating into our Snowflake account. We may have many CSP accounts connecting to a single Snowflake account, segregating our data sourcing and ingestion by geographic region or function.

Storage integration documentation is at `https://docs.snowflake.com/en/user-guide/data-load-s3-config-storage-integration.html#option-1-configuring-a-snowflake-storage-integration-to-access-amazon-s3`.

Step 1. Create IAM Role and Policy

Search for IAM (identity and access management) using the AWS Management Console and create a new policy and role. You will need your AWS account ID. Click your account name in the top-right corner of the screen. To show the name of the S3 bucket, I used btsdc-test-bucket. But S3 bucket names are globally unique, and all AWS accounts share the namespace; therefore, you must use a unique name. Then, update all references in the following code.

Note   Avoid using underscores in AWS names.

From the menu on the left, select Policy ➤ Create Policy. The IAM screen appears, looking similar to Figure 7-5.

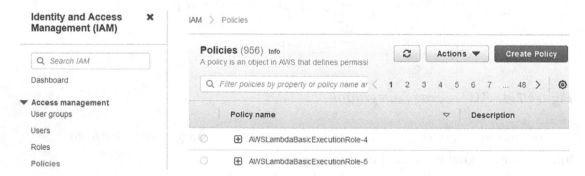

Figure 7-5.   *IAM policy (abridged)*

Listing 7-1 shows the JSON policy changes to be made. Simply cut and paste Listing 7-1 into the new policy. As S3 buckets are unique, rename btsdc-test-bucket to your chosen S3 bucket name to prevent a namespace clash.

Listing 7-1. Policy JSON

```
{
    "Version": "2012-10-17",
    "Statement": [
        {
            "Sid": "VisualEditor0",
            "Effect": "Allow",
            "Action": [
                "s3:PutObject",
                "s3:GetObject",
                "s3:DeleteObjectVersion",
                "s3:DeleteObject",
                "s3:GetObjectVersion"
            ],
            "Resource": "arn:aws:s3:::btsdc-test-bucket/*"
        },
        {
            "Sid": "VisualEditor1",
            "Effect": "Allow",
            "Action": "s3:ListBucket",
            "Resource": "arn:aws:s3:::*"
        }
    ]
}
```

When your policy has been amended, review the policy. Add your chosen textual descriptions, and then complete the configuration.

From the menu on the left, select Role ➤ Create Role. The IAM screen appears, looking similar to Figure 7-6.

Figure 7-6. *IAM role (abridged)*

In our example, we create a role called test_role, which generates a new ARN role appearing on the Summary page, as shown in Figure 7-7.

Figure 7-7. *Generated ARN (abridged)*

Copy the ARN for future use by our Snowflake storage integration created in step 2. If you chose a different role name, ensure the storage integration is updated to your role name; otherwise, authentication fails when attempting to access the S3 bucket. This is the affected line.

```
STORAGE_AWS_ROLE_ARN     = 'arn:aws:iam::616701129608:role/test_role'
```

Step 2. S3 Bucket

We may choose to segregate our data ingestion. One way to do this is by implementing separate S3 buckets. Using a browser, navigate to the AWS Management Console and search for S3.

1. Click Create Bucket using btsdc-test-bucket or the same S3 label used in the JSON section in step 1.

2.   Click Create Bucket, and verify bucket has been provisioned correctly.

3.   In the Permissions tab, edit the bucket policy, and if supplied by your organization, replace the policy with your local settings.

For those building out their organization infrastructure, a security scanning tool should be run to ensure the S3 bucket security policy complies with your organization's policy.

Step 3. Define Storage Integration

With the S3 bucket configured, let's turn our attention to Snowflake. Note the requirement to revert to the AWS account to establish the trust relationship. This example reuses the TEST database from Chapter 5. Given the custom nature of configuring S3, please treat the code as an example for modification to suit your own environment and rename btsdc-test-bucket to your S3 bucket name.

```
USE ROLE       accountadmin;
USE DATABASE TEST;
USE SCHEMA     public;

CREATE OR REPLACE STORAGE INTEGRATION test_integration
TYPE                       = EXTERNAL_STAGE
STORAGE_PROVIDER           = S3
ENABLED                    = TRUE
STORAGE_AWS_ROLE_ARN       = 'arn:aws:iam::616701129608:role/test_role'
STORAGE_ALLOWED_LOCATIONS = ( 's3://btsdc-test-bucket/' );
```

We may also specify multiple STORAGE_ALLOWED_LOCATIONS, including subdirectories, to segregate data sets in S3.

Unless there is a need, only run the storage integration once as the STORAGE_AWS_ EXTERNAL_ID changes each time, breaking the trust relationship.

Remember to grant usage to SYSADMIN.

```
GRANT USAGE ON INTEGRATION test_integration TO ROLE sysadmin;
```

Once our storage integration is created, we must retrieve the Amazon Resource Name (ARN). Execute the following command to display the Snowflake settings for the storage integration from which we use two, noted to establish the trust relationship in step 4.

```
DESC INTEGRATION test_integration;
```

Make a note of the following items from the command output.

STORAGE_AWS_IAM_USER_ARN ('arn:aws:iam::291942177718:user/6l9v-s-ukst4004')

STORAGE_AWS_EXTERNAL_ID ('VJ59634_SFCRole=3_6tGZNKWx80AtC+qFhlpF iMS1WYE=')

Step 4. Trust Relationship

Returning to AWS Management Console, we must edit our role trust relationship to add STORAGE_AWS_IAM_USER_ARN and STORAGE_AWS_EXTERNAL_ID values derived from step 3. Figure 7-8 illustrates how to edit the trust relationship.

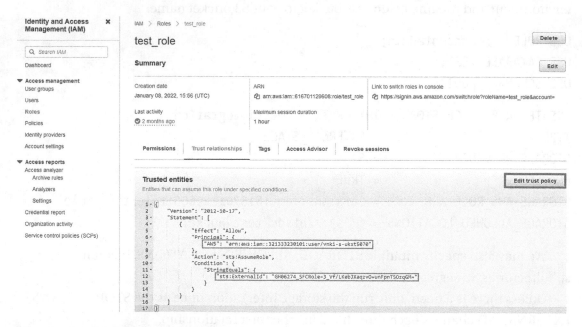

Figure 7-8. *Edit trust relationship*

Note   Every time storage integration is declared, the trust relationship MUST be updated.

Listing 7-2 shows parameters set in the JSON trust relationship.

Listing 7-2. Updated trust relationship

```
{
  "Version": "2012-10-17",
  "Statement": [
    {
      "Effect": "Allow",
      "Principal": {
        "AWS": "arn:aws:iam::291942177718:user/6l9v-s-ukst4004"
      },
      "Action": "sts:AssumeRole",
      "Condition": {
        "StringEquals": {
          "sts:ExternalId": "VJ59634_SFCRole=3_6tGZNKWx80AtC+qFhlpFiMS1WYE="
        }
      }
    }
  ]
}
```

Step 5. Define External Stage

With storage integration in place, let's move on to creating an external stage that acts as the interface between Snowflake and our external storage provider. In this example, our S3 bucket. Note the fully qualified stage path. Out of personal preference, I segregate system objects from application schema objects then I have a single location for all referenced STAGE objects in a database. Rename btsdc-test-bucket to your S3 bucket name.

```
USE ROLE      sysadmin;
USE DATABASE  TEST;
USE SCHEMA    public;

CREATE OR REPLACE STAGE TEST.public.test_stage
STORAGE_INTEGRATION = test_integration
DIRECTORY           = ( ENABLE = TRUE AUTO_REFRESH = TRUE )
```

```
ENCRYPTION              = ( TYPE = 'SNOWFLAKE_SSE' )
URL                     = 's3://btsdc-test-bucket/'
FILE_FORMAT             = TEST.public.test_pipe_format;
```

In the STAGE declaration are some non-obvious parameters to which I draw your attention because they are relevant for Chapter 9 but can be ignored for now.

- DIRECTORY specifies whether to add a directory table to the stage providing additional capability used later in this book, further explained at https://docs.snowflake.com/en/sql-reference/ sql/create-stage.html#directory-table-parameters- directorytableparams.

- ENCRYPTION specifies the type of encryption used for all files in the stage. We use SNOWFLAKE_SSE for server-side encryption and unstructured files. This is further explained at https:// docs.snowflake.com/en/sql-reference/sql/create-stage. html#internal-stage-parameters-internalstageparams.

- URL specifies the URL for the external location used to store files for loading and/or unloading, further explained at https:// docs.snowflake.com/en/sql-reference/sql/create-stage. html#external-stage-parameters-externalstageparams.

Typically, I would create a stage for each mapped S3 subfolder (if declared), providing physical segregation. Therefore, security for each source data set is to be loaded, and the same for data unloaded to S3 if permitted by our account security settings.

Step 6. Testing External Storage

With everything configured, we should now test our configuration to ensure our S3 integration to Snowflake works. To test, we must load a file into our S3 bucket. Any dummy data file will suffice.

Note   Ensure test files do not contain personally identifiable or otherwise sensitive information.

Using the AWS Management Console, navigate to your S3 bucket (mine is btsdc-test-bucket; your S3 bucket name will be different), then click Upload. The drag-and-drop option becomes available. Drop a file into the browser, and then click Upload. Your action should be successful.

All being well, we should be able to see the uploaded file via our stage.

```
LIST @TEST.public.test_stage;
```

Figure 7-9 shows our test file.

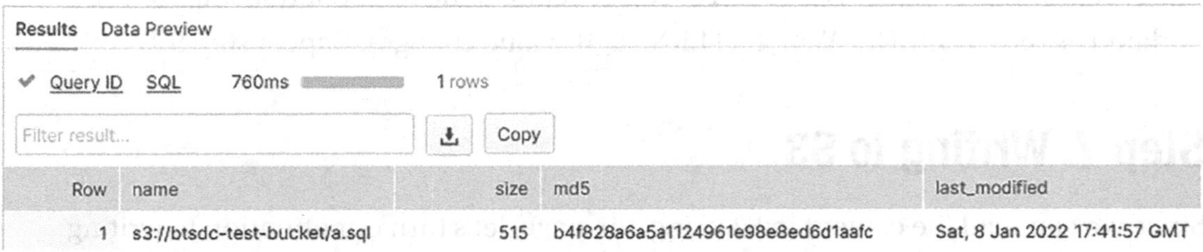

Row	name	size	md5	last_modified
1	s3://btsdc-test-bucket/a.sql	515	b4f828a6a5a1124961e98e8ed6d1aafc	Sat, 8 Jan 2022 17:41:57 GMT

Results   Data Preview

✔ Query ID   SQL   760ms   1 rows

Filter result...   Copy

Figure 7-9.   *Uploaded file in stage*

We can now SELECT from our file.

```
SELECT $1 FROM @TEST.public.test_stage;
```

Your results will differ from mine. And recalling our earlier discussion on file extensions, if your test file is a Microsoft Word document or similar, expect to see a corrupted result set as the file contents are not stored in plain text. Try it out by uploading a Microsoft Word document into your S3 bucket.

Note   Files in stages use positional notation for attributes. There is no metadata declared for mapping into Snowflake data types. Later in this book, we use external tables to overlay metadata.

Troubleshooting

An inevitable consequence of trying something for the first time is the probability of something failing. Here are some tips to assist with your troubleshooting.

When accessing external storage, you may see a SQL execution error: Error assuming AWS_ROLE. Please verify that the role and external ID are configured correctly in your AWS policy. This error implies a mismatch between the role declared in AWS and the role used in the storage integration. Refer to the preceding screenshots, check IAM Role ARN and JSON configurations along with SQL for extraneous space characters before and after single quote marks containing labels. Also, ensure the quote marks are correct because cutting and pasting between Microsoft Word documents and the Snowflake user interface can cause quotes to be misinterpreted.

Every time STORAGE INTEGRATION is redefined, the trust relationship must be updated as the STORAGE_AWS_EXTERNAL_ID value changes. Repeat step 5.

Step 7. Writing to S3

Since we can read file content in S3 using our stage, let's turn our attention to writing Snowflake data content out to S3, which is called *unloading*. Please also refer to Snowflake documentation found at `https://docs.snowflake.com/en/user-guide/data-unload-s3.html`.

First, we must grant entitlement to objects created in this chapter.

```
USE ROLE securityadmin;

GRANT USAGE ON FILE FORMAT TEST.public.test_pipe_format TO ROLE test_
object_role;
GRANT USAGE ON STAGE        TEST.public.test_stage      TO ROLE test_
object_role;
GRANT USAGE ON SCHEMA       TEST.public                 TO ROLE test_
object_role;
```

Then switch roles. Let's assume the declarations have been reused from Chapter 5.

```
USE ROLE      IDENTIFIER ( $test_object_role   );
USE DATABASE  IDENTIFIER ( $test_database       );
USE SCHEMA    IDENTIFIER ( $test_staging_schema );
USE WAREHOUSE IDENTIFIER ( $test_warehouse       );
```

Prove there is data available to unload.

```
SELECT * FROM TEST.test_owner.int_test_load;
```

And we can use the stage.

```
LIST @TEST.public.test_stage;
```

Now unload data to a file in the stage.

```
COPY INTO @TEST.public.test_stage/int_test_load FROM TEST.test_owner.int_
test_load;
```

Prove our file has been unloaded into the stage.

```
LIST @TEST.public.test_stage;
```

Figure 7-10 shows our unloaded file noting the file has been compressed using GZIP. The default behavior was called out previously when the file format was declared.

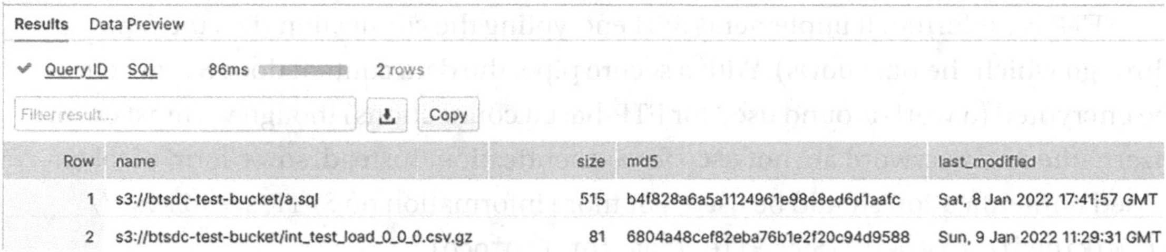

Results	Data Preview				

✔ Query ID	SQL	86ms		2 rows	

Filter result...		⤓	Copy		

Row	name	size	md5	last_modified
1	s3://btsdc-test-bucket/a.sql	515	b4f828a6a5a1124961e98e8ed6d1aafc	Sat, 8 Jan 2022 17:41:57 GMT
2	s3://btsdc-test-bucket/int_test_load_0_0_0.csv.gz	81	6804a48cef82eba76b1e2f20c94d9588	Sun, 9 Jan 2022 11:29:31 GMT

Figure 7-10. *Unloaded file in external stage*

Loading and Unloading Data

This section briefly describes ways to load files into our S3 bucket in preparation for ingestion into Snowflake. The following options are typically used in our organizations according to the level of maturity of our infrastructure and approach. But it should be obvious now that the Snowflake Data Cloud's success depends on lots of it, and the willing collaboration of data custodians to open their silos.

Referring to Chapter 2, you see how file-based transfers are typically implemented to support point-in-time data dumps from source to target representing the minimum capacity that can be delivered quickly and often on a tactical basis. For more sophisticated data transfers, we should look to API or possibly whole schema synchronization tooling, noting the challenges these approaches bring.

Notwithstanding, we may also want to write data from Snowflake into S3, as shown earlier, noting file format may affect how files are written.

AWS Management Console

We used the AWS Management Console while testing our STAGE to load files into S3. While useful for one-off loads or testing, we discourage using the AWS Management Console, which allows objects to be redefined. Just be aware the option is available.

Secure File Transfer Protocol (SFTP)

Back in the day, when IT development was much simpler, we used shell scripts and FTP to build point-to-point interfaces scheduled using cron for transferring data between systems. However, FTP is an insecure protocol because authentication is via username and password transferred in plain text, and data is unencrypted, therefore vulnerable to sniffing, spoofing, and brute force attacks.

SFTP is preferred. It implements SSH encrypting the connection (i.e., the pipe through which the data flows). With a secure pipe, the data content does not need to be encrypted (a work-around used for FTP-based connections) though we must ensure username and password are not used for authentication. Instead, some form of token-based authentication should be used. For more information on SFTP, see `https://en.wikipedia.org/wiki/SSH_File_Transfer_Protocol`.

As you might expect, AWS offers a suite of tools for implementing file transfer and integration with S3. Search for AWS Transfer Family; documentation is at `https://docs.aws.amazon.com/transfer/latest/userguide/what-is-aws-transfer-family.html`. Note your S3 bucket security profile may require a change to enable SFTP integration with a knock-on impact on your organization's security profile and security scanning outcome, do check first. Don't assume changes made will be acceptable to your cybersecurity team.

Managed File Transfer (MFT)

MFT is typically an on-prem tool that can connect with S3 buckets, ideal for file transfer integration with Snowflake. MFT provides a much wider suite of services than SFTP, including data protection at rest, scheduling, encryption, and desktop drag and drop integration. Many organizations provide MFT solutions and standard configurations, releasing developers to deliver functionality, not infrastructure. For more information on MFT, see `https://en.wikipedia.org/wiki/Managed_file_transfer`.

ELT/ETL

ELT (extract, load, and transform) loads high-volume data directly into a target system for later transformation. In contrast, ETL (extract, transform, and load) performs transformations before loading into a target system and is usually for low-volume data sets. A useful comparison is at `www.guru99.com/etl-vs-elt.html`.

This chapter so far has focused on S3 integration with Snowflake, while we can use either ELT or ETL tools to write to S3. Chapter 8 discusses a better integration pattern for ELT/ETL tooling.

SnowSQL

So far, our interaction with Snowflake has been via the user interface, but this is not the only tooling available to interact with Snowflake; documentation is at `https://docs.snowflake.com/en/user-guide/snowsql.html`. The following assumes you have installed SnowSQL.

SnowSQL supports operations not available via the user interface. For example, loading data from the desktop into Snowflake and attempting to use PUT results in an error: "SQL compilation error: The command is not supported from the UI: PUT."

In the Snowflake user interface, issue these commands, then manually construct your SnowSQL login noting the current_region needs amending to match this format: eu-west-2.aws.

```
SELECT current_account()||' '||current_region()||' '||current_user();
```

To log in to SnowSQL, open a command prompt and issue the constructed statement followed by the password.

```
snowsql -a vj59634.eu-west-2.aws -u andyc
```

As I prepared this chapter, I attempted to log in using an old version of SnowSQL, at which point the latest version of SnowSQL automatically downloaded and installed. Note the installation took a few seconds to begin, and several attempts to complete and post-login shows the latest version.

Direct UI Load

As Snowflake developers, we should already be aware of the capability to load data into our tables from the user interface. An example is shown in the next section.

Types of Stages

We briefly discussed stages in Chapter 3. Now that I've explained the external stages, let's dive into the remaining three types of stages, as shown in Figure 7-11.

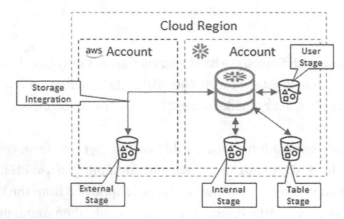

Figure 7-11. *Snowflake stages*

Internal stage types identified in Figure 7-9 are explained individually. Next, each stage resides in the Snowflake VPC, which is hosted on AWS in our example. We, therefore, incur storage charges for using internal stages as these resolve to S3 buckets in the underlying AWS account.

Named Stage

Internal stages are created similarly to external ones, though the options differ. Specifically, there is no STORAGE INTEGRATION clause, and internal stages may be either permanent or temporary. Suitable use cases for named stages include storing files for use by multiple users or data identified for load into multiple tables.

> **Note**   When referencing an internal stage, the syntax is identical to referencing an external stage. We use the @ symbol to prefix the STAGE name (e.g., @TEST. public.named_stage).

Creating an internal stage automatically creates an S3bucket in our Snowflake VPC.

```
USE ROLE      sysadmin;
USE DATABASE  TEST;
USE SCHEMA    public;

CREATE OR REPLACE STAGE TEST.public.named_stage
DIRECTORY         = ( ENABLE = TRUE )
ENCRYPTION        = ( TYPE = 'SNOWFLAKE_SSE' )
FILE_FORMAT       = TEST.public.test_pipe_format;
```

As yet, there are no files loaded into our named stage. To do so using SnowSQL, we use the PUT command. SnowSQL authentication may be username/password or token-based, depending on your organization. However, the PUT command can push a data file into a named stage, assuming a successful login.

To examine the contents of our named stage.

```
LIST @TEST.public.named_stage;
```

From SnowSQL (see the preceding), set our context as shown in Figure 7-12. Note the prompt changes to reflect the current context.

```
USE ROLE      sysadmin;
USE DATABASE  TEST;
USE SCHEMA    public;
```

Figure 7-12. *SnowSQL context*

Now upload a test file noting syntax for the file located in the drive A root directory.

```
PUT file:///a:/a.sql @TEST.public.named_stage;
```

Figure 7-13 illustrates the SnowSQL confirmation of file upload.

```
andyc#COMPUTE_WH@TEST.PUBLIC>PUT file:///a:/a.sql @TEST.public.named_stage;
+--------+-----------+-------------+-------------+-------------------+-------------------+----------+---------+
| source | target    | source_size | target_size | source_compression | target_compression | status   | message |
+--------+-----------+-------------+-------------+-------------------+-------------------+----------+---------+
| a.sql  | a.sql.gz  |         684 |         446 | NONE              | GZIP              | UPLOADED |         |
+--------+-----------+-------------+-------------+-------------------+-------------------+----------+---------+
1 Row(s) produced. Time Elapsed: 2.376s
```

Figure 7-13. *SnowSQL upload file*

Then confirm test file can be seen from the named stage.

```
LIST @TEST.public.named_stage;
```

Figure 7-14 illustrates the SnowSQL confirmation of file listing.

```
andyc#COMPUTE_WH@TEST.PUBLIC>LIST @TEST.public.named_stage;
+--------------------+------+----------------------------------+------------------------------------+
| name               | size | md5                              | last_modified                      |
+--------------------+------+----------------------------------+------------------------------------+
| named_stage/a.sql.gz | 446 | 467c3406633bccf91b1daef58615294e | Sun, 9 Jan 2022 16:17:10 GMT       |
+--------------------+------+----------------------------------+------------------------------------+
1 Row(s) produced. Time Elapsed: 0.163s
andyc#COMPUTE_WH@TEST.PUBLIC>
```

Figure 7-14. *SnowSQL stage contents*

Reverting to the user interface, we can now see the file uploaded to our named stage.

```
LIST @TEST.public.named_stage;
```

Note the file has been compressed using GZIP; hence examining the contents results in corrupted characters or an error since the character set appears corrupt.

To PUT the file uncompressed, either change the file format or override the default setting.

```
PUT file:///a:/a.sql @TEST.public.named_stage auto_compress=FALSE;
```

Using this query, we may also view information for both external and internal named stages.

```
SELECT * FROM information_schema.stages;
```

And finally, remove files uploaded into our named stage.

```
REMOVE @TEST.public.named_stage;
```

Figure 7-15 shows the response from the user interface when removing files.

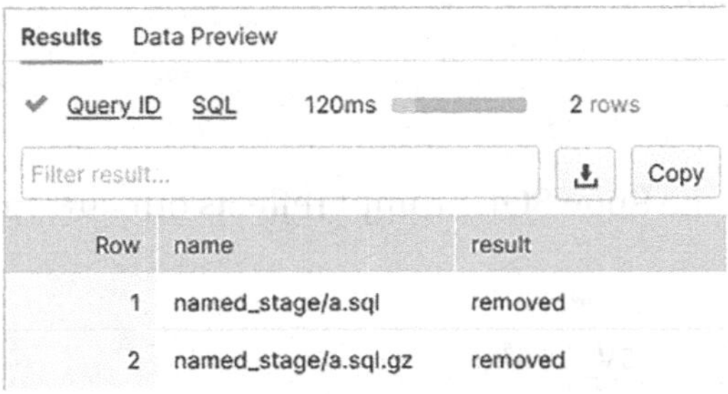

Figure 7-15. *Remove from named stage confirmation*

Table Stage

Every table in Snowflake has an associated table stage by default, which resolves to an underlying S3 bucket. We cannot rename or drop the associated table stage, and the file format cannot be associated with the stage.

Data upload can be done either through the user interface by directly referencing the table and uploading or using the PUT command from SnowSQL. Be aware of the file format used when uploading via a user interface. In our example, stripping a header record. The first row is removed automatically if your test data does not contain a header record.

Another interesting feature is to look at the user interface History tab, which shows two rows when uploading data. The first is a PUT statement, and the second is a COPY statement. You may find these useful when debugging your code. Furthermore, when running SnowSQL with the same user, commands issued in SnowSQL appear in the browser history, providing another means to gain information on executed queries. Just remember to refresh the history.

Note   Syntax when referencing a table stage differs. we use '@%' symbol to prefix the STAGE name (e.g., @%csv_test, or for fully qualified reference: '@"TEST"."PUBLIC".%"CSV_TEST"').

We cannot transform data when loading using a query into a table stage; therefore, direct load into a table stage is of limited use.

```
USE ROLE       sysadmin;
USE DATABASE   TEST;
USE SCHEMA     public;
```

With our context set, the SnowSQL prompt reflects our current context.

Create a test table.

```
CREATE OR REPLACE TABLE csv_test
(
id      NUMBER,
label   VARCHAR(30)
);
```

Create a dummy file (a.csv in the example) containing two records.

```
1000,ABC
1001,DEF
```

Upload file directly into table stage.

```
PUT file://a:\a.csv @%CSV_TEST auto_compress=FALSE;
```

To load the data into the table from its stage.

```
COPY INTO csv_test FROM @%CSV_TEST;
```

Prove test file a.csv has loaded. Two records should be returned, as shown in Figure 7-14.

```
SELECT * FROM csv_test;
```

Figure 7-16 illustrates the SnowSQL result set from our SQL command.

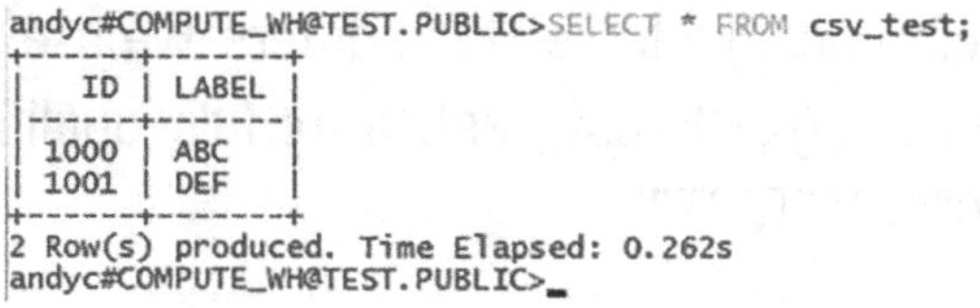

```
andyc#COMPUTE_WH@TEST.PUBLIC>SELECT * FROM csv_test;
+------+-------+
|   ID | LABEL |
|------+-------|
| 1000 | ABC   |
| 1001 | DEF   |
+------+-------+
2 Row(s) produced. Time Elapsed: 0.262s
andyc#COMPUTE_WH@TEST.PUBLIC>_
```

Figure 7-16. *SnowSQL stage contents*

And finally, remove files uploaded into our table stage.

```
REMOVE @%CSV_TEST;
```

You may recall that I mentioned uploading data via the user interface directly into a table. As you know, Snowflake has an immutable history, and the user interface has a History tab where every command is recorded. When uploading data via the user interface, the point and click commands are translated into PUT and COPY commands.

Here's an example from testing conducted while writing this chapter.

```
PUT 'file:///a.csv' '@"TEST"."PUBLIC".%"CSV_TEST"/ui1641748114927';

COPY INTO "TEST"."PUBLIC"."CSV_TEST" FROM @/ui1641748114927 FILE_FORMAT
= '"TEST"."PUBLIC"."TEST_PIPE_FORMAT"' ON_ERROR = 'ABORT_STATEMENT'
PURGE = TRUE;
```

User Stage

User stages are for internal use only, and direct access is strongly discouraged. Their inclusion in this chapter is for both completeness and interest.

Every Snowflake user has a user stage allocated by default that cannot be renamed or dropped. The file format cannot be declared for a user stage.

Note   Syntax when referencing a user stage differs. Use @~ to prefix the stage name (e.g., LIST @~;).

Not that we can use the information returned. You can see the user stages declared. The command executes in both the user interface and SnowSQL. Be aware that the result set largely contains records relating to the user interface tabs. There is one record for each, which looks like worksheet_data/d8498c76-7fdf-4c9b-81f4-e57d46543b66, plus a session record.

```
LIST @~;
```

Files can also be loaded into the user stage area via SnowSQL. Note that "dummy" can be any valid value.

```
PUT file://a:\a.csv @~/dummy auto_compress=FALSE;
```

And LISTed.

```
LIST @~;
LIST @~/dummy;
```

I do not recommend uploading files into the user stage because they are for the connected session only. Code is provided for example purposes only to illustrate the "art of the possible."

Cleanup

Apart from cleanup scripts to remove test code, you should also be mindful to periodically remove redundant files uploaded to internal stages. Remember, each file contributes toward storage costs.

```
USE ROLE       sysadmin;
USE DATABASE   TEST;
USE SCHEMA     public;

REMOVE @TEST.public.test_stage;
REMOVE @TEST.public.named_stage;
REMOVE @%CSV_TEST;

DROP STAGE       TEST.public.test_stage;
DROP STAGE       TEST.public.named_stage;
DROP FILE FORMAT TEST.public.test_pipe_format
DROP TABLE       csv_test;

USE ROLE     accountadmin;
DROP STORAGE INTEGRATION test_integration;
```

Summary

This chapter introduced the AWS account and provided information on configuring PrivateLink before introducing a new Snowflake object—file format.

I also introduced external storage concepts specifically focusing on AWS S3 and walked through the steps required to integrate a single AWS S3 bucket with Snowflake.

Interacting with the external stage content proved we can both read and write content, albeit with some default behavior affecting the outcome.

Interacting with the S3 bucket can be achieved in several ways. I introduced some techniques along with reference materials for further consideration. We build upon the principles as we progress through the next few chapters. I also introduced SnowSQL, which offers additional tooling, such as PUT and GET, for Snowflake's direct file load and unload.

For completeness, and because we are primarily dealing with stages in this chapter, we focused on understanding the remaining types of stages supported by Snowflake with a deep dive into each before cleaning up our code.

Next, let's focus on building data pipelines and integrating techniques from several chapters into useful tooling.

CHAPTER 8

Data Pipelines

Most of our data ingestion into Snowflake is via one of two routes: flat file submission loaded via S3 or directly landed using ELT/ETL tooling. We do not address ELT/ETL tooling specifically in this chapter but do call out entry points where relevant, instead of focusing on the Snowflake components and interactions.

You should be familiar with data historization and Slowly Changing Dimensions (SCD) concepts from previous chapters. I reuse the same pattern here but also recognize not every feed requires SCD2. However, for historical reporting purposes, it is most common to want to report temporally and be able to accurately re-create reports for any point in time that we hold data.

Once more, we call out how we address data corrections (at the source only, not in our repository) to preserve our system integrity and not allow data changes in two separate systems.

Please ensure your account was created for AWS with Business Critical Edition. The examples here are dependent upon Business Critical Edition features. Otherwise, please re-register using a different email address by selecting the Business Critical option at `https://signup.snowflake.com`.

A conceptual end-to-end data flow is shown in Figure 8-1. You have seen most of these components in previous chapters, but they are now collated into a simple view to be enriched throughout this chapter.

© Andrew Carruthers 2022
A. Carruthers, *Building the Snowflake Data Cloud*, https://doi.org/10.1007/978-1-4842-8593-0_8

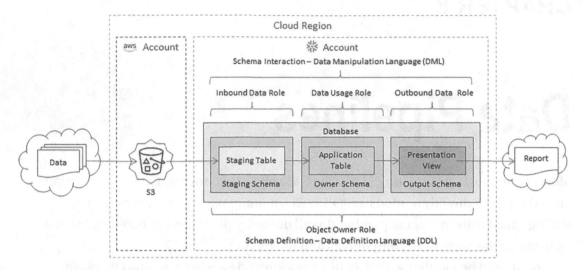

Figure 8-1. Conceptual data flow

I have referenced multiple schemas in diagrams to reinforce the message to segregate data for security and best practice. However, each chapter has a page budget. This chapter was proposed with 20 pages. I am far in excess, so I have developed everything using the SYSADMIN role, which is not recommended, and leave it for you to segregate into the proper roles. I trust you will forgive me.

What Is a Data Pipeline?

Data may be sourced from a variety of tools and delivered from data streams, low volume but high-frequency micro-batches, or high volume but less frequent data dumps. Data arrives continuously, with ever-increasing volumes, higher velocity, and from an expanding variety of sources. Our customers increasingly expect up-to-date data as freshness is critical and underpins decision-making. We also see a greater desire to hold historical data over longer periods for trend analysis. Traditional batch-based processing is being replaced with continuous feeds. Transactional boundaries become auto-ingestion, performance bottlenecks removed by serverless compute models.

We use the term *data pipeline* to describe a suite of related actions and operations to move data from one or more sources through a suite of transformation processes into a target destination for storage and subsequent consumption. Please also refer to the definition at `https://en.wikipedia.org/wiki/Pipeline_(computing)`.

In large organizations, we would typically land data in the AWS account S3 bucket, then ingest it into Snowflake using the external stage capability demonstrated in Chapter 7. Adopting an external stage approach enforces security options not available via other means. Another option utilizes named stages, noting any connections to upload files must be directly into Snowflake, a challenge we do not face when external stages are used, rendering named stages a far less attractive option. We would also prefer not to land data into table stages and certainly not into user stages. We want to make life a little easier for our operational support team. Our approach, therefore, focuses on external stages for the most part.

We have executed commands to create and merge data for loading and unloading files with discrete steps for each transition. We would not sit at our Snowflake console manually processing data in a real-world scenario. Instead, we would deploy components to automatically load and transform our data to publication, ready for consumption. Moving away from manual processing, Figure 8-2 illustrates data pipeline options and introduces new concepts for real-time data streaming and batch files.

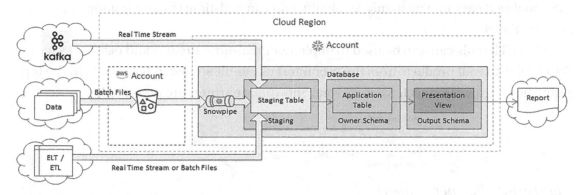

Figure 8-2. *Data pipeline options*

Batch Files

Based on a fixed schedule, batch files are typically produced by end-of-day processing from transactional source systems. Transactional systems may perform more than one operation on data throughout the working day. A single output file rarely captures all the individual operations carried out throughout the day but presents the most recent data state at the extract point. In other words, batch files ingested into Snowflake typically represent either a single end-of-day cut or perhaps several intra-day incremental cuts of data.

Source systems that feed into Snowflake are often constrained by both technology and design; therefore cannot easily make changes to facilitate data interchange. In an ideal world, data would be delivered in real time or as near real time as possible. Next, let's talk about some tooling.

Real-Time Stream

Apache Kafka is an open-source distributed event streaming platform used by thousands of companies for high-performance data pipelines, streaming analytics, data integration, and mission-critical applications. more information is at `https://kafka.apache.org`.

The Snowflake Connector for Kafka reads data from one or more Apache Kafka topics and loads the data into a Snowflake table. More information is at `https://docs.snowflake.com/en/user-guide/kafka-connector.html`.

Search for MSK – Amazon Managed Streaming for Apache Kafka from your AWS account. A fully managed, highly available, and secure Apache Kafka service. Amazon MSK makes it easy for you to migrate, build, and run real-time streaming applications on Apache Kafka.

ELT/ETL tools can also be used to implement real-time streams and batch files into Snowflake, to trial products from the user interface, as illustrated in Figure 8-3, in the top right click Partner Connect, where several Snowflake partners are presented.

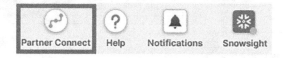

Figure 8-3. *Partner Connect*

Space does not permit a walk-through of every data pipeline option, and our focus returns to batch files. As we embark on a journey to develop our first data pipeline using Snowpipe, I must first introduce some new concepts.

Snowpipe

Snowpipe implements event-triggered data ingestion using AWS account components and built-in Snowflake capabilities. A full walk-through for AWS is found at `https://docs.snowflake.com/en/user-guide/data-load-snowpipe-auto-s3.html`.

This section relies upon the external stage built in Chapter 7. While Snowpipe can be configured to work with internal (named, table, and user) stages, the auto-ingest feature described next only works with external stages. We, therefore, focus on external stages for Snowpipe, and I leave the internal stage for you to investigate.

Note   Check the cost implications of using Snowpipe as the file ingestion and serverless compute charges soon mount up with many small files.

Figure 8-4 provides a Snowpipe overview showing internal and external stages. We use PUT from SnowSQL or file upload from the user interface into the internal stage but do not specify how data is landed in the S3 bucket for subsequent processing via the external stage.

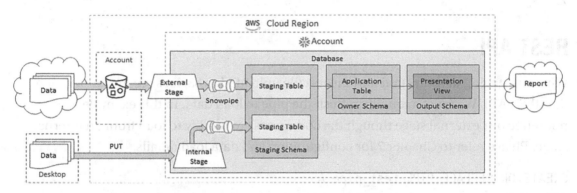

Figure 8-4. *Snowpipe overview*

Snowflake Pipe object is for a single target table and optionally has a file format. Pipes execute using serverless compute, which is a Snowflake provisioned resource. More information at https://docs.snowflake.com/en/user-guide/admin-serverless-billing.html.

Inheriting the following code from Chapter 7, an IAM role/profile, storage integration, and a trust relationship are assumed to be established. We begin by re-creating our external stage.

```
USE ROLE       sysadmin;
USE DATABASE   TEST;
USE WAREHOUSE  COMPUTE_WH;
USE SCHEMA     public;
```

```
CREATE OR REPLACE STAGE TEST.public.test_stage
STORAGE_INTEGRATION = test_integration
DIRECTORY           = ( ENABLE = TRUE AUTO_REFRESH = TRUE )
ENCRYPTION          = ( TYPE = 'SNOWFLAKE_SSE' )
URL                 = 's3://btsdc-test-bucket/'
FILE_FORMAT         = TEST.public.test_pipe_format;
```

Next, create a test staging table.

```
CREATE OR REPLACE TABLE pipe_load COPY GRANTS
(
id       NUMBER,
content  VARCHAR(255)
);
```

REST API

With our staging table created and the test file uploaded to S3, let's create a pipe using the REST API. We must manually refresh the pipe to load data. In this example, we reference our external stage though the code is easy to modify to load from an internal stage. Please refer to Chapter 7 for configuration and data load details.

```
CREATE OR REPLACE PIPE test_pipe AS
COPY INTO pipe_load FROM @TEST.public.test_stage
FILE_FORMAT = (TYPE = CSV);
```

And prove the pipe exists.

```
SHOW PIPES;
```

You should see results similar to those in Figure 8-5.

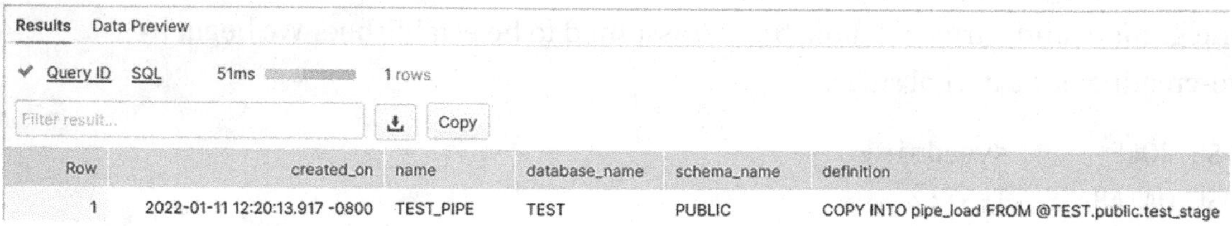

Results	Data Preview					
✔ Query ID SQL 51ms 1 rows						
Filter result...		⬆ Copy				
Row	created_on	name	database_name	schema_name	definition	
1	2022-01-11 12:20:13.917 -0800	TEST_PIPE	TEST	PUBLIC	COPY INTO pipe_load FROM @TEST.public.test_stage	

Figure 8-5. *SHOW PIPES output (part)*

So far, so good. You have learned how to create a PIPE object, but our test file has not been loaded. For example, you can try to query the data.

```
SELECT * FROM pipe_load;
```

We should not see any rows returned. We can check the pipe is running with this command.

```
SELECT system$pipe_status('TEST.public.test_pipe');
```

Where the output should be as in Figure 8-6.

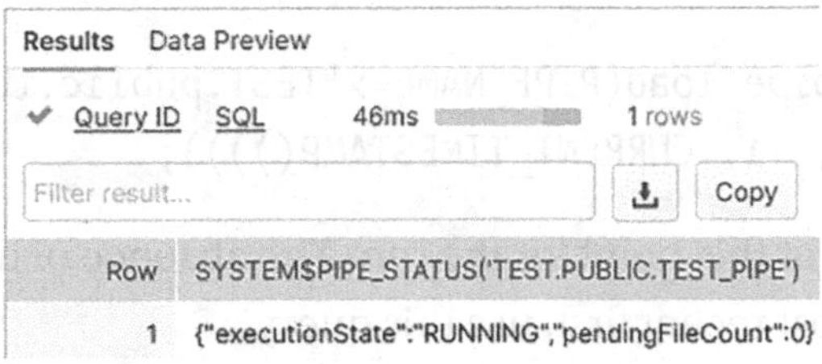

Figure 8-6. *Check pipe status*

With our pipe running, create a pipe_test.txt test file that contains the following.

```
id,content
1000,ABC
1001,DEF
```

Using the AWS Management Console, upload the pipe_test.txt test file into our btsdc-test-bucket S3 bucket. Note that your bucket name will differ. Do not forget to remove all other files, then prove the file has been uploaded.

```
LIST @TEST.public.test_stage;
```

To invoke the pipe, we must refresh. Note the response in Figure 8-7.

```
ALTER PIPE test_pipe REFRESH;
```

Figure 8-7. *ALTER PIPE response*

After a refresh, our pipe may take a few seconds to process the test file. If a huge test file was created, it might take longer. The following checks the progress.

```
SELECT *
FROM TABLE(validate_pipe_load(PIPE_NAME=>'TEST.public.test_pipe', START_
TIME=> DATEADD(hours, -1, CURRENT_TIMESTAMP())));
```

We may see an error at this point for the run. The absence of data indicates all is well, but to check all loads most recent first, use this query.

```
SELECT *
FROM TABLE(information_schema.copy_history(TABLE_NAME=>'PIPE_LOAD', START_
TIME=> DATEADD(hours, -1, CURRENT_TIMESTAMP())))
ORDER BY last_load_time DESC;
```

The result set in Figure 8-8 shows an error deliberately induced to illustrate what may be seen. Note the result set is only partial; further information is available in the user interface.

Row	FILE_NAME	STAGE_LOCATIO	LAST_LOAD_TIME	ROW_COUNT	ROW_PARSED	FILE_SIZE	FIRST_ERROR_ME	
1	pipe_test.txt	s3://btsdc-te...	2022-01-12 1...	2	2	32	NULL	r.public.test_stage
2	pipe_test.txt	s3://btsdc-te...	2022-01-12 1...	0	3	32	Numeric value...	

Figure 8-8. *Snowpipe copy history*

Finally, execute the following to clean up.

```
REMOVE @TEST.public.test_stage;
```

AUTO_INGEST

Having proven our pipe works with manual invocation, we turn our attention to automating Snowpipe by setting the AUTO_INGEST option. Figure 8-9 introduces a new AWS component, SQS enabling event notifications for our S3 bucket when new files are available to load. Snowflake documentation is at `https://docs.snowflake.com/en/user-guide/data-load-snowpipe-auto-s3.html#option-1-creating-a-new-s3-event-notification-to-automate-snowpipe`.

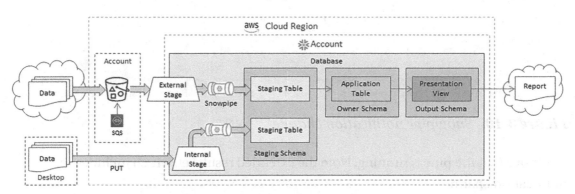

Figure 8-9. *SQS integration*

Set up Snowflake Objects

Create a test staging table for AUTO_INGEST.

```
CREATE OR REPLACE TABLE pipe_load_sqs COPY GRANTS
(
id       NUMBER,
content  VARCHAR(255)
);
```

Create a new pipe, test_pipe_sqs, with AUTO_INGEST = TRUE.

```
CREATE OR REPLACE PIPE test_pipe_sqs
AUTO_INGEST = TRUE
AS
COPY INTO pipe_load_sqs FROM @TEST.public.test_stage
FILE_FORMAT = (TYPE = CSV SKIP_HEADER = 1);
```

223

For example, redefining an existing pipe to change the file format retains the ARN; therefore, the SQS configuration does not need amending.

Check the pipe exists, taking note of the notification channel for the following configuration of AWS SQS.

```
SHOW PIPES;
SHOW PIPES LIKE 'test_pipe_sqs';
```

You should see the results in Figure 8-10.

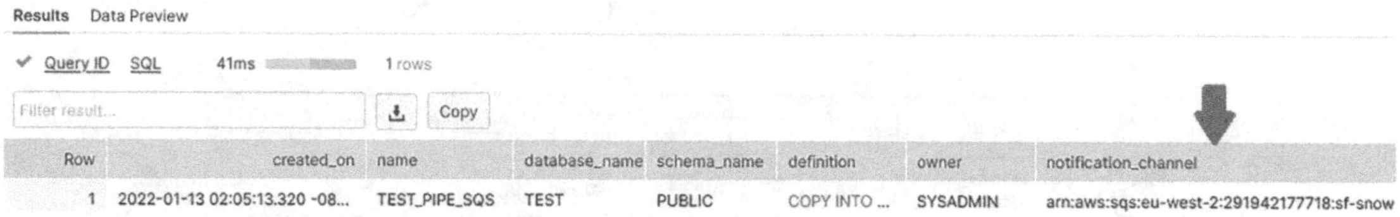

Figure 8-10. *Snowpipe notification channel*

Prove that the pipe is running. Note the extended result set, which includes the notification channel.

```
SELECT system$pipe_status('TEST.public.test_pipe_sqs');
```

Configure AWS SQS

Please read the Snowflake documentation on configuring event notifications before commencing work on this section. There are some limitations, specifically, a single event notification for the entire S3 bucket. Documentation is at `https://docs.snowflake.com/en/user-guide/data-load-snowpipe-auto-s3.html#step-4-configure-event-notifications`.

Each Snowpipe requires SQS configuring. The following information has been abstracted. The Snowflake documentation did not match the AWS fields listed when writing this book.

In the following walk-through, replace btsdc-test-bucket with your bucket name, remembering that S3 bucket names are global in scope.

Log in to the AWS Management Console.

1.   Select S3 ➤ btsdc-test-bucket ➤ Properties

2.   Scroll down to Event notifications ➤ Create event notification

3. General configuration ➤ Event name ➤ Snowpipe SQS

4. Ensure these options are checked.

    a. General configuration ➤ Event types ➤ All object create events

    b. General configuration ➤ Destination ➤ SQS queue

    c. General configuration ➤ Destination ➤ Enter SQS queue ARN

5. Populate the SQS queue with your pipe notification channel, which should look like this: "arn:aws:sqs:eu-west-2:291942177718:sf-snowpipe-AIDAUH6I4DO3MBH6RKJF6-pZC_wzWEcC6ATt5tvZpjyg".

6. Save the changes.

AWS documentation is at `https://docs.aws.amazon.com/AmazonS3/latest/userguide/enable-event-notifications.html`.

Test AUTO_INGEST

Our new pipe, test_pipe_sqs, should be running.

```
SELECT system$pipe_status('TEST.public.test_pipe_sqs');
```

With our pipe running, create a pipe_test_sqs.txt test file that contains the following.

```
id,content
1000,ABC
1001,DEF
```

Using the AWS Management Console, upload the pipe_test_sqs.txt test file into our btsdc-test-bucket S3 bucket, noting your bucket name will differ, not forgetting to remove all other files to prove the file can be seen via the external stage.

```
LIST @TEST.public.test_stage;
```

Note   Snowpipe typically runs within a minute of file receipt into S3.

Identical to the steps performed for our REST API Snowpipe walk-through, we can reuse the same SQL statements to check for progress and historical activity and ensure our test file has loaded into the target table.

```
SELECT *
FROM TABLE(validate_pipe_load(PIPE_NAME=>'TEST.public.test_pipe_sqs',
START_TIME=> DATEADD(hours, -1, CURRENT_TIMESTAMP())));

SELECT *
FROM TABLE(information_schema.copy_history(TABLE_NAME=>'PIPE_LOAD_SQS',
START_TIME=> DATEADD(hours, -1, CURRENT_TIMESTAMP())))
ORDER BY last_load_time DESC;

SELECT * FROM pipe_load_sqs;
```

Cleanup

Apart from cleanup scripts to remove our test code, we should also be mindful to periodically remove redundant files uploaded to external stages. Remember, each file contributes toward AWS account storage costs.

```
DROP PIPE test_pipe;
DROP PIPE test_pipe_sqs;

REMOVE @TEST.public.test_stage;
```

Further Considerations

- Snowpipe does not enforce load in file receipt order. If your requirement is for historized data preserving temporal integrity, then Snowpipe is not suitable. Snowpipe does not consider file upload to S3 timestamp and processes files in random order from those available.

- Snowpipe is one pipe per target table. A single pipe cannot load data into multiple target tables. With a 1:1 correlation of pipe to the table and SQS configuration, there could be a significant overhead to provisioning for large numbers of tables.

- Uploading identical files with the same name and content does not result in the file being ingested. Snowpipe maintains internal metadata for file tracking, and files are not removed from the stage.

- Not all CSP and storage combinations are supported. Please check valid combinations at `https://docs.snowflake.com/en/user-guide/data-load-snowpipe-intro.html#supported-cloud-storage-services`.

- Serverless compute is used to provision Snowpipe service; therefore, the billing model differs; more information is at `https://docs.snowflake.com/en/user-guide/data-load-snowpipe.html#loading-continuously-using-snowpipe`.

- AWS SNS may also be used to publish event notifications. More information is at `https://docs.snowflake.com/en/user-guide/data-load-snowpipe-auto-s3.html#option-2-configuring-amazon-sns-to-automate-snowpipe-using-sqs-notifications`. An explanation of the relationship between SNS and SQS is at `https://aws.amazon.com/blogs/aws/queues-and-notifications-now-best-friends/`.

- The Snowpipe troubleshooting guide is at `https://docs.snowflake.com/en/user-guide/data-load-snowpipe-ts.html#troubleshooting-snowpipe`.

Temporal Data Load

Snowpipe provides a simple, quick, and convenient way to rapidly ingest files into Snowflake, ideal for many micro-batch and continuous data ingestion use cases where temporal data management/SCD 2 is not required but noting Snowpipe does not remove consumed files.

This section is not prescriptive and offers an alternative way to ingest data into Snowflake, building upon the external stage created earlier. There are other ways to load data, and your use cases may significantly differ. I hope you find enough information of value to adapt to your own scenarios.

Recent changes to the stage syntax (for external stages only) optionally add a directory table to the stage. We previously included this line in our external stage definition, calling out the AUTO_REFRESH option, which manages the contents of the directory table. Note that directory content update is not immediate and may take a while to refresh after file upload into S3.

```
DIRECTORY          = ( ENABLE = TRUE AUTO_REFRESH = TRUE )
```

We extensively use the directory table features in Chapter 9. But, for now, we use the new capability in a limited manner to explain a temporal data load pattern where file receipt order determines ingestion.

Figure 8-11 depicts how Snowflake components can be configured to ingest data, noting the serial approach, which may be adequate for some use cases and is presented to show the "art of the possible."

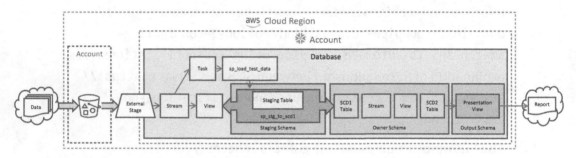

Figure 8-11. *Temporal data load schematic*

Streams

Streams can only be created on tables, external tables, and stage directory tables. Streams perform change data capture and are incredibly useful.

But there is a catch, and it is very subtle.

Suppose a stream is declared on a staging table, where we remove the contents in preparation for our next data load. In that case, the stream captures the events as DELETE operations, so the next run processes the deleted records in addition to the newly loaded data resulting in incorrect results. For this reason, we cannot use a stream when promoting data from the staging table (stg_content_test) through to the application table (scd1_content_test). But instead, we must use a stored procedure. If we give some thought to our design, we can create a generic stored procedure re-usable by many feeds, a single piece of code satisfying many feed ingestions, demonstrated next.

A stream can also generate an event since this fragment is later used to trigger a task. We use this feature to detect the presence of files loaded into S3.

```
WHEN system$stream_has_data ( '<your_stream_here>' )
```

According to Snowflake documentation, the directory table is only updated for new or updated files and not deleted files; see https://docs.snowflake.com/en/sql-reference/sql/create-stage.html#directory-table-parameters-directorytableparams.

During the preparation of this chapter, a time delay was observed between loading a file into S3 and the directory auto-refreshing, but the LIST command returned correct results immediately. Similar to Snowpipe, SQS can be configured to auto-refresh S3buckets. At the time of writing, this is a public preview feature. Instructions are at https://docs.snowflake.com/en/user-guide/data-load-dirtables-auto-s3.html.

To ensure that the S3 contents and directory table remain synchronized (or for the impatient), it may be necessary to force a refresh.

```
ALTER STAGE TEST.public.test_stage REFRESH;
```

Figure 8-12 shows the response when a deleted file has been removed from the directory table.

Figure 8-12. *Directory refresh*

Caution   Streams becomes stale when its offset is outside the data retention period for its source table; see `https://docs.snowflake.com/en/user-guide/streams.html`.

But we are getting ahead of ourselves and must focus on setting up our core structures and test cases without confusing matters, so ignore streams for this next section. We return to streams later.

Build Core Objects

First, create the content_test_20220115_130442.txt test file with the following data. The date_timestamp file suffix is explained later.

```
id,content,last_updated
1000,ABC,2022-01-15 13:04:42
1001,DEF,2022-01-15 13:04:42
1002,GHI,2022-01-15 13:04:42
```

Next, create a second test file named content_test_20220115_133505.txt. Populate that second file with the following data.

```
id,content,last_updated
1000,ABX,2022-01-15 13:35:05
1001,DEF,2022-01-15 13:04:42
1003,JKL,2022-01-15 13:35:05
```

The following differences now exist between the first file and the second.

```
1000 ABC -> ABX therefore last_updated has changed from original record
1001 No change
1002 Missing
1003 Is a new record
```

Upload both files into the S3 bucket, and then examine the content of our external stage where we should see both files.

```
LIST @TEST.public.test_stage;
```

Create tables to stage our files and then for our application.

```
CREATE OR REPLACE TABLE stg_content_test
(
id            NUMBER,
content       VARCHAR(30),
last_updated  TIMESTAMP_NTZ DEFAULT current_timestamp()::TIMESTAMP_NTZ
              NOT NULL
);

CREATE OR REPLACE TABLE scd1_content_test
```

```
(
id             NUMBER,
content        VARCHAR(30),
last_updated   TIMESTAMP_NTZ DEFAULT current_timestamp()::TIMESTAMP_NTZ
               NOT NULL
);
```

And create a stream on scd1_content_test.

```
CREATE OR REPLACE STREAM strm_scd1_content_test ON TABLE scd1_content_test;
```

Set up Test Cases

Before continuing, we must explain some assumptions. When ingesting temporal data, it is typical for submitted files to be suffixed with a date_timestamp. The usual convention for date_timestamp is YYYYMMDD_HH24MISS, making sorting easy for fairly obvious reasons. We might reasonably expect files to contain either every record from the source system (whole file submission) or only the changes made since the last file was submitted (delta file submission).

Our data pipeline must be configured according to the expected pattern. Here, we demonstrate the whole file submission pattern. In this example, we expect the id attribute to be unique in each file and never reused.

We have a lot of work to do, except for the end-state. We also add three more attributes: valid_from is the date_timestamp when the record was received into the system; valid_to is the date_timestamp when the record details changed, or a very long future date_timestamp defaulted to 99991231_235959; and current_flag is a Yes/No flag indicating the current record as a quick lookup.

Figure 8-13 shows our test files created side by side, along with the expected outcome from run 1, where we process our first file, content _test_20220115_130442.txt. Next, an Action column describes the expected outcome from run 2 where we process our second file, content _test_20220115_133505.txt.

content_test_20220115_130442.txt							Expected	content_test_20220115_133505.txt						
			Run 1									Run 2			
id	content	id	content	valid_from	valid_to	current_flag	Action	id	content	id	content	valid_from	valid_to	current_flag
1000	ABC	1000	ABC	20220114_220523	99991231_235959	Y	Update	1000	ABX	1000	ABC	20220114_220523	20220114_223139	N
										1000	ABX	20220114_223139	99991231_235959	Y
1001	DEF	1001	DEF	20220114_220523	99991231_235959	Y	No Action	1001	DEF	1001	DEF	20220114_220523	99991231_235959	Y
1002	GHI	1002	GHI	20220114_220523	99991231_235959	Y	Delete			1002	GHI	20220114_220523	20220114_223139	N
							Insert	1003	JKL	1003	JKL	20220114_223139	99991231_235959	Y

Figure 8-13. *Test file loads, actions, and outcomes*

In case you are wondering, this is an example of test-driven development, where we set out our test conditions, actions, and expected results before we begin developing our code. I strongly recommend this approach to software development. Chapter 13 explains how test cases validate code changes before deployment.

Although Figure 8-13 may be daunting, the content explains how SCD2 works.

- New records are those with an id attribute that does not exist in the most recently processed data set output. We expect an INSERT operation with valid_from date to the timestamp when data was received into the system, valid_to set to the default value, and current_flag set to Y.

- Absent records indicate a logical DELETE where valid_to is set to the timestamp of when the file was processed, current_flag set to N.

- Changed records result in a logical DELETE for the old record and an INSERT for the new record.

Build the Temporal Pipeline

Having established the test cases, we move on to write a stored procedure to both ingest and historize our data using the components developed so far and later introduce a task to implement the auto-ingestion process.

Tasks are the built-in Snowflake scheduling component. A full explanation of tasks is beyond the scope of this chapter. The page budget is under pressure through explaining in depth how components interact and are dependent, which is good preparation for later. Snowflake provides a far more detailed appraisal of tasks at `https://docs.snowflake.com/en/user-guide/tasks-intro.html`.

Our objectives are as follows.

- Identify each file in the external stage ordered by file name and date_timestamp, representing the timestamp recorded when the file was generated. We cannot rely upon the date_timestamp of when the file landed in S3 as this might result in files being processed out of sequence.

- Copy each file's content into the corresponding staging table. Ordinarily, we would hard-code the target table and filter our view, ensuring a 1:1 correlation between source and target; however, doing so would be of little value. Instead, we apply some logic to make our stored procedure more generic but at the cost of introducing an assumption. The file name and target table name are correlated.

- For each file content loaded into the staging table, determine the difference between the staging table content and the latest view of data in the application table, then merge the difference.

Let's unpack the most recent assumption a little more. Our example has two files containing full data sets but time separated: content_test_20220115_130442.txt and content_test_20220115_133505.txt. If we remove the leading underscore and date_timestamp suffix, we are left with content_test and this name, with a little imagination, becomes the target table.

Because we need to detect changes between our staging data and our latest held data, we must introduce new objects to enable automation. Figure 8-14 shows the addition of two stored procedures.

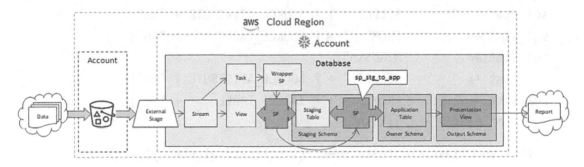

Figure 8-14. *Application objects*

Working backward, let's start with building our generic stored procedure (sp_stg_to_scd1) to ingest from the staging table to the application table. We need the stored procedure to embed it in the main ingestion stored procedure interfacing with the external stage.

```
CREATE OR REPLACE PROCEDURE sp_stg_to_scd1( P_SOURCE_DATABASE      STRING,
                                            P_SOURCE_TABLE         STRING,
                                            P_SOURCE_ATTRIBUTE     STRING,
                                            P_TARGET_TABLE         STRING,
                                            P_MATCH_ATTRIBUTE      STRING )
RETURNS STRING
LANGUAGE javascript
EXECUTE AS CALLER
AS
$$
   var sql_stmt     = "";
   var stmt         = "";
   var recset       = "";
   var result       = "";
   var update_cols  = "";
   var debug_string = '';

   sql_stmt  = "INSERT INTO " + P_TARGET_TABLE + "\n"
   sql_stmt += "SELECT *\n";
   sql_stmt += "FROM   " + P_SOURCE_TABLE     + "\n";
   sql_stmt += "WHERE  " + P_SOURCE_ATTRIBUTE + " IN\n";
   sql_stmt += "          (\n";
   sql_stmt += "          SELECT " + P_SOURCE_ATTRIBUTE + "\n";
   sql_stmt += "          FROM   " + P_SOURCE_TABLE     + "\n";
   sql_stmt += "          MINUS\n";
   sql_stmt += "          SELECT " + P_SOURCE_ATTRIBUTE + "\n";
   sql_stmt += "          FROM   " + P_TARGET_TABLE     + "\n";
   sql_stmt += "          );\n\n";

   stmt = snowflake.createStatement ({ sqlText:sql_stmt });
   debug_string += sql_stmt;

   try
   {
       recset = stmt.execute();
   }
   catch { result = sql_stmt; }
```

```
sql_stmt  = "SELECT column_name\n"
sql_stmt += "FROM    " + P_SOURCE_DATABASE + ".information_schema.
columns\n"
sql_stmt += "WHERE   table_name = :1\n"
sql_stmt += "AND     column_name NOT IN ( :2, :3 )\n"
sql_stmt += "ORDER BY ordinal_position ASC;\n\n"

stmt = snowflake.createStatement ({ sqlText:sql_stmt, binds:[ P_TARGET_
TABLE, P_SOURCE_ATTRIBUTE, P_MATCH_ATTRIBUTE ] });
sql_stmt      = sql_stmt.replace(":1", "'" + P_TARGET_TABLE      + "'");
sql_stmt      = sql_stmt.replace(":2", "'" + P_SOURCE_ATTRIBUTE + "'");
sql_stmt      = sql_stmt.replace(":3", "'" + P_MATCH_ATTRIBUTE  + "'");
debug_string += sql_stmt;

try
{
    recset = stmt.execute();
    while(recset.next())
    {
        update_cols += "tgt." + recset.getColumnValue(1) + " = stg." +
        recset.getColumnValue(1) + ",\n          "
    }
    update_cols = update_cols.substring(0, update_cols.length -9)
}
catch { result = sql_stmt; }

sql_stmt  = "UPDATE " + P_TARGET_TABLE + " tgt\n"
sql_stmt += "SET     " + update_cols    + ",\n";
sql_stmt += "        tgt." + P_MATCH_ATTRIBUTE + " = stg." + P_MATCH_
ATTRIBUTE +"\n";
sql_stmt += "FROM    " + P_SOURCE_TABLE + " stg\n";
sql_stmt += "WHERE   tgt." + P_SOURCE_ATTRIBUTE + "  = stg." + P_SOURCE_
ATTRIBUTE +"\n";
sql_stmt += "AND     tgt." + P_MATCH_ATTRIBUTE  + " != stg." + P_MATCH_
ATTRIBUTE +";\n\n";

stmt = snowflake.createStatement ({ sqlText:sql_stmt });
```

```
    debug_string += sql_stmt;

    try
    {
        recset = stmt.execute();
    }
    catch { result = sql_stmt; }

    sql_stmt  = "DELETE FROM " + P_TARGET_TABLE      + "\n"
    sql_stmt += "WHERE   " + P_SOURCE_ATTRIBUTE + " NOT IN\n";
    sql_stmt += "           (\n";
    sql_stmt += "            SELECT " + P_SOURCE_ATTRIBUTE + "\n";
    sql_stmt += "            FROM   " + P_SOURCE_TABLE      + "\n";
    sql_stmt += "           );\n\n";

    stmt = snowflake.createStatement ({ sqlText:sql_stmt });
    debug_string += sql_stmt;

    try
    {
        recset = stmt.execute();
    }
    catch { result = sql_stmt; }

    return debug_string;
//    return result;
$$;
```

Note the return value is the full list of generated SQL statements to assist with your debugging. Each SQL statement can be called independently.

Our test code provides a template to build our next stored procedure later in this chapter, but for now, we must prove sp_stg_to_scd1 works as expected, and the next code section provides the steps.

Load our content_test_20220115_130442.txt and content_test_20220115_133505.txt test files into S3, then load using the COPY command and stored procedure call and TRUNCATE to clear our staging table in between.

```
COPY INTO stg_content_test
FROM @TEST.public.test_stage/content_test_20220115_130442.txt
```

```
FILE_FORMAT = (TYPE = CSV SKIP_HEADER = 1)
PURGE       = TRUE;
```

Load first staged file into application table.

```
CALL sp_stg_to_scd1('TEST', 'STG_CONTENT_TEST', 'ID', 'SCD1_CONTENT_TEST',
'LAST_UPDATED');
```

Clear out the staging table in preparation for the second file load.

```
TRUNCATE TABLE stg_content_test;
```

Move the second staged file into a staging table.

```
COPY INTO stg_content_test
FROM @TEST.public.test_stage/content_test_20220115_133505.txt
FILE_FORMAT = (TYPE = CSV SKIP_HEADER = 1)
PURGE       = TRUE;
```

Merge second staged file into application table.

```
CALL sp_stg_to_scd1('TEST', 'STG_CONTENT_TEST', 'ID', 'SCD1_CONTENT_TEST',
'LAST_UPDATED');
```

Clear out the staging table in preparation for subsequent file load.

```
TRUNCATE TABLE stg_content_test;
```

Prove that the data is correct in table scd1_content_test, noting this is only an interim step. Additional processing has yet to be developed before we see the expected outcome shown in our test case.

```
SELECT * FROM scd1_content_test ORDER BY id ASC;
```

After both files have been run according to the instructions, the scd1_content_test table contains the current data view where only the latest records are available. For the curious, this is the Slowly Changing Dimension 1 pattern (SCD1); there is no history, only the latest view of data. Of note is the LAST_UPDATED attribute showing ID 1001 has the earliest date. Figure 8-15 shows the output from table scd1_content_test.

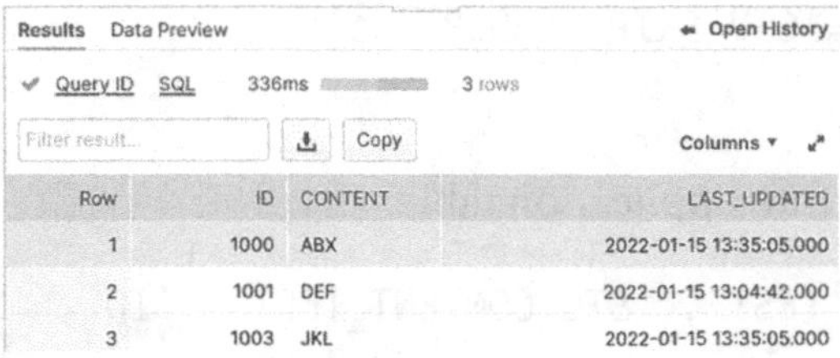

Figure 8-15. *Application results: scd1_content_test table*

Now that we have our staging to application stored procedure, we can build our main ingestion stored procedure interfacing with the external stage, then call sp_stg_to_scd1 to promote data, re-introduce our stream, which correctly captures deltas, and historize our data into SCD2 pattern.

Remember, our test files need uploading because the COPY command has PURGE = TRUE, meaning the files are removed from the STAGE after loading into the staging table.

The next step is to create our scd2_content_test table, where our historization will land. Note the decision attribute to assist in debugging and understanding; it is not essential for historization.

```
CREATE OR REPLACE TABLE scd2_content_test
(
id            NUMBER,
content       VARCHAR(30),
valid_from    TIMESTAMP_NTZ,
valid_to      TIMESTAMP_NTZ,
current_flag  VARCHAR(1),
decision      VARCHAR(100)
);
```

Next, let's implement a view joining both historization table scd2_content_test and the stream.

```
CREATE OR REPLACE VIEW v_content_test
AS
SELECT decision,
       id,
       content,
       valid_from,
```

```
        valid_to,
        current_flag,
        'I' AS dml_type
FROM    (
        SELECT 'New Record - Insert or Existing Record - Ignore' AS
        decision,
             id,
             content,
             last_updated         AS valid_from,
             LAG ( last_updated ) OVER ( PARTITION BY id ORDER BY last_
             updated DESC ) AS valid_to_raw,
             CASE
                WHEN valid_to_raw IS NULL
                     THEN '9999-12-31'::TIMESTAMP_NTZ
                     ELSE valid_to_raw
             END AS valid_to,
             CASE
                WHEN valid_to_raw IS NULL
                     THEN 'Y'
                     ELSE 'N'
             END AS current_flag,
             'I' AS dml_type
        FROM    (
             SELECT strm.id,
                    strm.content,
                    strm.last_updated
             FROM    strm_scd1_content_test   strm
             WHERE   strm.metadata$action   = 'INSERT'
             AND     strm.metadata$isupdate = 'FALSE'
             )
        )
UNION ALL
SELECT decision,
       id,
       content,
```

```
        valid_from,
        valid_to,
        current_flag,
        dml_type
FROM    (
        SELECT decision,
               id,
               content,
               valid_from,
               LAG ( valid_from ) OVER ( PARTITION BY id ORDER BY valid_from
               DESC ) AS valid_to_raw,
               valid_to,
               current_flag,
               dml_type
        FROM   (
               SELECT 'Existing Record - Insert' AS decision,
                      strm.id,
                      strm.content,
                      strm.last_updated           AS valid_from,
                      '9999-12-31'::TIMESTAMP_NTZ  AS valid_to,
                      'Y'                          AS current_flag,
                      'I' AS dml_type
               FROM   strm_scd1_content_test strm
               WHERE  strm.metadata$action   = 'INSERT'
               AND    strm.metadata$isupdate = 'TRUE'
               UNION ALL
               SELECT 'Existing Record - Delete',
                      tgt.id,
                      tgt.content,
                      tgt.valid_from,
                      current_timestamp(),
                      'N',
                      'D' AS dml_type
               FROM   scd2_content_test tgt
               WHERE  tgt.id IN
```

```
                 (
                 SELECT DISTINCT strm.id
                 FROM    strm_scd1_content_test strm
                 WHERE   strm.metadata$action   = 'INSERT'
                 AND     strm.metadata$isupdate = 'TRUE'
                 )
          AND    tgt.current_flag = 'Y'
          )
       )
UNION ALL
SELECT 'Missing Record - Delete',
      strm.id,
      strm.content,
      tgt.valid_from,
      current_timestamp()::TIMESTAMP_NTZ AS valid_to,
      NULL,
      'D' AS dml_type
FROM   scd2_content_test           tgt
INNER JOIN strm_scd1_content_test strm
   ON  tgt.id    = strm.id
WHERE   strm.metadata$action   = 'DELETE'
AND     strm.metadata$isupdate = 'FALSE'
AND     tgt.current_flag       = 'Y';
```

Test the Temporal Pipeline

As with all successful testing, we must ensure our baseline is known, and the cleanest way is to start from the top. Do not re-create storage integration or the trust relationship breaks; re-creating the stage is fine.

```
CREATE OR REPLACE STAGE TEST.public.test_stage
STORAGE_INTEGRATION = test_integration
DIRECTORY           = ( ENABLE = TRUE AUTO_REFRESH = TRUE )
ENCRYPTION          = ( TYPE = 'SNOWFLAKE_SSE' )
URL                 = 's3://btsdc-test-bucket/'
```

```
FILE_FORMAT            = TEST.public.test_pipe_format;
CREATE OR REPLACE STREAM strm_test_stage ON STAGE TEST.public.test_stage;
```

Upload files into S3 then refresh the stage directory.

```
ALTER STAGE TEST.public.test_stage REFRESH;
```

Clear out tables.

```
TRUNCATE TABLE stg_content_test;

TRUNCATE TABLE scd1_content_test;

CREATE OR REPLACE STREAM strm_scd1_content_test ON TABLE scd1_content_test;

TRUNCATE TABLE scd2_content_test;
```

Copy the first test file into the staging table.

```
COPY INTO stg_content_test
FROM @TEST.public.test_stage/content_test_20220115_130442.txt
FILE_FORMAT = (TYPE = CSV SKIP_HEADER = 1)
PURGE       = TRUE;
```

Promote to SCD1 table.

```
CALL sp_stg_to_scd1('TEST', 'STG_CONTENT_TEST', 'ID', 'SCD1_CONTENT_TEST',
'LAST_UPDATED');
```

Clear out staging table ready for next run.

```
TRUNCATE TABLE stg_content_test;
```

Merge SCD1 table into SCD2 table using v_content_test contents.

```
MERGE INTO scd2_content_test tgt
USING v_content_test strm
ON    tgt.id         = strm.id
AND   tgt.valid_from = strm.valid_from
AND   tgt.content    = strm.content
WHEN MATCHED AND strm.dml_type = 'U' THEN
UPDATE SET tgt.valid_to        = strm.valid_to,
```

```
            tgt.current_flag      = 'N',
            tgt.decision          = strm.decision
WHEN MATCHED AND strm.dml_type = 'D' THEN
UPDATE SET tgt.valid_to          = strm.valid_to,
            tgt.current_flag      = 'N',
            tgt.decision          = strm.decision
WHEN NOT MATCHED AND strm.dml_type = 'I' THEN
INSERT
(
tgt.id,
tgt.content,
tgt.valid_from,
tgt.valid_to,
tgt.current_flag,
tgt.decision
) VALUES (
strm.id,
strm.content,
current_timestamp(),
strm.valid_to,
strm.current_flag,
strm.decision
);
```

Copy the second test file into the staging table.

```
COPY INTO stg_content_test
FROM @TEST.public.test_stage/content_test_20220115_133505.txt
FILE_FORMAT = (TYPE = CSV SKIP_HEADER = 1)
PURGE       = TRUE;
```

Promote to the SCD1 table.

```
CALL sp_stg_to_scd1('TEST', 'STG_CONTENT_TEST', 'ID', 'SCD1_CONTENT_TEST',
'LAST_UPDATED');
```

Clear out the staging table to be ready for the next run.

```
TRUNCATE TABLE stg_content_test;
```

Merge SCD1 table into SCD2 table using v_content_test contents.

```
MERGE INTO scd2_content_test tgt
USING v_content_test strm
ON    tgt.id          = strm.id
AND   tgt.valid_from = strm.valid_from
AND   tgt.content    = strm.content
WHEN MATCHED AND strm.dml_type = 'U' THEN
UPDATE SET tgt.valid_to        = strm.valid_to,
           tgt.current_flag    = 'N',
           tgt.decision        = strm.decision
WHEN MATCHED AND strm.dml_type = 'D' THEN
UPDATE SET tgt.valid_to        = strm.valid_to,
           tgt.current_flag    = 'N',
           tgt.decision        = strm.decision
WHEN NOT MATCHED AND strm.dml_type = 'I' THEN
INSERT
(
tgt.id,
tgt.content,
tgt.valid_from,
tgt.valid_to,
tgt.current_flag,
tgt.decision
) VALUES (
strm.id,
strm.content,
current_timestamp(),
strm.valid_to,
strm.current_flag,
strm.decision
);
```

Now check the SCD2 table contents match the expected results.

```
SELECT * FROM scd2_content_test ORDER BY id ASC;
```

That's a lot of code and testing. I have added information into v_content_test and scd2_content_test, which I hope you find useful in explaining how the delta records are derived. Please also see the accompanying script with additional queries to assist in debugging.

But we are not finished yet. We still have to build our wrapper stored procedure to knit everything together and automate using a task.

Automate the Temporal Pipeline

The wrapper stored procedure that brings everything together follows. I leave it for you to review and understand as this chapter is far too long already. Note the hard-coding in to fix the database name and match column. This is intentionally left for you to figure out a way of dynamically generating, but I would add parameters to sp_load_test_data and then replace them in the CALL statement. Note the parameters MUST be in UPPERCASE when referenced. Likewise, the MERGE statement attributes need further consideration where code generation and lookup tables are both valid approaches.

```
CREATE OR REPLACE PROCEDURE sp_load_test_data() RETURNS STRING
LANGUAGE javascript
EXECUTE AS CALLER
AS
$$
   var sql_stmt  = "";
   var stmt      = "";
   var recset    = "";
   var result    = "";

   var debug_string   = '';

   var path_to_file   = "";
   var table_name     = "";

   sql_stmt  = "SELECT path_to_file,\n"
   sql_stmt += "        table_name\n"
   sql_stmt += "FROM    v_strm_test_stage\n"
   sql_stmt += "WHERE   metadata$action = 'INSERT'\n"
   sql_stmt += "ORDER BY path_to_file ASC;\n\n";
```

```
stmt = snowflake.createStatement ({ sqlText:sql_stmt });

debug_string = sql_stmt;

try
{
    recset = stmt.execute();
    while(recset.next())
    {
        path_to_file    = recset.getColumnValue(1);
        table_name      = recset.getColumnValue(2);

        sql_stmt  = "COPY INTO stg_" + table_name + "\n"
        sql_stmt += "FROM " + path_to_file +"\n"
        sql_stmt += "FILE_FORMAT = (TYPE = CSV SKIP_HEADER = 1)\n"
        sql_stmt += "PURGE        = TRUE;\n\n";

        debug_string = debug_string + sql_stmt;

        stmt = snowflake.createStatement ({ sqlText:sql_stmt });

        try
        {
            stmt.execute();
            result = "Success";
        }
        catch { result = sql_stmt; }

        sql_stmt  = "CALL sp_stg_to_scd1('TEST', 'STG_" + table_name +
        "', 'ID', 'SCD1_" + table_name + "', 'LAST_UPDATED');\n\n";

        debug_string = debug_string + sql_stmt;

        stmt = snowflake.createStatement ({ sqlText:sql_stmt });

        try
        {
            stmt.execute();
            result = "Success";
        }
```

```
catch { result = sql_stmt; }

sql_stmt  = "MERGE INTO scd2_" + table_name + " tgt\n"
sql_stmt += "USING v_" + table_name + " strm\n"
sql_stmt += "ON    tgt.id         = strm.id\n"
sql_stmt += "AND    tgt.valid_from = strm.valid_from\n"
sql_stmt += "AND    tgt.content   = strm.content\n"
sql_stmt += "WHEN MATCHED AND strm.dml_type = 'U' THEN\n"
sql_stmt += "UPDATE SET tgt.valid_to       = strm.valid_to,\n"
sql_stmt += "              tgt.current_flag    = 'N',\n"
sql_stmt += "              tgt.decision        = strm.decision\n"
sql_stmt += "WHEN MATCHED AND strm.dml_type = 'D' THEN\n"
sql_stmt += "UPDATE SET tgt.valid_to       = strm.valid_to,\n"
sql_stmt += "              tgt.current_flag    = 'N',\n"
sql_stmt += "              tgt.decision        = strm.decision\n"
sql_stmt += "WHEN NOT MATCHED AND strm.dml_type = 'I' THEN\n"
sql_stmt += "INSERT\n"
sql_stmt += "(\n"
sql_stmt += "tgt.id,\n"
sql_stmt += "tgt.content,\n"
sql_stmt += "tgt.valid_from,\n"
sql_stmt += "tgt.valid_to,\n"
sql_stmt += "tgt.current_flag,\n"
sql_stmt += "tgt.decision\n"
sql_stmt += ") VALUES (\n"
sql_stmt += "strm.id,\n"
sql_stmt += "strm.content,\n"
sql_stmt += "current_timestamp(),\n"
sql_stmt += "strm.valid_to,\n"
sql_stmt += "strm.current_flag,\n"
sql_stmt += "strm.decision\n"
sql_stmt += ");\n\n";

debug_string = debug_string + sql_stmt;

stmt = snowflake.createStatement ({ sqlText:sql_stmt });
```

```
            try
            {
                stmt.execute();
                result = "Success";
            }
            catch { result = sql_stmt; }

            sql_stmt  = "TRUNCATE TABLE stg_" + table_name + ";\n\n";

            debug_string = debug_string + sql_stmt;

            stmt = snowflake.createStatement ({ sqlText:sql_stmt });

            try
            {
                stmt.execute();
                result = "Success";
            }
            catch { result = sql_stmt; }
        }
    }
    catch { result = sql_stmt; }
    return debug_string;
//    return result;
$$;
```

The v_strm_test_stage stage directory view also needs further consideration because the path_to_file is hard-coded.

```
CREATE OR REPLACE VIEW v_strm_test_stage COPY GRANTS
AS
SELECT '@TEST.public.test_stage/'||relative_
path                                AS path_to_file,
       SUBSTR ( relative_path, 1, REGEXP_INSTR ( relative_path, '_20' ) - 1 )
       AS table_name,
       size,
       last_modified,
       metadata$action
FROM   strm_test_stage;
```

After repeating our test setup, run sp_load_test_data.

```
CALL sp_load_test_data();
```

Note   Repeatedly loading the same test file results in the SCD2 table containing multiple records. This is due to the content not changing. It is expected, if unwelcome, behavior.

Finally, we create the task to call the wrapper stored procedure, noting the use of system$stream_has_data, which detects the presence of new data in our stage directory table. If no new files exist, execution is ignored, and the task does not run.

```
CREATE OR REPLACE TASK task_load_test_data
WAREHOUSE = COMPUTE_WH
SCHEDULE  = '1 minute'
WHEN system$stream_has_data ( 'strm_test_stage' )
AS
CALL sp_load_test_data();
```

Before we can execute tasks, our role must be entitled, and we must use the ACCOUNTADMIN role.

We normally create a separate role to execute tasks, but space does not permit full production configuration, so more thought should be given to your setup. For expediency, we GRANT TO SYSADMIN, but this does not represent best practice.

```
USE ROLE accountadmin;
```

```
GRANT EXECUTE TASK ON ACCOUNT TO ROLE sysadmin;
```

Switch back to SYSADMIN.

```
USE ROLE sysadmin;
```

On creation and by default, tasks are suspended to enable.

```
ALTER TASK task_load_test_data RESUME;
```

To see which tasks have been declared.

```
SHOW tasks;
```

And to check when a task was last run.

```
SELECT timestampdiff ( second, current_timestamp, scheduled_time ) as
next_run,
       scheduled_time,
       current_timestamp,
       name,
       state
FROM   TABLE ( information_schema.task_history())
WHERE  state = 'SCHEDULED'
ORDER BY completed_time DESC;
```

Finally, to suspend a task.

```
ALTER TASK task_load_test_data SUSPEND;
```

Note   Tasks can quickly consume credits, ensuring only essential tasks are left running. Also consider adding resource monitors.

Cleanup

By now, you know how to clean up most objects, but to drop a JavaScript stored procedure, the declared prototype must match.

```
DROP PROCEDURE sp_stg_to_app( VARCHAR, VARCHAR, VARCHAR, VARCHAR, VARCHAR );
```

External Tables

If working through this chapter sequentially, ensure the task is suspended as this interferes with this external table section as we reuse one of the test files.

```
ALTER TASK task_load_test_data SUSPEND;
```

We might also consider external tables as an ingestion pattern. They are easy to configure and use. Documentation is at `https://docs.snowflake.com/en/sql-reference/sql/create-external-table.html`.

After suspending our task, upload files to S3, refresh, and check contents are visible.

```
ALTER STAGE TEST.public.test_stage REFRESH;

LIST @TEST.public.test_stage;

SELECT $1 FROM @TEST.public.test_stage/content_test_20220115_130442.txt;
```

Reusing our external stage and the first test file, the syntax for mapping a file to an external table is as follows.

```
CREATE OR REPLACE EXTERNAL TABLE ext_content_test_20220115_130442
(
id           VARCHAR AS (value:c1::varchar),
content      VARCHAR AS (value:c2::varchar),
last_updated VARCHAR AS (value:c3::varchar)
)
WITH LOCATION = @TEST.public.test_stage/
FILE_FORMAT = (TYPE = CSV SKIP_HEADER = 1)
PATTERN     = content_test_20220115_130442.txt;
```

Parquet, Avro, and ORC file formats metadata can be inferred. The documentation is at `https://docs.snowflake.com/en/sql-reference/functions/infer_schema.html`.

In line with Snowpipe auto-ingest, external tables can be refreshed using SQS. Documentation is at `https://docs.snowflake.com/en/user-guide/tables-external-s3.html`. Otherwise, refresh the external table.

```
ALTER EXTERNAL TABLE ext_content_test_20220115_130442 REFRESH;
```

To see which external tables are declared.

```
SHOW EXTERNAL TABLES;
```

Display content and associated file information.

```
SELECT $1, metadata$filename FROM @TEST.public.test_stage/;
```

Finally, the following extracts the content from our external table.

```
SELECT * FROM ext_content_test_20220115_130442;
```

This simple example can easily be extended to several files of the same format.

I prefer not to use external tables due to additional management overhead in declaring and maintaining, but I offer this example as another ingestion pattern.

Cleanup

Apart from cleanup scripts to remove out test code, we should also be mindful to periodically remove redundant files uploaded to internal stages remembering each file contributes toward storage costs.

```
USE ROLE       sysadmin;
USE DATABASE   TEST;
USE SCHEMA     public;

DROP TASK      task_load_test_data;
DROP DATABASE test;

USE ROLE       accountadmin;


DROP STORAGE INTEGRATION test_integration;
REVOKE EXECUTE TASK ON ACCOUNT FROM ROLE sysadmin;
```

Summary

We began this chapter by explaining data pipelines and illustrating the difference between batch and real-time data streams. We then dived into Snowpipe, Snowflake's out-of-the-box data ingestion tool showcasing both modes of operation: manual invocation via REST API and automated invocation via AUTO_INGEST.

Recognizing Snowpipe does not process files in receipt order, we investigated an alternative approach. Temporal data load guarantees files are loaded in order. Introducing streams, tasks, and stored procedures by providing a worked example with the expectation the code samples provide a platform for experimentation and enhancement. Note logging and correct homing of components are left for you to develop. The diagrams show a correct physical separation of components in a target state for production deployment rather than the code walk-through, which is delivered using the SYSADMIN role for expediency (and page count).

Personally, this has been the hardest chapter to write by far. Technically complex, demanding, and over the page budget. But I want to give you a decent starting point. I hope you agree.

We are halfway there and now focus on extending data pipelines through data validation, enrichment, and presentation.

CHAPTER 9

Data Presentation

This chapter follows directly from Chapter 8, where we built our sample data pipeline. I now introduce reasoning and logic before building out components to implement data-driven dynamic validations, then identify how to enrich data. We then dive into data masking, row-level security, and object tagging.

The whole purpose of Chapter 2 is to underpin everything we do in this book. There is no point in creating another data silo, but we do not have space to cover everything in this chapter, so I must defer some related topics until later.

Figure 9-1 is becoming difficult to read but illustrates the reference architecture for validation, enrichment, and presentation components.

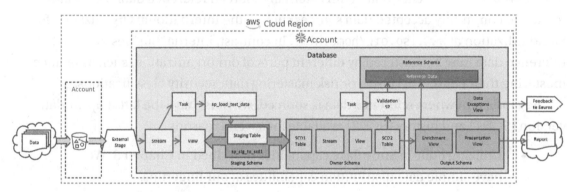

Figure 9-1. *Conceptual data flow*

I have referenced multiple schemas in diagrams to reinforce the message to segregate data for security and best practices. However, I have developed everything using the SYSADMIN role, which is not recommended. I leave it for you to segregate into the proper roles.

© Andrew Carruthers 2022
A. Carruthers, *Building the Snowflake Data Cloud*, https://doi.org/10.1007/978-1-4842-8593-0_9

Reference Data

What do we mean by *reference data*? Why do we need it? How do we source it? What is reference data?

Reference data consists of commonly accepted industry standard definitions. Country names and currency codes are examples of reference data. They are typically static dimensions, although occasionally, we see the birth of a new country or the adoption of a new currency.

We might think there is a single "golden source" or "system of record" in our organizations. but in fact, many applications maintain their own local reference data, and only a few create (or "master") reference data for use in other departments. Mastering reference data means the application is the (hopefully) single authoritative source for creating reference data in a specific business domain. And very few organizations have the foresight to centralize their master reference data into a single source, like Snowflake, distributing reference data to all consumers.

There is a difference between externally sourced reference data and internal business-focused reference data. With externally sourced reference data, we should reference commonly accepted data sources such as the International Organization for Standardization at `www.iso.org/home.html`. In contrast, internal business-focused reference data may be mastered by different parts of our organization, such as finance mastering their chart of accounts or risk mastering data security classifications.

Regardless of where reference data is sourced, we must have the latest up-to-date version (SCD1) and history (SCD2) available for several reasons.

- Utilizing built-in Snowflake capabilities described in Chapter 3. We have simple data distribution capabilities at our fingertips.

- We must also ensure data received into Snowflake conforms to our reference data and mark— but not exclude—failing records accordingly.

- Where record enrichment is required, we rely upon accurate reference data to enhance our customer reporting experience.

Without common reference data, we are certain to encounter data mismatches. We also find people making manual adjustments to correct data in their local system but never feeding back to the golden source. The endless cycle of poor-quality data issues continually repeats.

The manual approach to correct data is labor-intensive, costly, and error-prone because humans make mistakes. Our objective must always be to simplify, automate and remove cottage industries wherever possible. Companies need their brightest and smartest people to add real value and not be tied up with endless administration and data corrections.

Having explained why, how, and what for reference data, let's now look at practical use cases found in all organizations. I assume you have already created a reference schema and objects and loaded your reference data.

Validation

At the simplest level, we define validation as checking against reference data to ensure conformance. Equally, checks can be in a data set, for example, to ensure an employee is not their own manager.

We need to keep things simple, but as you might imagine, with software development, complexity rears its head at every opportunity, and designing generic processes which can be applied across many feeds is non-trivial.

Validations do not exclude data from flowing through our data pipeline. They simply mark records and attributes as failing validation rules, which may be considered data quality rules. Over time, we collate counts of records failing applied data quality rules leading to metrics such as confidence scores and common or repeat offenders.

Figure 9-2 illustrates what we are building.

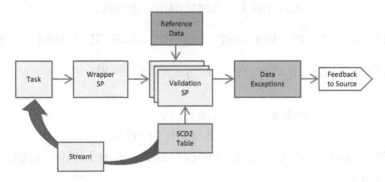

Figure 9-2. *Validation overview*

Data lands in our SCD2 table against which a stream has been declared detecting data landing, enabling the task to begin executing the wrapper stored procedure. We expect to apply several validation routines expressed via JavaScript stored procedures, each implementing a business rule applied via a wrapper stored procedure where each routine runs serially. Data quality exceptions are saved to a table with a view on top, making data available for extraction and feedback to the source. Naturally, we cannot be prescriptive in how data quality exceptions are propagated, but remember the core principle of not allowing corrections in our system. it is for each organization to determine their feedback pattern.

In our example, we hard-code a few rules, but later you see how to dynamically generate rules on the fly.

We reuse code from Chapter 8. Assume our storage integration is configured along with dependent objects to populate scd2_content_test, including the task_load_test_ data task.

First, let's establish our baseline.

```
USE ROLE       sysadmin;
USE DATABASE   TEST;
USE WAREHOUSE  COMPUTE_WH;
USE SCHEMA     public;

CREATE OR REPLACE STAGE TEST.public.test_stage
STORAGE_INTEGRATION = test_integration
DIRECTORY          = ( ENABLE = TRUE AUTO_REFRESH = TRUE )
ENCRYPTION         = ( TYPE = 'SNOWFLAKE_SSE' )
URL                = 's3://btsdc-test-bucket/'
FILE_FORMAT        = TEST.public.test_pipe_format;

CREATE OR REPLACE STREAM strm_test_stage ON STAGE TEST.public.test_stage;

CREATE OR REPLACE VIEW v_strm_test_stage COPY GRANTS
AS
SELECT '@TEST.public.test_stage/'||relative_
path                              AS path_to_file,
       SUBSTR ( relative_path, 1, REGEXP_INSTR ( relative_path, '_20' ) - 1
       ) AS table_name,
       size,
       last_modified,
       metadata$action
FROM   strm_test_stage;
```

```
ALTER STAGE TEST.public.test_stage REFRESH;

CREATE OR REPLACE TABLE stg_content_test
(
id            NUMBER,
content       VARCHAR(30),
last_updated  TIMESTAMP_NTZ DEFAULT current_timestamp()::TIMESTAMP_NTZ
              NOT NULL
);

CREATE OR REPLACE TABLE scd1_content_test
(
id            NUMBER,
content       VARCHAR(30),
last_updated  TIMESTAMP_NTZ DEFAULT current_timestamp()::TIMESTAMP_NTZ
              NOT NULL
);

CREATE OR REPLACE STREAM strm_scd1_content_test ON TABLE scd1_content_test;

CREATE OR REPLACE TABLE scd2_content_test
(
id            NUMBER,
content       VARCHAR(30),
valid_from    TIMESTAMP_NTZ,
valid_to      TIMESTAMP_NTZ,
current_flag  VARCHAR(1),
decision      VARCHAR(100)
);
```

I assume JavaScript sp_stg_to_scd1 and sp_load_test_data stored procedures, v_content_test, and task_load_test_data exist, with task task_load_test_data suspended.

Upload Test Files into S3

Create a target table to store data quality exceptions generated by our validation routines.

```
CREATE OR REPLACE TABLE data_quality_exception
(
data_quality_exception_id        NUMBER        NOT NULL,
validation_routine               VARCHAR(255)  NOT NULL,
data_quality_exception_code_id   NUMBER        NOT NULL,
source_object_name               VARCHAR(255)  NOT NULL,
source_attribute_name            VARCHAR(255)  NOT NULL,
source_record_pk_info            VARCHAR(255)  NOT NULL,
insert_timestamp                 TIMESTAMP_NTZ DEFAULT current_
timestamp()::TIMESTAMP_NTZ NOT NULL
);
```

Create a local reference table holding the list of exception codes. We would usually create referential integrity, too, even though Snowflake does not enforce referential integrity constraints.

```
CREATE OR REPLACE TABLE data_quality_exception_code
(
data_quality_exception_code_id            NUMBER        NOT NULL,
data_quality_exception_code_name          VARCHAR(255)  NOT NULL,
data_quality_exception_code_description VARCHAR(255)  NOT NULL,
insert_timestamp                          TIMESTAMP_NTZ DEFAULT current_
                                          timestamp()::TIMESTAMP_NTZ NOT NULL
);
```

Create Sequences to uniquely identify records.

```
CREATE OR REPLACE SEQUENCE seq_data_quality_exception_id      START
WITH 10000;
CREATE OR REPLACE SEQUENCE seq_data_quality_exception_code_id START
WITH 10000;
```

Create a data_quality_exception_code record.

```
INSERT INTO data_quality_exception_code  VALUES (seq_data_quality_
exception_code_id.NEXTVAL, 'ATTRIBUTE_IS_NULL', 'Attribute is declared as
NOT NULL but NULL value found', current_timestamp());
```

Create a simple validation stored procedure sp_is_attribute_null, noting the embedded use of current_flag to select the latest records.

```
CREATE OR REPLACE PROCEDURE sp_is_attribute_null( P_ROUTINE          STRING,
                                                  P_DQ_EXEP_CODE     STRING,
                                                  P_SOURCE_OBJECT    STRING,
                                                  P_SOURCE_ATTRIBUTE STRING,
                                                  P_SOURCE_PK_INFO   STRING
) RETURNS STRING
LANGUAGE javascript
EXECUTE AS CALLER
AS
$$
    var sql_stmt  = "";
    var stmt      = "";
    var result    = "";

    sql_stmt  = "INSERT INTO data_quality_exception ( data_quality_
    exception_id, validation_routine, data_quality_exception_code_id,
    source_object_name, source_attribute_name, source_record_pk_info,
    insert_timestamp )\n"
    sql_stmt += "SELECT seq_data_quality_exception_id.NEXTVAL,\n"
    sql_stmt += "        :1,\n"
    sql_stmt += "        ( SELECT data_quality_exception_code_id FROM data_
                         quality_exception_code WHERE data_quality_exception_
                         code_name = :2 ),\n"
    sql_stmt += "        :3,\n"
    sql_stmt += "        :4,\n"
    sql_stmt += "        :5,\n"
    sql_stmt += "        current_timestamp()::TIMESTAMP_NTZ\n"
    sql_stmt += "FROM    " + P_SOURCE_OBJECT    + "\n"
    sql_stmt += "WHERE   " + P_SOURCE_ATTRIBUTE + " IS NULL\n"
    sql_stmt += "AND     current_flag = 'Y';\n\n";

    stmt = snowflake.createStatement ({ sqlText:sql_stmt, binds:[P_ROUTINE,
    P_DQ_EXEP_CODE, P_SOURCE_OBJECT, P_SOURCE_ATTRIBUTE, P_SOURCE_PK_INFO] });
```

```
    try
    {
        result = stmt.execute();
        result = "Number of rows found: " + stmt.getNumRowsAffected();
    }
    catch { result = sql_stmt; }
    return result;
$$;
```

Recognizing we have not loaded any data into our tables, run a test. The result should be 'Number of rows found: 0'.

```
CALL sp_is_attribute_null ( 'Test',
                            'ATTRIBUTE_IS_NULL',
                            'scd2_content_test',
                            'content',
                            'id');
```

We should now test for all possible scenarios, so load up data into scd2_content_test, then update the content attribute to be NULL. Re-run the test harness. If we have set our test up correctly, we can expect records in data_quality_exception.

```
SELECT * FROM data_quality_exception;
```

Before re-running our test, we must remove previous records from data_quality_exception, always reset the baseline before re-testing and work methodically. There are no shortcuts in software development. Part of the learning is to approach our subject in a logical, consistent, repeatable manner. We need this discipline when developing a continuous integration framework, and for the curious, these repeat steps will be useful as plug-in test cases later.

```
TRUNCATE TABLE data_quality_exception;
```

When we are confident, our validation routine works as expected. We can create a new stored procedure sp_validate_test_data to wrap validation procedures. Note that there are two calls. You may wish to add control logic to ensure the first stored procedure execution outcome is properly returned in the event of failure.

```
CREATE OR REPLACE PROCEDURE sp_validate_test_data() RETURNS STRING
LANGUAGE javascript
EXECUTE AS CALLER
AS
$$
   var stmt      = "";
   var result    = "";

   stmt = snowflake.createStatement ({ sqlText: "CALL sp_is_attribute_
   null(?,?,?,?,?);",
                                       binds:['Test', 'ATTRIBUTE_IS_NULL',
                                       'scd2_content_test', 'id', 'id'] });

   try
   {
       result = stmt.execute();
       result = "SUCCESS";
   }
   catch { result = sql_stmt; }

   stmt = snowflake.createStatement ({ sqlText: "CALL sp_is_attribute_
   null(?,?,?,?,?);",
                                       binds:['Test', 'ATTRIBUTE_IS_NULL',
                                       'scd2_content_test', 'content',
                                       'id'] });

   try
   {
       result = stmt.execute();
       result = "SUCCESS";
   }
   catch { result = sql_stmt; }
   return result;
$$;
```

And test our wrapper.

```
CALL sp_validate_test_data();
```

Remember to grant entitlement to execute tasks to sysadmin.

```
USE ROLE accountadmin;
```

```
GRANT EXECUTE TASK ON ACCOUNT TO ROLE sysadmin;
```

And now, create a task triggered by the presence of new data in strm_scd2_content_test with our new sp_validate_test_data stored procedure.

```
USE ROLE sysadmin;
```

Create a stream on table strm_scd2_content_test. we use this to trigger our validation wrapper task.

```
CREATE OR REPLACE STREAM strm_scd2_content_test ON TABLE scd2_content_test;
```

Create a new task to run validations.

```
CREATE OR REPLACE TASK task_validate_test_data
WAREHOUSE = COMPUTE_WH
SCHEDULE  = '1 minute'
WHEN system$stream_has_data ( 'strm_scd2_content_test' )
AS
CALL sp_validate_test_data();
```

Set task to run.

```
ALTER TASK task_validate_test_data RESUME;
```

Check task status.

```
SHOW tasks;
```

Identify next scheduled run time.

```
SELECT timestampdiff ( second, current_timestamp, scheduled_time ) as
next_run,
      scheduled_time,
      current_timestamp,
      name,
      state
FROM   TABLE ( information_schema.task_history())
```

```
WHERE  state = 'SCHEDULED'
ORDER BY completed_time DESC;
```

By now, you should be comfortable with testing the data pipeline from resetting the test case baseline, loading files from S3, ingesting them into the SCD2 table, and running validations. Once your testing is complete, suspend the task.

```
ALTER TASK task_load_test_data SUSPEND;
```

While a simple example has been provided, more complex validations should be considered. Examples include lookups to reference data to ensure attributes match known, approved values and internal data consistency checks, such as ensuring an employee is not their own manager. The possibilities are almost endless, and the more complete our data validations are, the better.

Having worked through examples of how to both identify and record data quality exceptions, let's turn our attention to enriching data.

Data Quality Exceptions

Having run our validation routines, the presence of data quality exceptions generates metrics we can use for several purposes. We can determine the quality of data in our source systems, provide a measure of data quality improvements over time, and give consumers confidence in our data's fidelity.

Once we have generated data quality exceptions, we must decide what to do with the data. Most organizations do not have a clearly defined feedback route to the source. Still, the generation of data quality exceptions exposes gaps or issues with the supplied data. From previous discussions, you know correction in a data warehouse is the wrong action to allow.

Ideally, we would automatically feedback data to the source for correction, leading to a virtuous circle where the number of data quality exceptions reduces to near zero over time. An initial objective should be set to reduce the time to correct data between submission and validation cycles.

Options to identify data quality exceptions include provisioning database views and associated reporting screens, automated email notifications, and periodic extract to flat files in a shared directory. You may have your own preferred option.

I do not propose a means to manage, accept, or clear data quality exceptions for our data warehouse. The operational management is left for you to determine.

Enrichment

Data enrichment is the process by which we add attributes into a data set to make the data more easily understood, business-focused, and useful to the end consumer. Here we focus on the data itself and not the metadata definitions, which are addressed later in this chapter with object tagging. Figure 9-3 introduces enrichment views into our data pipeline.

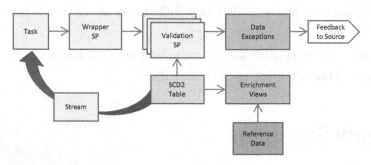

Figure 9-3. *Enrichment overview*

Often our business colleagues have poorly articulated expectations and make assumptions over the artifacts we deliver to them. As software developers, we must always consider the business outcomes of our work. Everything we do has a cost and should (but not always) deliver tangible business benefits.

A practical example of data enrichment is to expand an ISO country code from the three-letter alpha code (e.g., USA) to the entire country name, (e.g., United States of America), along with any other attributes required by our users or we proactively think they will find useful. See `https://en.wikipedia.org/wiki/List_of_ISO_3166_country_codes` here for further information.

There is nothing particularly clever about this approach. Every reporting data set has joins to reference data. Hence we are not diving into any particular design pattern or code sample. Instead, we reiterate the point of not excluding records by using INNER JOINs. We prefer to use OUTER JOINs with default values for missing attributes to preserve the principle of preserving all records throughout the data lifecycle. Our operational support teams have better things to do than chase a single missing record.

Dynamic Data Masking

In our data sets, we find a wide variety of content, some of which is highly sensitive for various reasons. We may also have a requirement to mask production data for use in lower environments such as user acceptance testing or development where the use of production data is prohibited. Although cloning databases is (usually) trivial in Snowflake, to do so when generating a development environment may expose data. Snowflake documentation introduces the subject of data masking at `https://docs.` `snowflake.com/en/user-guide/security-column-ddm-intro.html`.

Data masking may also be applied to external tables but not at the point of creation, only by altering after that, as the metadata required does not exist until after creation. Further restrictions on unmasking policy usage apply. Please refer to the documentation at `https://docs.snowflake.com/en/user-guide/security-column-intro.` `html#label-security-column-intro-cond-cols`.

Dynamic data masking determines what the user sees and should be considered as horizontal filtering of attribute values. In other words, data masking does not filter the rows returned by a query, only the visible content displayed.

Note   Apply data masking selectively. There is a performance impact where it has been excessively implemented.

Simple Masking

this section dives into implementing data masking with hands-on demonstration. First, we must establish a maskingadmin administrative role to administer our tags.

```
USE ROLE securityadmin;

CREATE OR REPLACE ROLE maskingadmin;

GRANT CREATE MASKING POLICY ON SCHEMA TEST.public TO ROLE maskingadmin;
```

Grant entitlement to use the schema we allow masking policies to be created. In this example, we use the TEST database and PUBLIC schema.

```
GRANT USAGE ON DATABASE  TEST            TO ROLE maskingadmin;
GRANT USAGE ON WAREHOUSE compute_wh  TO ROLE maskingadmin;
GRANT USAGE ON SCHEMA    TEST.public TO ROLE maskingadmin;
```

Grant maskingadmin to your user.

```
GRANT ROLE maskingadmin TO USER <YOUR_USER>;
```

Because APPLY MASKING is set at the account level, we must use the ACCOUNTADMIN role.

```
USE ROLE accountadmin;
```

```
GRANT APPLY MASKING POLICY ON ACCOUNT TO ROLE maskingadmin;
```

Set context.

```
USE ROLE       sysadmin;
USE DATABASE   TEST;
USE WAREHOUSE  COMPUTE_WH;
USE SCHEMA     public;
```

Create the masking_test table.

```
CREATE OR REPLACE TABLE masking_test
(
user_email         VARCHAR(30)  NOT NULL,
user_email_status VARCHAR(30)  NOT NULL
);
```

Create a stream on the masking_test table.

```
CREATE OR REPLACE STREAM strm_masking_test ON TABLE masking_test;
```

Create two rows in the masking_test table.

```
INSERT INTO masking_test
VALUES
('user_1@masking_test.com', 'Public' ),
('user_2@masking_test.com', 'Private');
```

Prove we have two rows in our masking_test table.

```
SELECT * FROM masking_test;
```

Create a View on table masking_test;

```
CREATE OR REPLACE VIEW TEST.public.v_masking_test
AS
SELECT * FROM masking_test;
```

Grant entitlement to maskingadmin on masking_test table and v_masking_test view.

```
GRANT SELECT ON masking_test      TO ROLE maskingadmin;
GRANT SELECT ON v_masking_test     TO ROLE maskingadmin;
GRANT SELECT ON strm_masking_test TO ROLE maskingadmin;
```

At this point no masking policies exist.

```
SHOW masking policies;
SHOW masking policies IN ACCOUNT;
```

Switch to the maskingadmin role and set schema context.

```
USE ROLE       maskingadmin;
USE DATABASE   TEST;
USE WAREHOUSE  COMPUTE_WH;
USE SCHEMA     TEST.public;
```

Prove we can see unmasked data in masking_test.

```
SELECT * FROM masking_test;
```

Figure 9-4 shows the expected output from our query.

Figure 9-4. *Unmasked raw data*

This step is only required if the whole test case has been re-run from the start and requires some explanation. If the table is re-created *after* the masking policy has been applied to an attribute, then it is not possible to re-create the masking policy. The behavior is inconsistent with expectations, and the next few lines are a work-around.

```
USE ROLE accountadmin;

DROP MASKING POLICY IF EXISTS dq_code_mask;
```

Revert to the maskingadmin role.

```
USE ROLE maskingadmin;
```

Create a masking policy.

```
CREATE OR REPLACE MASKING POLICY dq_code_mask AS ( P_PARAM STRING ) RETURNS
STRING ->
CASE
    WHEN current_role() IN ('SYSADMIN') THEN P_PARAM
    ELSE '*********'
END;
```

Check that the masking policy has been created.

```
SHOW masking policies LIKE 'dq_code_mask';
```

Apply masking policy to an attribute on the masking_test table.

```
ALTER TABLE TEST.public.masking_test
MODIFY COLUMN user_email SET MASKING POLICY dq_code_mask;
```

Check masking_test.user_email is now displayed as '*********'.

```
SELECT * FROM masking_test;
```

Figure 9-5 shows the expected output from our query.

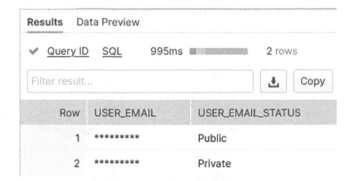

Figure 9-5. *Masked user_email*

This simple demonstration shows the effect of adding a masking policy to a single table. But for our data pipelines, we also need to cascade the masking policy to other objects in our pipeline. Let's see what happens when we create a view on top of our table.

```
SELECT * FROM v_masking_test;
```

The result set is identical to Figure 9-5.
And now a stream.

```
SELECT * FROM strm_masking_test;
```

Figure 9-6 shows the expected output from our query.

Row	USER_EMAIL	USER_EMAIL_STATUS	METADATA$ACTION	METADATA$ISUPDATE	METADATA$ROW_ID
1	*********	Public	INSERT	FALSE	b8498fadf820a29890cf7653aee57d29cad71b5b
2	*********	Private	INSERT	FALSE	e8cf9af3d6c3a69d07bc889448da93e660db96b5

Figure 9-6. *Masked stream output*

For both the view and stream, the data remains masked.
Change role to SYSADMIN and ensure masking_test.user_email is now displayed unmasked.

```
USE ROLE sysadmin;
```

```
SELECT * FROM masking_test;
```

Figure 9-7 shows the expected output from our query.

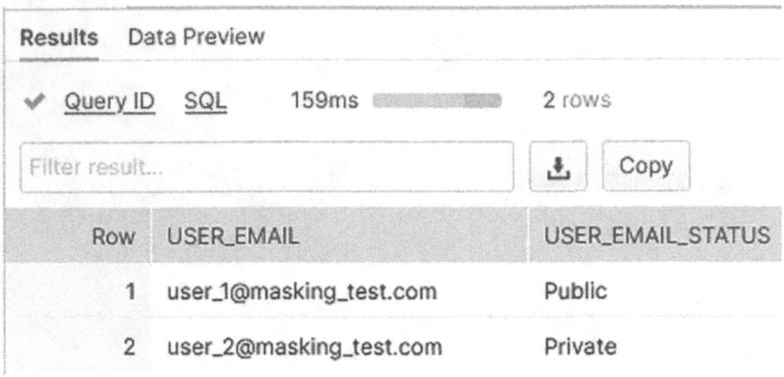

Figure 9-7. *Unmasked raw data*

Reset role to maskingadmin.

```
USE ROLE maskingadmin;
```

Unset the masking policy for masking_test.user_email.

```
ALTER TABLE TEST.public.masking_test
MODIFY COLUMN user_email UNSET MASKING POLICY;
```

And drop the masking policy.

```
DROP MASKING POLICY dq_code_mask;
```

Conditional Masking

So far, our test case has used the user's current role to determine whether an attribute should be masked or not. A more sophisticated approach is to extend the masking policy to refer to a related attribute, as explained at https://docs.snowflake.com/en/user-guide/security-column-intro.html#label-security-column-intro-cond-cols.

In simple terms, we can now apply data masking by either role or attribute value, reusing our test case. We now rely upon the value of attribute user_email_status to determine masking outcome.

Note the masking policy prototype has been extended to two parameters with data types.

```
CREATE OR REPLACE MASKING POLICY dq_code_mask AS ( user_email VARCHAR,
user_email_status VARCHAR ) RETURNS STRING ->
```

```
CASE
  WHEN current_role() IN ('SYSADMIN') THEN user_email
  WHEN user_email_status = 'Public'   THEN user_email
  ELSE '*********'
END;
```

Likewise, setting the masking policy for an attribute has also been extended.

```
ALTER TABLE TEST.public.masking_test
MODIFY COLUMN user_email
SET MASKING POLICY dq_code_mask
USING (user_email, user_email_status);
```

The remainder of our test cases are identical to the preceding one.

```
SELECT * FROM masking_test;
```

Figure 9-8 shows the expected output from our query with masking applied according to the user_email_status.

Figure 9-8. *Conditional data masking*

```
SELECT * FROM v_masking_test;
```

The result set is identical to Figure 9-8.

And now a stream.

```
SELECT * FROM strm_masking_test;
```

Figure 9-9 shows the expected output from our query.

Row	USER_EMAIL	USER_EMAIL_STATUS	METADATA$ACTION	METADATA$ISUPDATE	METADATA$ROW_ID
1	user_1@masking_test.com	Public	INSERT	FALSE	b8498fadf820a29890cf7653aee57d29cad71b5b
2	*********	Private	INSERT	FALSE	e8cf9af3d6c3a69d07bc889448da93e660db96b5

Figure 9-9. *Masked stream output*

For both view and stream, the data remains masked.

Revert to the sysadmin role and re-test.

```
USE ROLE sysadmin;
```

```
SELECT * FROM masking_test;
```

Figure 9-10 shows the expected output from our query.

Row	USER_EMAIL	USER_EMAIL_STATUS
1	user_1@masking_test.com	Public
2	user_2@masking_test.com	Private

Figure 9-10. *Unmasked raw data*

Cleanup also remains the same.

```
USE ROLE maskingadmin;
```

```
ALTER TABLE TEST.public.masking_test
MODIFY COLUMN user_email UNSET MASKING POLICY;
```

```
DROP MASKING POLICY dq_code_mask;
```

Data Masking Considerations

With a simplistic test case, data masking is readily demonstrated, but we are in the business of building data pipelines and must consider how roles and data masking interact. Let's assume our cybersecurity colleagues impose a policy that states: All confidential attributes must be masked from the point of ingestion into Snowflake, through all internal processing, until the ultimate point of consumption where confidential attributes must be unmasked. In other words, the security policy states specific attributes that persisted in Snowflake must have data masking applied and only be visible after leaving Snowflake.

The security policy would be easily achieved if we had a single role that handles every step from data ingestion to data consumption. A single masking policy with a single role would suffice. But, our data lives in a highly secure environment, with multiple actors with specific roles operating on our data throughout its lifecycle. It is incumbent upon us to ensure security policies are enforced. Figure 9-11 articulates a real-world scenario where roles and interactions pose challenges in preserving data masking as data transits our pipeline. Let's categorize our roles as follows.

- Ingestion: The role that lands data into our staging table must not have entitlement to read the application table or presentation view.

- Persistence: The role that moves data through our pipeline from staging table to presentation view must not have entitlement to read the presentation view.

- Consumption: The role that presents data to the consuming service must not have entitlement to read either the staging table or application table.

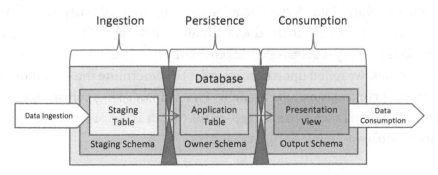

Figure 9-11. *Real-world data masking*

The challenge with implementing data masking relates to the context required to allow each actor to read data unmasked while saving it into tables in the data pipeline where masking policies are applied. The answer is to carefully consider the masking policy declaration and only include the minimum required context allowing clear text to be returned.

Masking policies are agnostic and can be applied to any attribute. The context contained in the masking policy determines what the invoker sees. Therefore a few masking policies can be declared and applied ubiquitously across our data model according to the boundary at which they sit and the objects they protect.

Making changes to existing masking policies comes at a price. The masking policy after redefinition must then be re-tested, and the application testing process is both time-consuming and expensive.

Another use for data masking is to obfuscate production data rendering the content suitable for use in a testing environment. If the rendered values are acceptable and do not remove required data, then data masking policies may work very well.

The key to success with data masking is to apply masking policies selectively. Every SELECT statement invokes the masking policy applied to chosen attributes, therefore incurring additional processing overhead to unmask or partially mask attributes. It is recommended that only those attributes determined to be highly confidential, highly restricted, or commercially sensitive be masked. Not forgetting to conduct testing with real-world scenarios to ensure masking is applied correctly, then load test using production-like data volumes.

Advanced Topics

External tokenization is an extension to data masking where data can be tokenized before loading into Snowflake. As tokenization relies upon third-party tooling, we do not discuss it further. Instead, refer to the documentation at `https://docs.snowflake.com/en/user-guide/security-column-ext-token.html`.

In our test cases, we relied upon current_role() to determine the execution context, a broader range of context functions available for use in masking policies is explained at `https://docs.snowflake.com/en/user-guide/security-column-advanced.html#advanced-column-level-security-topics`.

Row-Level Security (RLS)

While data masking determines what the user sees in terms of attribute content displayed, row-level security (RLS) applies filters that determine the actual rows returned based upon a role. Snowflake also refers to it as row access policies. Documentation is at `https://docs.snowflake.com/en/user-guide/security-row.html`.

From Chapter 5, you know object level entitlement is granted to roles, and with any given role, if we have entitlement to SELECT from an object, then we have access to all the data in the table.

In contrast, RLS applies filters to the object content, restricting returned data to only those rows matching the RLS definition. Figure 9-12 illustrates the difference in approach between data masking and RLS. The actual data is of no real interest. But for the curious, it derived from lathe tool cutting speeds for different materials. The effects are what we need to know.

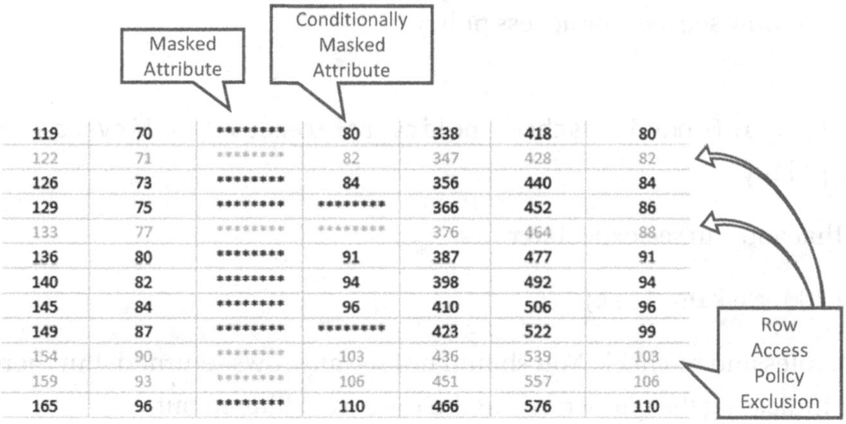

Figure 9-12. *Data masking and row-level security*

Having explained RLS and the impact, let's look at a simple worked example. First, we must either create or entitle an existing role, and for simplicity, we reuse maskingadmin as the capabilities of RLS are compatible with data masking.

```
USE ROLE accountadmin;
```

```
GRANT APPLY ROW ACCESS POLICY ON ACCOUNT TO ROLE maskingadmin;
```

We rely upon previous entitlement grants to maskingadmin hereafter and expect masking_test table to exist.

Switch back to the sysadmin role and create a row access policy.

```
USE ROLE      sysadmin;
USE DATABASE  TEST;
USE WAREHOUSE COMPUTE_WH;
USE SCHEMA    public;

CREATE OR REPLACE ROW ACCESS POLICY sysadmin_policy
AS ( user_email VARCHAR ) RETURNS BOOLEAN ->
'SYSADMIN' = current_role();
```

Now apply the row access policy to masking_test.user_email.

```
ALTER TABLE TEST.public.masking_test
ADD ROW ACCESS POLICY sysadmin_policy ON ( user_email );
```

You should now see the row access policy.

```
SELECT *
FROM   TABLE ( information_schema.policy_references ( policy_name =>
'sysadmin_policy' ));
```

Check that you can see expected rows.

```
SELECT * FROM masking_test;
```

Change roles and re-check. You should not see any rows returned, but there should be no error indicating the query ran, but the rows were filtered out.

```
USE ROLE      maskingadmin;
```

```
SELECT * FROM masking_test;
```

Change back.

```
USE ROLE      sysadmin;
```

Before a row access policy can be dropped, we must remove it from all assigned tables.

```
ALTER TABLE TEST.public.masking_test DROP ROW ACCESS POLICY
sysadmin_policy;

DROP ROW ACCESS POLICY sysadmin_policy;
```

What did we do? We created a simple row access policy that prevents access to data except when SYSADMIN is the current role, even if any other role has SELECT access to the table. And yes, this is a trivial example used to illustrate how row access policy statements are used, also to provide a template for extension into a more useful pattern as at `https://docs.snowflake.com/en/user-guide/security-row-using.html`.

Object Tagging

This section addresses a complex subject with far-reaching consequences necessitating explanation. As you become aware, our data must be protected as it transits through our systems. I have showcased both design techniques and implementation patterns adhering to best practices.

But at some point, we need a way of categorizing our data by object and attribute, making data discovery (relatively) easy. We can implement data discovery via external tooling such as Collibra, exposing the data model and metadata. This approach remains valid while the Collibra catalog and Snowflake information schemas are aligned.

Another way to enable data discovery is to tag each object and attribute with one or more tags. Supporting documentation is at `https://docs.snowflake.com/en/user-guide/object-tagging.html`.

Tag Uniqueness

We assume that organizations want to use an external tool such as Collibra to implement object tagging, and some general rules must be applied to ensure consistency across our estate.

Object tags are not a unique feature of Snowflake. However, their value to an organization is only realized if tags are unique across the organization regardless of the location and system where they are applied. Without enforced uniqueness, the likelihood of duplicate tags enables data to be misidentified and potentially disclosed inappropriately.

It is strongly recommended that tag creation and maintenance be centralized into a single team whose primary role is the maintenance of a unique organization tag taxonomy, the golden source of all tags used in the organization. Please be aware of the implications of changing or reorganizing tags when managed externally to Snowflake. The same changes must be reflected in our data warehouse, and lineage must be preserved.

Note   Snowflake limits the number of tags in an account to 10,000 and limits the number of allowed values for a single tag to 20.

Tag Taxonomy

We might decide to define our taxonomies according to the line of business, function, or capability with a root node from which all other usages down to the leaf are represented by a colon delimiter. By way of example, let's look at privacy data as this domain has a small set of tags noting the ones presented here are for demonstration purposes only derived from `www.investopedia.com/terms/p/personally-identifiable-information-pii.asp`. Feel free to reuse but check that the definitions match your organization's requirements.

Our taxonomy begins with the top-level PII (Personally Identifiable Information) with two subcategories for sensitive and non-sensitive attributes. For each of the two subcategories, we select a few identifiers, the full lists are more extensive, but we need a representative sample only.

For sensitive information, we use full name and Social Security number; for non-sensitive information, we use gender and date of birth. Our taxonomy may look like the following. Underscores are used to separate fields.

> PII
>
> PII_S_FullName
>
> PII_S_SSN
>
> PII_N_Gender
>
> PII_N_DoB

To make the tags concise, we may use abbreviations. Naturally, we would use a short name and full description along with SCD2 historization when entered in our cataloging tool for management purposes.

Single Tag Value Implementation

With our tags identified, we can start to build out our sample implementation creating a new role for object tagging—tagadmin. We assume familiarity with the following code. and for further information, refer to the Snowflake documentation at `https://docs.snowflake.com/en/user-guide/object-tagging.html#object-tagging`.

```
USE ROLE securityadmin;

CREATE OR REPLACE ROLE tagadmin;

GRANT USAGE ON DATABASE  TEST          TO ROLE tagadmin;
GRANT USAGE ON WAREHOUSE compute_wh  TO ROLE tagadmin;
GRANT USAGE ON SCHEMA    TEST.public TO ROLE tagadmin;
```

Tags are applied in a schema.

```
GRANT CREATE TAG ON SCHEMA TEST.public TO ROLE tagadmin;
```

Assign to your user.

```
GRANT ROLE tagadmin TO USER <YOUR_USER>;
```

Entitle tagadmin to apply tags to the account.

```
USE ROLE accountadmin;

GRANT APPLY TAG ON ACCOUNT TO ROLE tagadmin;
```

We must set the context before attempting to work with object tags.
Create a test table and load sample data.

```
USE ROLE      sysadmin;

CREATE OR REPLACE TABLE pii_test
(
id                    NUMBER,
```

```
full_name                VARCHAR(255),
social_security_number VARCHAR(255),
gender                   VARCHAR(255),
date_of_birth            TIMESTAMP_NTZ
);
```

```
CREATE OR REPLACE SEQUENCE seq_pii_test_id     START WITH 10000;
```

```
INSERT INTO pii_test VALUES (seq_pii_test_id.NEXTVAL, 'John Doe',
'12345678', 'Male', current_timestamp()), (seq_pii_test_id.NEXTVAL, 'Jane
Doe', '23456789', 'Female', current_timestamp());
```

See documentation for allowable entitlement options at https://docs.snowflake.com/en/user-guide/security-access-control-privileges.html#tag-privileges.

```
GRANT SELECT ON pii_test TO ROLE tagadmin;
```

Switch to the tagadmin role, noting we have segregated the object creation and data from the tagging. This is another example of separating capabilities via RBAC.

```
USE ROLE       tagadmin;
USE DATABASE   TEST;
USE WAREHOUSE COMPUTE_WH;
USE SCHEMA     public;
```

Create sample tags. The documentation is at https://docs.snowflake.com/en/sql-reference/sql/create-tag.html#create-tag.

```
CREATE OR REPLACE TAG PII             COMMENT = 'Personally Identifiable
Information';
CREATE OR REPLACE TAG PII_S_FullName COMMENT = 'Personally Identifiable
Information -> Sensitive -> Full Name';
CREATE OR REPLACE TAG PII_S_SSN      COMMENT = 'Personally Identifiable
Information -> Sensitive -> Social Security Number';
CREATE OR REPLACE TAG PII_N_Gender   COMMENT = 'Personally Identifiable
Information -> Non-Sensitive -> Gender';
CREATE OR REPLACE TAG PII_N_DoB      COMMENT = 'Personally Identifiable
Information -> Non-Sensitive -> Date of Birth';
```

The following displays tags, noting context.

```
SHOW tags;
```

Ensure we can retrieve data from pii_test.

```
SELECT * FROM pii_test;
```

With tags declared, we can now assign to pii_test.

```
ALTER TABLE pii_test SET TAG PII = 'Personally Identifiable Information';

ALTER TABLE pii_test MODIFY COLUMN full_name              SET TAG PII_S_
FullName = 'Personally Identifiable Information -> Sensitive -> Full Name';
ALTER TABLE pii_test MODIFY COLUMN social_security_number SET TAG PII_S_
SSN      = 'Personally Identifiable Information -> Sensitive -> Social
Security Number';
ALTER TABLE pii_test MODIFY COLUMN gender                 SET TAG PII_N_
Gender   = 'Personally Identifiable Information -> Non-Sensitive ->
Gender';
ALTER TABLE pii_test MODIFY COLUMN date_of_birth          SET TAG PII_N_
DoB      = 'Personally Identifiable Information -> Non-Sensitive -> Date
of Birth';
```

Fetch the tags associated with pii_test.

```
SELECT *
FROM   TABLE ( TEST.information_schema.tag_references( 'pii_test',
'TABLE' ));
```

Fetch the tags associated with the full_name attribute. Note that the TABLE tag is inherited. Tags applied to an object are automatically applied to the attributes.

```
SELECT *
FROM   TABLE ( TEST.information_schema.tag_references( 'pii_test.full_
name', 'COLUMN' ));
```

We can also fetch the label associated with each tag.

```
SELECT system$get_tag ( 'PII', 'pii_test', 'TABLE' );
SELECT system$get_tag ( 'PII_S_FullName', 'pii_test.full_name', 'COLUMN' );
```

For a named table we can also use this query.

```
SELECT *
FROM   TABLE ( information_schema.tag_references_all_columns ( 'pii_test',
'TABLE' ));
```

Multiple Tag Value Implementation

We can also assign multiple values to a tag noting the limitation of 20 values per tag. Note that at the time of writing this book, this feature is in public preview as documented at https://docs.snowflake.com/en/sql-reference/sql/alter-tag.html#specifying-allowed-values.

```
ALTER TAG PII ADD ALLOWED_VALUES 'Personally Identifiable Information',
'PII Admin: pii@your_org.xyz';
```

The following displays the allowed values for a tag.

```
SELECT system$get_tag_allowed_values ( 'TEST.public.PII' );
```

The following unsets them.

```
ALTER TAG PII UNSET ALLOWED_VALUES;
```

Tag Identification

So far, our work has focused on the bottom-up build of object tagging capability. Our users are less interested in programmatic constructs. They need a way to quickly identify tags of interest by searching and filtering all available tags in our data warehouse, selecting, and then applying their criteria to drill down into the information required. And there are challenges with dynamically driving table functions. The parameters do not accept attributes passed in a SQL query. Let's turn our attention to a top-down view of object tagging, and for this, we must extend our Account Usage store knowledge.

Please refer to Chapter 6 for details on accessing the Account Usage store.

To satisfy our users' need to identify available tags, and noting latency for tags and tag_references are stated to be 2 hours at https://docs.snowflake.com/en/sql-reference/account-usage.html#account-usage-views. Run the below query:

```
SELECT *
FROM    snowflake.account_usage.tags
WHERE   deleted IS NULL
ORDER BY tag_id;
```

The following identifies where objects have been tagged.

```
SELECT *
FROM    snowflake.account_usage.tag_references
ORDER BY tag_name, domain, object_id;
```

With these queries, we can create a view for end user access. I leave this as an exercise for you to consolidate your knowledge.

Tag Cleanup

Here's how we unset tags.

```
ALTER TABLE pii_test UNSET TAG PII;

ALTER TABLE pii_test MODIFY COLUMN full_name                UNSET TAG PII_S_
                                                            FullName;
ALTER TABLE pii_test MODIFY COLUMN social_security_number UNSET TAG PII_S_SSN;
ALTER TABLE pii_test MODIFY COLUMN gender                   UNSET TAG
                                                            PII_N_Gender;
ALTER TABLE pii_test MODIFY COLUMN date_of_birth            UNSET TAG
                                                            PII_N_DoB;
```

And drop tags.

```
DROP TAG PII;
DROP TAG PII_S_FullName;
DROP TAG PII_S_SSN;
DROP TAG PII_N_Gender;
DROP TAG PII_N_DoB;
```

Data Presentation

Outbound data from our data warehouse is the point at which our users interact to select data sets, and reporting tooling delivers business value. Several options are discussed in Chapter 13, but first, a note of warning. Data modeling can be a fractious subject to be approached with a degree of sensitivity.

I offer some general comments without wishing to take sides or inflame the debate. Utilizing outbound data may include aggregations, summarizations, and filters, which may lend weight to a presentation layer design depending upon the use cases. Also, we cannot ignore data governance because, as you see in Chapter 12, the ability to control access and make visible who can see what is most important.

We do not describe or propose specific outbound data patterns as space does not permit and defer to Chapter 13 for a more in-depth discussion.

Data Quality Exceptions

We have identified data quality exceptions and the need to propagate back to the source. In Snowflake, a secure view provides a simple mechanism to extract data from individual sources on demand. We do not anticipate anything more than a response file generated per feed ingestion on a 1:1 basis plus a user interface where aggregated or summary metrics with trends over time are provisioned.

Tag-Based Query

Adopting tag-based queries is more promising for self-service users. However, a high degree of complexity awaits the unwary. While technically not difficult to provision automatic serving up of data, the ever-changing governance aspects in multi-jurisdiction organizations make this approach impossible even where a comprehensive suite of tags is deployed across multiple dimensions, never mind the tag maintenance aspect.

We envisage a tag-based query as a tool used internally in an organization to identify objects and attributes of interest. With object tagging, primary use cases include privacy impact assessments or data subject access requests. Naturally, you may identify your own use cases, and those suggested are not considered exclusive.

Cleanup

Apart from cleanup scripts to remove our test code, we should also be mindful to periodically remove redundant files uploaded to internal stages remembering each file contributes toward storage costs.

```
USE ROLE       sysadmin;
USE DATABASE   TEST;
USE SCHEMA     public;

DROP TASK       task_load_test_data;
DROP DATABASE test;

USE ROLE       accountadmin;


DROP STORAGE INTEGRATION test_integration;
REVOKE EXECUTE TASK ON ACCOUNT FROM ROLE sysadmin;
```

Summary

This chapter introduced reference data and its importance in validating data quality. I then explained how to extend data pipelines by providing a validation that showcases how we can implement parameter-driven stored procedures to make them generic and reusable across all tables and views.

Our discussion moved on to handling the data quality exceptions generated by validation routines and enriching our data by joining reference data.

Dynamic data masking proved to be an interesting and surprisingly complex topic. The chapter provided insight, a test case, and a basis for further investigation and self-learning.

We then investigated row-level security and had a hands-on practical examination of RLS before moving to object tagging. This evolving subject-provoking system design enables users to interact with tags without programming knowledge.

Working through this chapter has proven personally to be very interesting. I learned a lot, including that most topics are evolving, with some features not yet generally available. The best advice is to evaluate each section in conjunction with the documentation when ready to build capability and refactor accordingly.

Having extended data pipelines into end-to-end delivery, let's look at structured, semi-structured, and unstructured data.

CHAPTER 10

Semi-Structured and Unstructured Data

This chapter is the one I have been looking forward to the most. Although all the chapters have been very enjoyable to write, setting the right tone and content is technically challenging. This one is set to stretch our boundaries the most.

Note While writing this chapter, some differences were observed between the Snowflake documentation and AWS Management Console layout; hence some sections have textual descriptions.

What is semi-structured and unstructured data?

Semi-structured data is typically a document format where all the relationships are described in the document. In other words, a single document containing semi-structured data is self-contained, self-describing, and not reliant upon external references to be complete. While the content is usually human-readable, we often use external tools to display the content in an easily understood format. Don't worry. All will become clear when we dive into one of the semi-structured formats (JSON).

Unstructured data is all around us. We rarely think of our daily interaction with media in general as unstructured data, but it is. We use unstructured data whenever we watch TV, read a book, or open an email. Learning how to interact with the emerging— and for some of us, the very exciting—realm of unstructured data is where the "art of the possible" has become a reality. This chapter demonstrates how to load images into Snowflake, allowing programmatic content interrogation. I can't wait to get into this part. I never thought I would get excited over invoices.

Figure 10-1, used with the kind permission of Snowflake Inc., illustrates the three layers of data management available in Snowflake. I assume structured data is well

A. Carruthers, *Building the Snowflake Data Cloud*, https://doi.org/10.1007/978-1-4842-8593-0_10

known and understood by most readers; therefore, I do not dwell on what is considered common knowledge.

EVOLUTION OF DATA MANAGEMENT

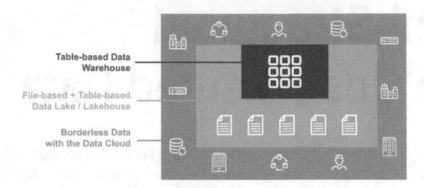

Figure 10-1. Evolution of data management

There are plenty of books available on structured data. Instead, we start with semi-structured data, JSON, and other familiar formats. I hope to offer fresh insights, tools, and techniques to address content before moving on to unstructured data, where the various data formats bring their own challenges. I offer ways to extract the content into Snowflake, taking advantage of the latest (at the time of writing) features and providing a framework for extending unstructured support according to your needs.

As Saurin Shah, a Snowflake product manager, says, "It doesn't matter how we store the data. It matters how data is stored and consumed." Integrating semi-structured and unstructured data into Snowflake has significant potential to unlock and transform value for our organizations, particularly when enriching with our relational data. Imagine the possibilities that become a reality if we can readily access content locked up inside our documents, PDFs, and images of various formats and then add real-time streaming data.

Semi-Structured Data

This chapter focuses on JSON. Snowflake supplies the tutorial at `https://docs.snowflake.com/en/user-guide/json-basics-tutorial.html`. Snowflake also supports other semi-structured file formats, including Avro, ORC, Parquet, and XML to a lesser degree. Returning to JSON, you may find the tool at `https://jsonformatter.org` useful.

AWS allows a single SQS queue on the S3 bucket. Using AWS Management Console, create a new S3 bucket, btsdc-json-bucket, copying settings from the existing btsdc-test-bucket.

Begin by creating a new storage integration named json_integration.

```
USE ROLE      accountadmin;
USE DATABASE TEST;
USE SCHEMA    public;

CREATE OR REPLACE STORAGE INTEGRATION json_integration
TYPE                    = EXTERNAL_STAGE
STORAGE_PROVIDER        = S3
ENABLED                 = TRUE
STORAGE_AWS_ROLE_ARN    = 'arn:aws:iam::616701129608:role/test_role'
STORAGE_ALLOWED_LOCATIONS = ( 's3://btsdc-json-bucket/' );

USE ROLE securityadmin;

GRANT USAGE ON INTEGRATION json_integration TO ROLE sysadmin;

DESC INTEGRATION json_integration;
```

Make a note of your values as they will differ. The following are mine.

STORAGE_AWS_IAM_USER_ARN:
arn:aws:iam::321333230101:user/vnki-s-ukst5070

STORAGE_AWS_EXTERNAL_ID: GH06274_SFCRole=3_Vf/
LKebJXaqzvO+unFpnT5OzqGM=

In AWS Management Console, we must edit our role trust relationship to add the preceding STORAGE_AWS_IAM_USER_ARN and STORAGE_AWS_EXTERNAL_ID values. Navigate to IAM ➤ Roles. In the drop-down list, select S3, and then "Allow S3 to call AWS services on your behalf," and click Next.

Click Edit policy and modify your JSON as follows.

```
{
  "Version": "2012-10-17",
  "Statement": [
    {
      "Effect": "Allow",
```

```
    "Principal": {
      "AWS": "arn:aws:iam::321333230101:user/vnki-s-ukst5070"
    },
    "Action": "sts:AssumeRole",
    "Condition": {
      "StringEquals": {
        "sts:ExternalId": "GHO6274_SFCRole=3_Vf/
        LKebJXaqzvO+unFpnT5OzqGM="
      }
    }
  }
]
}
```

Update test_policy to include a new bucket.

```
{
    "Version": "2012-10-17",
    "Statement": [
        {
            "Sid": "VisualEditor0",
            "Effect": "Allow",
            "Action": [
                "s3:PutObject",
                "s3:GetObject",
                "s3:DeleteObjectVersion",
                "s3:DeleteObject",
                "s3:GetObjectVersion"
            ],
            "Resource": [ "arn:aws:s3:::btsdc-test-bucket/*",
                          "arn:aws:s3:::btsdc-ingest-bucket/*",
                          "arn:aws:s3:::btsdc-json-bucket/*"
            ]
        },
        {
```

```
        "Sid": "VisualEditor1",
        "Effect": "Allow",
        "Action": "s3:ListBucket",
        "Resource": "arn:aws:s3:::*"
      }
  ]
}
```

When we upload files into our S3 buckets, we should be able to select contents.

```
SELECT $1 FROM @TEST.public.json_stage/json_test.json;
```

Having resolved AWS security dependencies for our new S3 bucket and storage integration, we can rely upon Snowflake structures built in previous chapters; therefore, assume familiarity and set our context accordingly.

```
USE ROLE       sysadmin;
USE DATABASE   TEST;
USE WAREHOUSE  COMPUTE_WH;
USE SCHEMA     public;
```

To automate, we create a new stage to address two parameters: DIRECTORY and ENCRYPTION. These were previously introduced but not explained.

```
CREATE OR REPLACE STAGE TEST.public.json_stage
STORAGE_INTEGRATION = json_integration
DIRECTORY           = ( ENABLE = TRUE AUTO_REFRESH = TRUE )
ENCRYPTION          = ( TYPE = 'SNOWFLAKE_SSE' )
URL                 = 's3://btsdc-json-bucket/'
FILE_FORMAT         = TEST.public.test_pipe_format;
```

ENCRYPTION addresses where in the data upload process our files are encrypted, documentation notes we should specify server-side encryption (SNOWFLAKE_SSE) for unstructured data files; see https://docs.snowflake.com/en/sql-reference/sql/create-stage.html#internal-stage-parameters-internalstageparams.

DIRECTORY has several parameters, all documented at https://docs.snowflake.com/en/sql-reference/sql/create-stage.html#directory-table-parameters-directorytableparams. At the time of writing, the Snowflake documentation does not explain how AUTO_REFRESH is enabled, this document explains how SQS should be configured; see `https://docs.snowflake.com/en/user-guide/data-load-dirtables-auto-s3.html#step-2-configure-event-notifications`. First, we must identify the ARN of the SQS queue for the directory table in the directory_notification_channel field.

```
DESC STAGE TEST.public.json_stage;

SELECT "property_value"
FROM   TABLE ( RESULT_SCAN ( last_query_id()))
WHERE  "property" = 'DIRECTORY_NOTIFICATION_CHANNEL';
```

In our example. your results will differ. an example response is.

```
arn:aws:sqs:eu-west-2:321333230101:sf-snowpipe-AIDAUVUHT7IKQ7JAXV7BV-NcSl4ctKOyISNW9XBbvhOQ
```

Switching to AWS Management Console, navigate to S3 ➤ btsdc-json-bucket ➤ Properties ➤ Event notification. Create an S3_json_directory_update event notification. Check the event notifications, as shown in Figure 10-2.

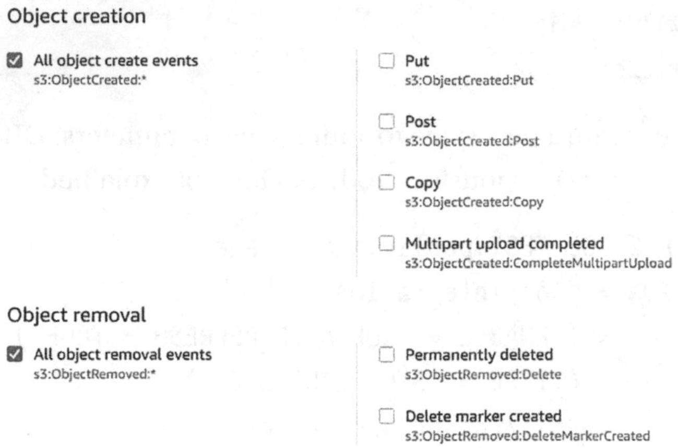

Figure 10-2. *Setting event notification*

At the foot of the page, select the SQS queue, and enter the SQS queue ARN as shown in Figure 10-3.

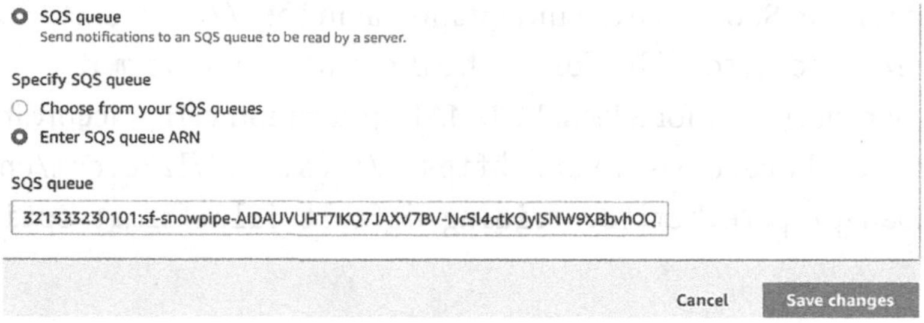

Figure 10-3. *Setting SQS queue*

Then save the changes.

If SQS is not configured, we may need to manually refresh our stage directory.

```
ALTER STAGE TEST.public.json_stage REFRESH;
```

File Format

In previous chapters, we encountered the FILE FORMAT object, but we did not pay much attention to them. We used simple declarations to skip the first record, since this is typically a header record used to convey attribute names that already exist in our staging tables.

But there are many other uses for FILE FORMAT objects, and some relate to semi-structured data where we might use a FILE FORMAT to interact with a stage, including setting the TYPE to JSON, as this example illustrates.

```
CREATE OR REPLACE FILE FORMAT TEST.public.test_json_format
TYPE = JSON;
```

When writing this book, XML is only supported as a Public Preview feature.

Other options include setting the compression type. While Snowflake automatically detects several compression types by default, some are not yet natively supported; see https://docs.snowflake.com/en/user-guide/intro-summary-loading. html#compression-of-staged-files. This example shows how to manually set the compression type.

```
CREATE OR REPLACE FILE FORMAT TEST.public.test_json_format_brotli
TYPE        = JSON
COMPRESSION = Brotli;
```

Please also refer to Snowflake documentation at `https://docs.snowflake.com/en/sql-reference/sql/create-file-format.html#create-file-format`.

The order of precedence for a FILE FORMAT application varies according to where FILE FORMAT is declared. Please refer to `https://docs.snowflake.com/en/user-guide/data-load-prepare.html#overriding-default-file-format-options`.

JSON

JSON is short-hand for JavaScript Object Notation, a human-readable text format for storing and transporting information. JSON content is self-describing and, due to the nested nature of relationships in parent-child form, easy to understand.

This JSON fragment expands upon the same data set in Chapter 9 used in the object tagging discussion.

```
{
  "employee":[
  {
    "firstName": "John","lastName": "Doe",
    "gender": "Male",
    "socialSecurityNumber": "12345678",
    "dateOfBirth": "12-Oct-2001",
    "address": {
        "streetAddress": "43 Kingston Drive","city": "Doncaster","state":
        "South Yorkshire",
        "postalCode": "DN9 4BS"
    },
    "phoneNumbers": [
      { "type": "home", "number": "9012345678" } ]
  },
  {
    "firstName": "Jane",
    "lastName": "Jones",
    "gender": "Female",
    "socialSecurityNumber": "23456789",
    "dateOfBirth": "23-Jan-2005",
    "address": {
```

```
        "streetAddress": "22 St James Drive","city": "Leeds",
        "state": "West Yorkshire","postalCode": "DN9 4BX"
    },
    "phoneNumbers": [{ "type": "home", "number": "8901234567" },{
    "type": "work", "number": "7890123456" }]
  }]
}
```

Immediately we can read the JSON structure and recognize meaningful information, but what is less clear (this example has deliberately been structured to illustrate the point) is the built-in structured nature of JSON. To view in a more readable format, please navigate to https://jsonformatter.org, then type or cut and paste the preceding into the browser from the accompanying file, where a cleaner, easier-to-read version is presented in the right-hand pane.

Please experiment with changing the content and then "validate" your changes. Note the use of curly braces ({ }) that hold objects resulting in key/value pairs and square brackets ([ ]) that hold lists and arrays. These concepts are important to grasp for later use in materialized views explained next.

Variant Data Type

Now that you have a basic grasp of JSON, let's prepare a file containing the preceding or reformatted content from https://jsonformatter.org and load it into S3. For this demonstration and using your favorite text editor, please create json_test.json and upload it into S3.

Another way to view JSON, assuming your browser opens JSON files, is to associate the file type with your browser when attempting to open json_test.json. You should see a stylized representation of the file content and tabs offering different views. In Firefox, these look like Figure 10-4. Note the partial screenshot shown.

Figure 10-4. *Firefox JSON sample display*

With our sample file json_test.json uploaded to S3, we can see the json_test.json file contents.

```
SELECT $1 FROM @TEST.public.json_stage/json_test.json;
```

For troubleshooting external stage issues, please see Chapter 7, where a step-by-step configuration guide is available.

For our semi-structured data, we use the VARIANT data type. There are others, including OBJECT and ARRAY. All are described at https://docs.snowflake.com/en/ sql-reference/data-types-semistructured.html#semi-structured-data-types.

Note   VARIANT is limited to 16 MB in size.

Declare our staging table to ingest the json_test.json file.

```
CREATE OR REPLACE TABLE stg_json_test
(
json_test    VARIANT
);
```

Now load json_test.json into stg_json_test using FILE FORMAT declared earlier.

```
COPY INTO stg_json_test FROM @TEST.public.json_stage/json_test.json
FILE_FORMAT = test_json_format;
```

Confirm data has loaded as expected.

```
SELECT * FROM stg_json_test;
```

We should see our JSON data set as shown in Figure 10-5. when you click the record, the pop-up window displays nicely formatted JSON, as you saw in the right-hand pane at `https://jsonformatter.org`.

Figure 10-5. Firefox JSON sample display

Handling Large Files

What if our JSON file exceeds the maximum variant size of 16 MB? What options do we have to load larger files?

Depending upon how our JSON is constructed and remembering earlier comments regarding square brackets, which hold lists and arrays. One option is to strip the outer array by redefining the file format, truncate the staging table, and then reload. This follows Snowflake's recommendation at `https://docs.snowflake.com/en/user-guide/data-load-considerations-prepare.html#semi-structured-data-size-limitations`.

```
CREATE OR REPLACE FILE FORMAT TEST.public.test_json_format
TYPE              = JSON
STRIP_OUTER_ARRAY = TRUE
TRUNCATE TABLE stg_json_test;
```

```
COPY INTO stg_json_test FROM @TEST.public.json_stage/json_test.json
FILE_FORMAT = test_json_format;
```

But this approach does not work for JSON without an outer array to strip. We need to consider other options.

While researching this chapter, I came across a technique worthy of consideration. I have not tested or endorsed the next approach. I simply offer it as a novel way forward; see `https://medium.com/snowflake/breaking-the-16-mb-limit-kinda-fe8d36658b`.

Other approaches to splitting large JSON files include utilizing an ELT tool, which may offer built-in capabilities, or developing a custom process using a third-party tool to pre-process JSON into manageable chunks, as jq purports to do; see `https://stedolan.github.io/jq/`. Also, consider using an external table to access your JSON file and pre-processing data before loading it into Snowflake.

Regardless of the approach chosen, remember the S3 bucket where your JSON file lands using SQS could invoke a Lambda function to pre-process, then copy to the S3 bucket where Snowpipe or other auto-ingestion processing occurs. Unfortunately, a fuller investigation is beyond the scope of this book.

Materialized Views

Having successfully landed our test file into our staging table, we must consider how to access the JSON content. At this point, we now discuss the merits of directly accessing JSON content and creating materialized views that overlay the JSON structure.

We aim to normalize the JSON structure into a relational format to facilitate joining with our structured data content. We may choose to implement a suite of three normal forms of materialized views which would extend the pattern explained here, which stops at a single denormalized materialized view.

If we consider the structure of our sample JSON document and attempt to directly access elements, we soon identify reasons to consider adopting a more sophisticated approach. Referencing individual elements in an array does not provide a scalable delivery pattern.

```
SELECT  json_test:employee[0].firstName::STRING,
        json_test:employee[1].firstName::STRING
FROM    stg_json_test;
```

To explain notation used in the code sample: The canonical reference to attribute tables then attribute:top level array[n].leaf node::datatype. Note the difference in the field delimiters. Figure 10-6 shows the output. As we expect, both attributes are in the same record.

Results   Data Preview

✔ Query ID   SQL   25ms ▬▬▬▬▬   1 rows

| Filter result... | | 📥 | Copy |

Row	JSON_TEST:EMPLOYEE[0].FIRSTNAME::STRING	JSON_TEST:EMPLOYEE[1].FIRSTNAME::STRING
1	John	Jane

Figure 10-6. *Firefox JSON sample display*

We need tools to make JSON content accessible by turning the semi-structured format into a more familiar relational format. In other words, extract each nested section embedded in the JSON document into its constituent structured component, ready for consumption. We now walk through how to do this using Snowflake functions as mechanisms to turn an array into a nested result set.

- LATERAL (documentation at `https://docs.snowflake.com/en/sql-reference/constructs/join-lateral.html`)

- FLATTEN() (documentation at `https://docs.snowflake.com/en/sql-reference/functions/flatten.html#flatten`)

```
SELECT e.value:firstName::STRING
FROM   stg_json_test,
       LATERAL FLATTEN ( json_test:employee ) e;
```

Figure 10-7 shows the effect of using both LATERAL and FLATTEN. We can extend the attribute list using the same pattern to all attributes at the same level in the JSON record, which we do next.

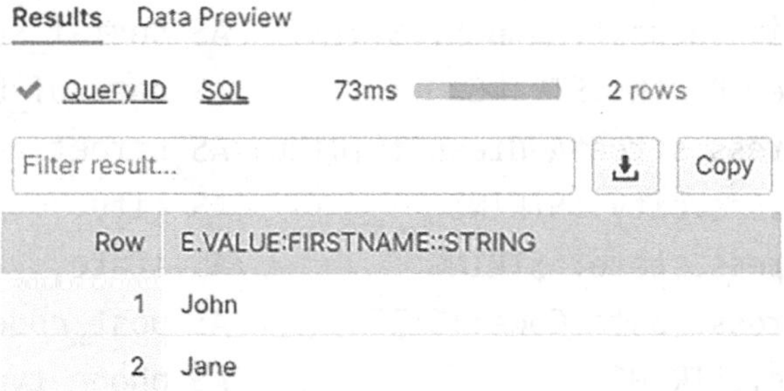

Results   Data Preview

✔ Query ID   SQL   73ms ▬▬▬▬▬   2 rows

| Filter result... | | 📥 | Copy |

Row	E.VALUE:FIRSTNAME::STRING
1	John
2	Jane

Figure 10-7. *LATERAL and FLATTEN output*

By referencing nested object attributes we can flatten the address object too.

```
SELECT  e.value:firstName::STRING,
        e.value:lastName::STRING,
        e.value:gender::STRING,
        e.value:socialSecurityNumber::STRING,
        e.value:dateOfBirth::STRING,
        e.value:address.streetAddress::STRING,
        e.value:address.city::STRING,
        e.value:address.state::STRING,
        e.value:address.postalCode::STRING,
        p.value:type::STRING,
        p.value:number::STRING
FROM    stg_json_test,
        LATERAL FLATTEN ( json_test:employee )                 e,
        LATERAL FLATTEN ( e.value:phoneNumbers, OUTER => TRUE ) p;
```

Note   Nested records in square brackets should be lateral flattened.

With the preceding query, we can create a view and/or a materialized view.

```
CREATE OR REPLACE MATERIALIZED VIEW mv_stg_json_test
AS
SELECT  e.value:firstName::STRING              AS first_name,
        e.value:lastName::STRING               AS last_name,
        e.value:gender::STRING                 AS gender,
        e.value:socialSecurityNumber::STRING   AS social_security_number,
        e.value:dateOfBirth::STRING            AS date_of_birth,
        e.value:address.streetAddress::STRING  AS street_address,
        e.value:address.city::STRING           AS city,
        e.value:address.state::STRING          AS state,
        e.value:address.postalCode::STRING     AS post_code,
        p.value:type::STRING                   AS phone_type,
        p.value:number::STRING                 AS phone_number
FROM    stg_json_test,
```

```
    LATERAL FLATTEN ( json_test:employee )                        e,
    LATERAL FLATTEN ( e.value:phoneNumbers, OUTER => TRUE ) p;
```

The following identifies materialized views in the current schema.

```
SHOW MATERIALIZED VIEWS;
```

The following demonstrates there are records in the new materialized view.

```
SELECT * FROM mv_stg_json_test;
```

Direct Query or Materialized View

Now we have built code to perform both direct query access against our staged data and a materialized view. We should consider the relative merits of both approaches.

With direct query access, we can easily address new elements as they appear in the source JSON document when they arrive but at the cost of explicit attribute referencing. As you can see, the syntax is somewhat convoluted, labor intensive to implement, does not itself to integration with other Snowflake objects and JSON does not support a data type any data, dates, timestamps, and numbers are treated as strings. The advantages of direct query access are no additional disk storage charges and end-user access to the raw JSON.

The materialized view approach requires up-front configuration. It is relatively inflexible in the event of changes to the underlying JSON document structure and incurs additional storage costs. The advantages materialized views offer include.

- Abstraction from the JSON pathing by renaming attributes into more user-friendly conventions

- Improved performance when accessing deeply buried information in the JSON

- Cleaner syntax when integrating with other Snowflake objects and the possibility of micro-partition pruning improves performance due to both micro-partition and cluster key definition

- Materialized views used for summarizations and aggregations and subject to RBAC

- Conversion from JSON date and timestamp strings into their corresponding Snowflake data type provides an opportunity for performant temporal queries.

- Materialized views can be used for query rewrite even if not referenced in the SQL

The general use cases for direct query access are when results change often, results are not used often, and queries are not resource-intensive. The general use cases for materialized views are when results don't change often, results are often used, and queries are resource intensive.

In summary, our preferred approach is to retain our raw JSON and build materialized views on top, providing the best of both approaches but at the cost of additional (cheap) storage for the materialized view. Note that materialized views can also be used on external tables.

MV Limitations

While materialized views offer significant benefits over direct query access, there are some limitations: One or more materialized views can be declared but only against a single table. For the best performance, the cluster key should differ from the underlying table. The number of cluster columns should be kept low, with the cluster key defined as lowest to highest cardinality. Temporal functions such as current_timestamp() cannot be used in materialized view declaration. For replicated materialized views, only the definition is carried across to the secondary, not the results, to maintain data.

```
SET AUTO_REFRESH_MATERIALIZED_VIEWS_ON_SECONDARY = TRUE;
```

Serverless compute is used for maintenance and therefore incurs both storage and compute costs for automatic management as the materialized view stores the results of the declaring query. Costs are a function of the amount of data that changes in each base table, the number of materialized views created on each table with compute costs calculated in 1-second increments.

Note   The Time Travel feature is not supported in materialized views.

Automation

In Chapter 8, we implemented Snowpipe, recognizing that materialized views by their nature are limited to the current version of data only. We may wish to historize ingested JSON documents.

We can now take advantage of the stage directory, which provides a much richer and more natural SQL interface than the LIST command.

```
LIST @TEST.public.json_stage;

SELECT * FROM DIRECTORY ( @TEST.public.json_stage );
```

We can build a stream on top of the stage directory for change data capture.

```
CREATE OR REPLACE STREAM strm_test_public_json_stage ON STAGE TEST.public.
json_stage;
```

And query the stage.

```
SELECT * FROM strm_test_public_json_stage;
```

However, we can do more with a stage directory where Snowflake supplies several file functions documented at https://docs.snowflake.com/en/sql-reference/ functions-file.html#file-functions.

Combining the file functions with stream attributes, we can build a view. Note that metadata$row_id is always blank and therefore omitted from view declaration.

```
CREATE OR REPLACE SECURE VIEW v_strm_test_public_json_stage
AS
SELECT '@TEST.public.json_stage'                        AS stage_name,
       get_stage_location    ( @TEST.public.json_stage ) AS stage_location,
       relative_path,
       get_absolute_path     ( @TEST.public.json_stage,
                               relative_path )           AS absolute_path,
       get_presigned_url     ( @TEST.public.json_stage,
                               relative_path )           AS presigned_url,
       build_scoped_file_url ( @TEST.public.json_stage,
                               relative_path )           AS scoped_file_url,
       build_stage_file_url  ( @TEST.public.json_stage,
                               relative_path )           AS stage_file_url,
```

```
        size,
        last_modified,
        md5,
        etag,
        file_url,
        metadata$action,
        metadata$isupdate
FROM    strm_test_public_json_stage;
```

We can now query the view noting the contents are empty because creation time was after directory refresh time; therefore, to test, we must reload files into S3 and then refresh.

```
SELECT * FROM v_strm_test_public_json_stage;
```

And finally, add a task to detect the presence of new data in our stream and trigger a stored procedure (not shown) to historize.

```
 CREATE OR REPLACE TASK task_load_json_data
WAREHOUSE = COMPUTE_WH
SCHEDULE  = '1 minute'
WHEN system$stream_has_data ( 'strm_test_public_json_stage' )
AS
CALL sp_load_json_data();
```

Note the task has not resumed because the stored procedure has not been declared; this is an exercise for you to complete.

Unstructured Data

With a little thought, we can leverage AWS to pre-process our documents in S3 and deliver content ready for ingestion in JSON format. Figure 10-8 outlines the steps required: (1) Files land in a new S3 bucket; (2) SQS detects file presence and calls (3) Lambda, which detects the file type and converts to JSON; (4) file is written to existing S3 bucket mapped via external stage.

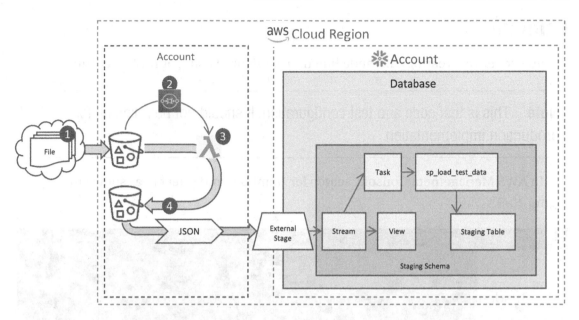

Figure 10-8. *Integrated approach*

Note   If using AWS Free Tier, be aware that Textract usage limitations of 100 invocations before charges apply.

Once a JSON file has landed in S3, we can treat the content as semi-structured data and ingest it according to the preceding pattern.

If our file exceeds 16 MB, we may also pre-process it to split into multiple source files or strip the outer array according to need.

File Preparation

In AWS Management Console, navigate to S3 and create a new S3 bucket: btsdc-ingest-bucket. We require a second bucket to prevent circular writes when the Lambda function is triggered.

For testing purposes, we use a simple PNG file created from a screenshot of an invoice.

We will later upload the test files into btsdc-ingest-bucket. The Lambda function is triggered, and files are copied to btsdc-test-bucket.

Lambda

We now focus on creating a Lambda function configured using our AWS account.

Note   This is test code and test configuration. It should not be relied on for production implementation.

In AWS Management Console, search for "lambda" and select it, as shown in Figure 10-9.

Figure 10-9. *AWS Lambda*

Click "Create function" and ensure "Author from scratch" is selected. Populate the "Basic information" fields, as shown in Figure 10-10.

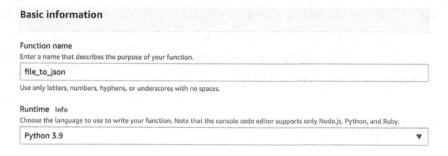

Figure 10-10. *AWS Lambda basic information*

Click the "Create function" button. You are presented with more configuration options, as shown in Figure 10-11. Note the Copy ARN, Add trigger and Upload from buttons.

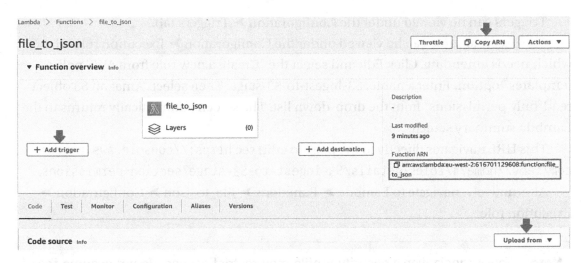

Figure 10-11. *AWS Lambda basic information*

A basic Lambda function is available in the Code source ➤ lambda_function tab. Note that the Deploy button is grayed out. Whenever we change the Python code, the Deploy button is enabled.

Click "Add trigger" and populate S3. Select your ingestion bucket name, and the "All object create events" event type. Select the "Recursive invocation" checkbox and click Add. See Figure 10-12.

Trigger configuration

> 🪣 **S3**
> aws   storage

Bucket
Please select the S3 bucket that serves as the event source. The bucket must be in the same region as the function.

> btsdc-ingest-bucket

Event type
Select the events that you want to have trigger the Lambda function. You can optionally set up a prefix or suffix for an event. However, for each bucket, individual events cannot have multiple configurations with overlapping prefixes or suffixes that could match the same object key.

> All object create events

Figure 10-12. *AWS Lambda basic information*

If an error is received during your testing, it may be due to prior event registration. Navigate to S3 ➤ Properties ➤ Event notification and remove any events from file_to_json, and then retry.

Triggers can be viewed under the Configuration ➤ Triggers tab.

The execution role can be viewed under the Configuration ➤ Execution role tab, which needs amending. Click Edit and select the "Create a new role from AWS policy templates" option. Enter a name: S3-ingest-to-S3-stage. Then select "Amazon S3 object read-only permissions" from the drop-down list. The screen automatically returns to the Lambda summary screen.

This URL navigates directly to the page to edit; see `https://console.aws.amazon.com/iamv2/home?#/roles/details/S3-ingest-to-S3-stage?section=permissions`.

Alternatively, navigate to Lambda ➤ Functions ➤ file_to_json ➤ Configuration ➤ Execution role.

Note   Your organization's security profile may restrict access, do not assume the following suggested policy meets requirements.

Click the policy name beginning with AWSLambdaS3ExecutionRole-nnnnnnn. Then navigate to the Edit policy ➤ JSON tab and paste the following. The permissions apply to all S3 buckets and those for Amazon Textract.

```
{
    "Version": "2012-10-17",
    "Statement": [
        {
            "Effect": "Allow",
            "Action": [
                "s3:GetObject",
                "s3:PutObject",
                "s3:DeleteObject"
            ],
            "Resource": "arn:aws:s3:::*"
        },
        {
            "Effect": "Allow",
            "Action": [
                "textract:DetectDocumentText",
                "textract:AnalyzeDocument",
```

```
        "textract:StartDocumentTextDetection",
        "textract:StartDocumentAnalysis",
        "textract:GetDocumentTextDetection",
        "textract:GetDocumentAnalysis"
      ],
      "Resource": "*"
    }
  ]
}
```

Click "Review policy" and then the "Save changes" button, which returns to the Lambda function overview.

This code copies a file from the source S3 bucket into the destination bucket. Cut and paste it into the Code tab to test if the policy works correctly.

Note   destination_bucket_name must be changed as in AWS. S3 bucket names are global in scope across all AWS accounts.

The following code is intended to be cut and pasted, Python is indenting sensitive, and the code format incorrectly implies a line break after the "# Specify source bucket" and "# Write copy statement" statements.

Note   The AWS Lambda console implements alignment markers that must be adhered to; otherwise, an indentation error occurs.

```python
import time
import json
import boto3

s3_client=boto3.client('s3')

# lambda function to copy file from one s3 bucket to another s3 bucket
def lambda_handler(event, context):

    # Specify source bucket
```

```
source_bucket_name=event['Records'][0]['s3']['bucket']['name']

# Get file that has been uploaded
file_name=event['Records'][0]['s3']['object']['key']

# Specify destination bucket - this must be changed as AWS S3 bucket
names are global in scope
destination_bucket_name='btsdc-test-bucket'

# Specify from where file needs to be copied
copy_object={'Bucket':source_bucket_name,'Key':file_name}

# Write copy statement
s3_client.copy_object(CopySource=copy_object,Bucket=destination_bucket_
name,Key=file_name)

return {
    'statusCode': 200,
    'body': json.dumps('File has been Successfully Copied')
}
```

Then click Deploy.

Upload files to btsdc-ingest-bucket and check they appear in btsdc-test-bucket. We may also check the file has been copied across correctly by downloading it from the target S3 bucket.

Delete files from both btsdc-ingest-bucket and btsdc-test-bucket.

Extracting File Content

This section has the potential to expand into a separate book itself. One challenge with authoring code samples is keeping content to manageable sections while conveying enough information to inform and equip the audience. Cue one epic fail!

Let's start by investigating Amazon Textract, a file processing utility. Its documentation is at https://boto3.amazonaws.com/v1/documentation/api/latest/reference/services/textract.html#Textract.Client.analyze_document. Note that Textract has two modes of operation: synchronous for small files and asynchronous for larger files. In this chapter, we investigate Synchronous operation using textract.analyze_

document. Documentation is at https://docs.aws.amazon.com/textract/latest/dg/
analyzing-document-text.html while integrating with our S3 copy code.

I confess. I did not author all of the following code. It is based upon code supplied
and used with the kind permission of Saurin Shah of Snowflake Inc., extended to
showcase the "art of the possible" with apologies for removing comments and blank
lines to reduce the page count. The accompanying file has a complete version.

Amend our Lambda function by cutting and pasting it into the body, then deploy.
Testing should be straightforward. Upload a PNG or JPG file into your S3 ingest
bucket, and assuming all is well, a JSON file containing all the text appears in the S3
output bucket.

```
import time
import pathlib
import json
import boto3

s3_client          = boto3.client('s3')
json_return_array = [ ]

def textract_analyze_document(source_bucket_name, file_name):
    doc_client = boto3.client('textract')
    response = doc_client.analyze_document(Document={'S3Object': { 'Bucket':
    source_bucket_name, 'Name': file_name}}, FeatureTypes=["TABLES"])
    blocks = response['Blocks']
    key_map = {}
    value_map = {}
    block_map = {}
    for block in blocks:
        block_id = block['Id']
        block_map[block_id] = block
        if block['BlockType'] == "KEY_VALUE_SET":
            if 'KEY' in block['EntityTypes']:
                key_map[block_id] = block
            else:
                value_map[block_id] = block
        elif block['BlockType'] == "LINE":
            key_map[block_id] = block
```

```python
        return key_map, value_map, block_map

def find_value_block(key_block, value_map):
    value_block = {}
    for relationship in key_block['Relationships']:
        if relationship['Type'] == 'VALUE':
            for value_id in relationship['Ids']:
                value_block = value_map[value_id]
    return value_block

def get_kv_relationship(key_map, value_map, block_map):
    kvs = {}
    for block_id, key_block in key_map.items():
        value_block = find_value_block(key_block, value_map)
        key = get_text(key_block, block_map)
        val = get_text(value_block, block_map)
        kvs[key] = val
    return kvs

def get_text(result, blocks_map):
    text = ''
    if 'Relationships' in result:
        for relationship in result['Relationships']:
            if relationship['Type'] == 'CHILD':
                for child_id in relationship['Ids']:
                    word = blocks_map[child_id]
                    if word['BlockType'] == 'WORD':
                        text += word['Text'] + ' '
                    if word['BlockType'] == 'SELECTION_ELEMENT':
                        if word['SelectionStatus'] == 'SELECTED':
                            text += 'X '
    return text

def lambda_handler(event, context):
    source_bucket_name = event['Records'][0]['s3']['bucket']['name']
    file_name = event['Records'][0]['s3']['object']['key']
    target_bucket_name='btsdc-test-bucket'
```

```
file_extension = pathlib.Path(file_name).suffix
file_extension = file_extension.lower()
copy_object = {'Bucket':source_bucket_name, 'Key':file_name}
if file_extension == '.png' or file_extension == '.jpg':
    key_map, value_map, block_map = textract_analyze_document(source_
    bucket_name, file_name)
    kvs = get_kv_relationship(key_map, value_map, block_map)
    json_return_array.append(kvs)
    json_return_string = json.dumps({"data" : json_return_array})
    s3_client.put_object(Bucket = target_bucket_name, Body = json_return_
    string, Key = file_name + '.json')
    s3_client.copy_object(CopySource = copy_object, Bucket = target_
    bucket_name, Key=file_name)
    s3_client.delete_object(Bucket = source_bucket_name, Key=file_name)
return {
    'statusCode': 200,
    'body': json.dumps('File has been Successfully Copied')
}
```

Mapping to JSON

With our PNG document converted to JSON, we can use the semi-structured data integration pattern articulated at the start of the chapter to ingest content into Snowflake.

```
 LIST @TEST.public.json_stage;
```

```
SELECT $1 FROM @TEST.public.json_stage/test_json.png.json;
```

There is one caveat. During testing, we encountered the following error.

Invalid UTF8 detected in string ' "0xA3 ": "," File 'test_json.png.json', line 38, character 1, row 37, column "TRANSIENT_STAGE_TABLE"["$1":1].

The fix was to delete the offending line (38) in JSON and then reload and retry.

You may also find it necessary to load your file into https://jsonformatter.org and save the output before uploading.

With data accessible, I leave it to you for further processing.

Troubleshooting

If files have not been copied across, navigate to Lambda ➤ Functions ➤ file_to_json
➤ Monitor ➤ Logs. Click the "View logs in CloudWatch" option, which opens a new
browser tab showing all available log streams in date order, the most recent first. Refresh,
and then click the most recent log where information similar to what is shown in
Figure 10-13 is displayed. Note the example shown is for two Lambda invocations, the
first failed, and the second most recent invocation succeeded.

Figure 10-13. *CloudWatch sample output*

Further information can be found by opening up each section where the Lambda
error line and reason appear, as shown in Figure 10-14.

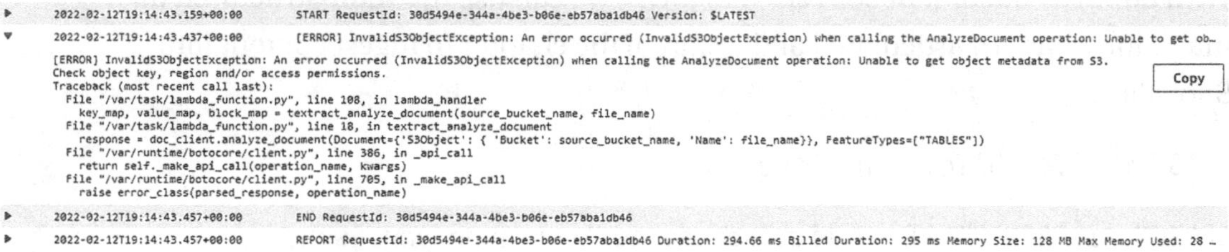

Figure 10-14. *CloudWatch error example*

SQL Integration

The previous section demonstrated how to automatically process files on receipt and
convert them to JSON. This chapter builds all required Snowflake capability to stage and
interact with JSON. Let's now focus on directly accessing the content in files held on S3
using SQL. To do so, we must create an external function to call a remote service, in our

example, an AWS Python Lambda, using API integration for which documentation is at `https://docs.snowflake.com/en/sql-reference/sql/create-api-integration.html#for-amazon-api-gateway` and `https://docs.snowflake.com/en/sql-reference/external-functions-creating-aws.html`. Note the very helpful embedded YouTube tutorial.

The remote service must expose an HTTPS endpoint, both accept and return a single JSON record. Figure 10-15 illustrates the proposed approach where (1) represents our file delivered into S3; (2) is our new Lambda function, which extracts the file content; (3) maps the remote service endpoint into Snowflake; (4) external function and SQL call to extract file content.

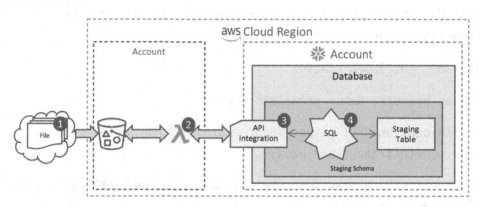

Figure 10-15. *Remote service and API integration*

Tracking Template

We populate the following template copied from `https://docs.snowflake.com/en/sql-reference/external-functions-creating-aws-planning.html#preparing-to-use-the-aws-management-console`. You may also wish to add this template to your support documentation for each external function created.

Note   Using the template saves you time and prevents troubleshooting and rework. I found out the hard way!

```
==========================================================================
================ Tracking Worksheet: AWS Management Console ================
==========================================================================
```

****** Step 1: Information about the Lambda Function (remote service) *****

Your AWS Account ID: 616701129608

Lambda Function Name: parse_image

******** Step 2: Information about the API Gateway (proxy Service) ********

New IAM Role Name: Snowflake_External_Function

New IAM Role ARN: arn:aws:iam::616701129608:role/Snowflake_External_
Function

Snowflake VPC ID (optional): com.amazonaws.vpce.eu-west-2.vpce-
svc-0839061a5300e5ac1

New API Name: get_image_content

API Gateway Resource Name: snowflake_proxy

Resource Invocation URL: https://8qdfb0w8fh.execute-api.eu-west-2.
amazonaws.com/test/snowflake_proxy

Method Request ARN: arn:aws:execute-api:eu-west-2:616701129608:8qdfb0w8
fh/*/POST/snowflake_proxy

API Key (optional): uqkOKvci6OajRNCYXvohX9WTY8HpxGzt5vj9eDHV

*** Step 3: Information about the API integration and External Function ***

API Integration Name: document_integration

API_AWS_IAM_USER_ARN: arn:aws:iam::321333230101:user/vnki-s-ukst5070

API_AWS_EXTERNAL_ID: GH06274_SFCRole=3_8GebOvugHdYOQcaNVeO5Ki/H2+Q=

External Function Name: get_image_data

Create a Python Lambda Function

Repeating the steps from Figure 10-9 and Figure 10-10, create a new Lambda function called parse_image. Select Python 3.9 as the runtime, then leave the default execution role as "Create a new role with basic Lambda permissions" before clicking the "Create function" button. After a minute or so, our new function appears similar to Figure 10-11.

Cut and paste the sample code from https://docs.snowflake.com/en/sql-reference/external-functions-creating-aws-sample-synchronous.html#sample-synchronous-lambda-function (as follows) and then deploy. This proves that end-to-end configuration works and is replaced later in this chapter.

```python
import json

def lambda_handler(event, context):

    # 200 is the HTTP status code for "ok".
    status_code = 200

    # The return value will contain an array of arrays (one inner array per
    input row).
    array_of_rows_to_return = [ ]

    try:
        # From the input parameter named "event", get the body, which
        contains
        # the input rows.
        event_body = event["body"]

        # Convert the input from a JSON string into a JSON object.
        payload = json.loads(event_body)
        # This is basically an array of arrays. The inner array
        contains the
        # row number, and a value for each parameter passed to the
        function.
        rows = payload["data"]

        # For each input row in the JSON object...
        for row in rows:
            # Read the input row number (the output row number will be
            the same).
            row_number - row[0]

            # Read the first input parameter's value. For example, this
            can be a
```

```python
        # numeric value or a string, or it can be a compound
        value such as
        # a JSON structure.
        input_value_1 = row[1]

        # Read the second input parameter's value.
        input_value_2 = row[2]

        # Compose the output based on the input. This simple example
        # merely echoes the input by collecting the values into an
        array that
        # will be treated as a single VARIANT value.
        output_value = ["Echoing inputs:", input_value_1, input_
        value_2]

        # Put the returned row number and the returned value into
        an array.
        row_to_return = [row_number, output_value]

        # ... and add that array to the main array.
        array_of_rows_to_return.append(row_to_return)

    json_compatible_string_to_return = json.dumps({"data" : array_of_
    rows_to_return})

except Exception as err:
    # 400 implies some type of error.
    status_code = 400
    # Tell caller what this function could not handle.
    json_compatible_string_to_return = event_body

# Return the return value and HTTP status code.
return {
    'statusCode': status_code,
    'body': json_compatible_string_to_return
}
```

Now test the Lambda function using the supplied test case, cut and paste into the Test tab. Then click the Test button, after which you can see the execution result. Use CloudWatch to investigate.

```
{
  "body":
    "{ \"data\": [ [ 0, 43, \"page\" ], [ 1, 42, \"life, the universe, and
    everything\" ] ] }"
}
```

You should see "Execution result: succeeded."

Create AWS IAM Role

We now use AWS Management Console to configure the IAM role for Snowflake use. Click Roles ➤ Create role. Our role is for another AWS account which is a trusted entity. Select "AWS account" and use your account number recorded in the tracking template (mine is 616701129608). Then click Next ➤ Next. Add a meaningful name such as Snowflake_External_Function, and then click Create role.

Once created, click the new role and record the ARN. (Mine is arn:aws:iam::616701129608:role/Snowflake_External_Function.)

Get Snowflake VPC Information

We must now retrieve the Snowflake VPC information, which relates to the account provisioned by Snowflake, and record information in our tracking template. To do this, from the Snowflake user interface, issue the following commands.

```
USE ROLE accountadmin;

SELECT system$get_privatelink_config();
```

Assuming no private links have been configured, and after reformatting at https://jsonformatter.org, the returned JSON document should look like the following.

```
{
  "privatelink-account-name": "gh06274eu-west-2.privatelink",
```

```
    "privatelink-vpce-id": "com.amazonaws.vpce.eu-west-2.vpce-
    svc-0839061a5300e5ac1",
    "privatelink-account-url": "gh06274eu-west-2.privatelink.
    snowflakecomputing.com",
    "regionless-privatelink-account-url": "ewaomog-bx93955.privatelink.
    snowflakecomputing.com",
    "privatelink_ocsp-url": "ocsp.gh06274eu-west-2.privatelink.
    snowflakecomputing.com",
    "privatelink-connection-urls": "[]"
}
```

From which we add privatelink-vpce-id: com.amazonaws.vpce.eu-west-2.vpce-svc-0839061a5300e5ac1 to our tracking template. Note that yours will differ.

Create API Gateway Endpoint

In AWS Management Console, search for "API Gateway" and select the service. From the options presented, find Rest API and click the Build button. Then select New API, at which point you see the Settings screen.

Enter the API name: get_image_content. Optionally, you can populate the Description field. Leave the Endpoint Type set to Regional, and then click Create API.

In the Methods screen, click Actions and select Create Resource. Set the resource name to snowflake_proxy before clicking the "Create resource" button.

Next, click Actions and select Create Method. From the drop-down list, select POST, and click the tick mark. Leave Lambda Function selected as the integration type. Select the "Use Lambda Proxy integration" option. In the Lambda Function dialog box, enter parse_image and click Save. In the pop-up box, click OK. The screen now presents a schematic showing request/response. This can be ignored.

From the Actions drop-down, select Deploy API action. Set the Deployment Stage to [New Stage], set the stage name to "test", and then click Deploy.

The newly created "test" stage can now be selected and the tree opens. Click POST and record the Invoke URL (in my example, https://8qdfbOw8fh.execute-api.eu-west-2.amazonaws.com/test/snowflake_proxy). Record this in the tracking template. Note that yours will differ.

Secure API Gateway Proxy

Click Amazon API Gateway ➤ get_image_content ➤ POST ➤ Method Request. Next, click the Edit symbol in Authorization, and in the drop-down list, select AWS IAM. Click the tick mark to save. From the Actions drop-down list, select the Deploy API action. Set the deployment stage to "test" and click Deploy.

Then click Method Execution and retrieve the ARN; in my example, it is arn:aws:execute-api:eu-west-2:616701129608:8qdfb0w8fh/*/POST/snowflake_proxy. Record this in the tracking template. Note thatyours will differ.

On the left of the screen, select Resource Policy and paste the example resource policy from the documentation at https://docs.snowflake.com/en/sql-reference/external-functions-creating-aws-ui-proxy-service.html#secure-your-amazon-api-gateway-endpoint.

Replace <12-digit-number> with your AWS account ID, in our example this is 616701129608. Replace <external_function_role> with Snowflake_External_Function. Then replace <method_request_ARN> with your method execution ARN.

The following code uses my settings; yours will differ.

```
{
    "Version": "2012-10-17",
    "Statement":
    [
        {
        "Effect": "Allow",
        "Principal":
            {
            "AWS": "arn:aws:sts::616701129608:assumed-role/Snowflake_
            External_Function/snowflake"
            },
        "Action": "execute-api:Invoke",
        "Resource": "arn:aws:execute-api:eu-west-2:616701129608:8qdfb0w8
        fh/*/POST/snowflake_proxy"
        }
    ]
}
```

Click the Save button before clicking get_image_content in the header bar. Click Actions ➤ Deploy API. In the deployment stage, select "test" and then click Deploy.

Snowflake API Integration

In this section, I have substituted values from my AWS account; yours will differ and must be replaced.

Create API Key

This section is included for completeness only. You may not need an API key, but if your organization requires API keys, here is how to create them. Documentation is at `https://docs.snowflake.com/en/sql-reference/external-functions-security.html#using-the-api-key-option-in-create-api-integration`.

From AWS Management Console, navigate to API Gateway ➤ get_image_content, and on the left-hand panel, select API Keys ➤ Actions ➤ Create API key.

In the Name dialog, enter get_image_content_key. Leave the API key set to Auto Generate, and optionally add a description before clicking Save.

To see the generated key, navigate to API Gateway ➤ get_image_content. On the left-hand panel, select API Keys ➤ get_image_content_key ➤ API key ➤ show. My key looks like this: uqkOKvci6OajRNCYXvohX9WTY8HpxGzt5vj9eDHV. Add this to the tracking template.

Navigate to API Gateway ➤ get_image_content and on the left hand panel select Resources ➤ Method Request ➤ Settings ➤ API Key Required. Ensure it is set to "true". If not, edit and click the checkbox then Actions ➤ Deploy API.

AWS documentation is at `https://docs.aws.amazon.com/apigateway/latest/developerguide/api-gateway-setup-api-key-with-console.html`.

Create API Integration

An API integration stores information about an HTTPS proxy service. We create our API integration to access the Lambda function created.

```
USE ROLE accountadmin;
```

The use of the AWS API key is not covered in the online video tutorial, hence the preceding section. The API key section is commented out in the following command, which serves as a placeholder. Your API key will differ from mine.

Define our API integration—document_integration.

```
CREATE OR REPLACE API INTEGRATION document_integration
API_PROVIDER         = aws_api_gateway
API_AWS_ROLE_ARN     = 'arn:aws:iam::616701129608:role/Snowflake_External_
Function'
//API_KEY            = 'uqkOKvci6OajRNCYXvohX9WTY8HpxGzt5vj9eDHV'
ENABLED              = TRUE
API_ALLOWED_PREFIXES = ('https://8qdfbOw8fh.execute-api.eu-west-2.
amazonaws.com/test/snowflake_proxy');
```

As we might expect, when creating a new integration, the trust relationship must be established, for which we need some information.

```
DESC INTEGRATION document_integration;
```

Record these settings in the tracking template. Yours will differ.

API_AWS_IAM_USER_ARN: arn:aws:iam::321333230101:user/vnki-s-ukst5070

API_AWS_EXTERNAL_ID: GH06274_SFCRole=3_8GebOvugHdY0QcaNV eO5Ki/H2+Q=

Establish Trust Relationship

In AWS Management Console, navigate to IAM and select the Snowflake_External_ Function role created. Select the Trust relationships tab and click "Edit trust policy". Replace AWS ARN with your API_AWS_IAM_USER_ARN. Expand the Condition square brackets to include the following text shown in curly braces {"StringEquals": {"sts:ExternalId": " GH06274_SFCRole=3_8GebOvugHdY0QcaNVeO5Ki/H2+Q=}}: and replace with your API_AWS_EXTERNAL_ID. Mine is shown.

```
{
        "Version": "2012-10-17",
        "Statement": [
                {
                        "Effect": "Allow",
```

```
                    "Principal": {
                            "AWS": "arn:aws:iam::321333230101:user/
                            vnki-s-ukst5070"
                    },
                    "Action": "sts:AssumeRole",
                    "Condition": {"StringEquals": {"sts:ExternalId": "
                    GHO6274_SFCRole=3_8GebOvugHdYOQcaNVeO5Ki/H2+Q=}}
                }
        ]
}
```

Then click the "Update policy" button. If we later update the API integration, the trust relationship will also need updating.

Create Snowflake External Function

After updating the trust relationship, we can create our get_image_data external function.

```
CREATE OR REPLACE EXTERNAL FUNCTION get_image_data ( n INTEGER, v VARCHAR )
RETURNS VARIANT
API_INTEGRATION = document_integration
AS 'https://8qdfbOw8fh.execute-api.eu-west-2.amazonaws.com/test/
snowflake_proxy';
```

Calling External Function

We can test our function by calling with appropriate values for INTEGER and VARCHAR parameters.

```
SELECT get_image_data ( 1, 'name' );
```

For our next section.

```
GRANT USAGE ON FUNCTION get_image_data(INTEGER, VARCHAR) TO ROLE sysadmin;
```

The successful response is to echo the input parameters, as shown in Figure 10-16.

Figure 10-16. *Successful Lambda invocation*

Deploy Lambda Image

This section modifies parse_image to scrape content from an image file.

Prove we can execute the existing external function as sysadmin, returning the same results shown in Figure 10-16.

```
USE ROLE        sysadmin;
USE DATABASE    TEST;
USE WAREHOUSE   COMPUTE_WH;
USE SCHEMA      public;

SELECT get_image_data ( 1, 'name' );
```

In AWS Management Console, navigate to parse_image. If you have made changes to the template code, consider saving a copy before deploying this code. Note that spaces and comments have been removed to reduce page count.

```
import time
import json
import boto3
from urllib.request import urlopen

def textract_analyze_document(contents):
    doc_client = boto3.client('textract')
    response = doc_client.analyze_document(Document={'Bytes': contents},
    FeatureTypes=["FORMS"])
    blocks = response['Blocks']
    key_map = {}
    value_map = {}
```

```python
    block_map = {}
    for block in blocks:
        block_id = block['Id']
        block_map[block_id] = block
        if block['BlockType'] == "KEY_VALUE_SET":
            if 'KEY' in block['EntityTypes']:
                key_map[block_id] = block
            else:
                value_map[block_id] = block
        elif block['BlockType'] == "LINE":
            key_map[block_id] = block
    return key_map, value_map, block_map

def find_value_block(key_block, value_map):
    value_block = {}
    for relationship in key_block['Relationships']:
        if relationship['Type'] == 'VALUE':
            for value_id in relationship['Ids']:
                value_block = value_map[value_id]
    return value_block

def get_kv_relationship(key_map, value_map, block_map):
    kvs = {}
    for block_id, key_block in key_map.items():
        value_block = find_value_block(key_block, value_map)
        key = get_text(key_block, block_map)
        val = get_text(value_block, block_map)
        kvs[key] = val
    return kvs

def get_text(result, blocks_map):
    text = ''
    if 'Relationships' in result:
        for relationship in result['Relationships']:
            if relationship['Type'] == 'CHILD':
                for child_id in relationship['Ids']:
                    word = blocks_map[child_id]
```

```
            if word['BlockType'] == 'WORD':
                text += word['Text'] + ' '
            if word['BlockType'] == 'SELECTION_ELEMENT':
                if word['SelectionStatus'] == 'SELECTED':
                    text += 'X '
    return text

def lambda_handler(event, context):
    array_of_rows_to_return = [ ]
    event_body = event["body"]
    payload = json.loads(event_body)
    rows = payload["data"]
    for row in rows:
        row_number    = row[0]
        param_1       = row[1]
        presigned_url = row[2]
        contents = urlopen(presigned_url).read()
        key_map, value_map, block_map = textract_analyze_document(contents)
        kvs = get_kv_relationship(key_map, value_map, block_map)
        row_to_return = [row_number, kvs]
        array_of_rows_to_return.append(row_to_return)
        json_compatible_string_to_return = json.dumps({"data" : array_of_
        rows_to_return})

    return {
        'statusCode': 200,
        'body': json_compatible_string_to_return
    }
```

Assign IAM Policy

We must assign entitlement for parse_image to access S3 bucket btsdc-json-bucket and use Textract. We implemented the same entitlement previously. First, edit test_policy.

```
{
    "Version": "2012-10-17",
    "Statement": [
```

```
    {
        "Sid": "VisualEditor0",
        "Effect": "Allow",
        "Action": [
            "s3:PutObject",
            "s3:GetObject",
            "s3:DeleteObjectVersion",
            "s3:DeleteObject",
            "s3:GetObjectVersion",
            "s3:ListBucket"
        ],
        "Resource": "arn:aws:s3:::*"
    },
    {
        "Sid": "VisualEditor1",
        "Effect": "Allow",
        "Action": [
            "textract:DetectDocumentText",
            "textract:StartDocumentTextDetection",
            "textract:StartDocumentAnalysis",
            "textract:AnalyzeDocument",
            "textract:GetDocumentTextDetection",
            "textract:GetDocumentAnalysis"
        ],
        "Resource": "*"
    }
    ]
}
```

Click the "Review policy" button and then save the changes. You may be prompted to remove an old version of test_policy.

Next, attach policy. Navigate to IAM ➤ Roles ➤ Snowflake_External_Function ➤ Add permissions ➤ Attach policies. You may also find a default role created for parse_image called something like parse_image-role-eu6asv26 to which test_policy should be attached.

Set the checkbox for test_policy and then click the "Attach policies" button. Do not forget to check that the policy has been attached. Navigate to IAM ➤ Roles ➤ Snowflake_ External_Function and check that test_policy appears under the "Permissions policies" section.

Invoke Lambda Image

Let's reuse btsdc-json-bucket; your S3 bucket name will differ. Upload a PNG file to your corresponding S3 bucket and from the Snowflake workspace refreshing stage if required and ensure the file is visible

```
ALTER STAGE TEST.public.json_stage REFRESH;

SELECT * FROM DIRECTORY ( @TEST.public.json_stage );

SELECT * FROM v_strm_test_public_json_stage;
```

There is a limitation with get_presigned_url function insofar as this must be a fully qualified literal stage name declared as an object and not a quoted string passed through. For more information, see the documentation at https://docs.snowflake. com/en/sql-reference/functions/get_presigned_url.html#get-presigned-url.

Using the get_presigned_url function, we now invoke parse_image. Replace <your_ file_here> with the file loaded into btsdc-json-bucket.

```
SELECT get_image_data(0, get_presigned_url(@TEST.public.json_stage, get_
relative_path(@TEST.public.json_stage, 's3://btsdc-json-bucket/<your_file_
here>.png')));
```

Figure 10-17 shows expected results, yours will differ, and the address has been deliberately blanked out.

Figure 10-17. *Sample PNG to JSON*

The JSON document returns fields and values, but due to the way in which Textract operates, not every key/value pair is returned. A full explanation is beyond the scope of this chapter, but briefly, text in images is bound by boxes before extract, and it is the content of each box that is returned. Examine the returned JSON for information. Note thattrailing spaces are significant for matching specific fields.

In the following code sample, replace <your_string> with the text to search for and <your_file_here> with the file loaded into btsdc-json-bucket.

```
SELECT parse_json(get_image_data(0, get_presigned_url(@TEST.public.json_
stage, relative_path))):"<your_string>"::string
FROM   v_strm_test_public_json_stage
WHERE  relative_path = '<your_file_here>.png';
```

Automation

Our next step is to automate document JSON extract and parsing, an exercise left for you to complete using your knowledge of streams and tasks.

Monitoring

In AWS Management Console, navigate to CloudWatch ➤ Log groups, where you see the /aws/lambda/parse_image log group. Click it to show the latest logs.

Note that upstream errors (i.e., those which do not cause Lambda to be invoked) are not recorded. Only Lambda invocations create log entries that may not be successful.

Troubleshooting

Inevitably there will be times when we need to fix our code, and during the creation of this example, I made several, hence my note to use the tracking template. One "gotcha" stands out. In the Method Request" setting, I failed to redeploy Actions ➤ Deploy API. I easily missed it. After which, I encountered a further error relating to the API key. See https://docs.snowflake.com/en/sql-reference/external-functions-creating-aws-troubleshooting.html#request-failed-for-external-function-ext-func-with-remote-service-error-403-message-forbidden to which the answer was to disable the API key from the Method Request.

The last error encountered related to Lambda not having an entitlement to access Textract, specifically analyze_document, as identified in the CloudWatch logs. The solution was to add a policy containing appropriate entitlement to the Lambda function.

The following document offers solutions to common issues: `https://docs.snowflake.com/en/sql-reference/external-functions-creating-aws-troubleshooting.html#troubleshooting-external-functions-for-aws`.

And for AWS-specific issues, see `https://docs.snowflake.com/en/sql-reference/external-functions-creating-aws-troubleshooting.html#label-external-functions-creating-aws-troubleshooting`.

Cleanup

Apart from cleanup scripts to remove our test code, we should also be mindful to periodically remove redundant files uploaded to stages remembering each file contributes toward storage costs.

```
USE ROLE        sysadmin;
USE DATABASE    TEST;
USE SCHEMA      public;

DROP TASK       task_load_test_data;
DROP DATABASE   test;

USE ROLE        accountadmin;

DROP STORAGE INTEGRATION json_integration;
DROP API INTEGRATION document_integration;
REVOKE EXECUTE TASK ON ACCOUNT FROM ROLE sysadmin;
```

Summary

This chapter introduced semi-structured data, focusing on JSON, and loading a sample file into the Snowflake VARIANT data type. We discussed techniques to handle source files of greater than 16 MB. Note that the maximum size a VARIANT can accept is 16 MB.

Noting the self-contained and complex nature of semi-structured data, we utilized materialized views to normalize into a more easily understood format leading to natural integration with our relational format data model. A summary of use cases for the direct query of materialized view creation was also provided. Materialized view limitations were discussed before touching upon automation of semi-structured data ingestion.

Moving on to unstructured data, we dived into unfamiliar territory, looking inside AWS Management Console, configuring a Python lambda function to automatically invoke a Lambda function, and copying and processing an image file into JSON format, thus demonstrating the capability to convert an unstructured image file into a semi-structured format.

Our next example delivered a "hello world" style Lambda function as an external procedure accessible via Snowflake. Using SQL proved that end-to-end connectivity is not a trivial undertaking. We then adapted our working Lambda function to scrape content out of an image document, proving we can convert an image file into a JSON document, albeit with some limitations imposed by Amazon Textract, and then extract specific key/value pairs using SQL. I left automation as an exercise for you to implement using streams and tasks.

Having worked through semi-structured and unstructured data, we now look at the query optimizer and performance tuning.

PART IV

Management

## CHAPTER 11

Query Optimizer Basics

Performance tuning is often regarded as a "black" art, available to those with experience gained over many years of hands-on work. This section offers insights into how Snowflake works "under the hood" and provides tools to better understand behavior and identify options to improve performance.

Chapter 3 discusses micro-partitions and cluster keys for large-volume data sets that are critical to performance. This chapter delves into the fundamentals driving performance, starting with clustering. We cannot proactively propose solutions without understanding how clustering is implemented and a few techniques to optimally cluster our tables. I provide summary guidance, or "rules of thumb," to facilitate design or later remediation.

By understanding how Snowflake processes queries to enable efficient data access, we can choose the most appropriate storage and warehouse strategy to reduce costs. Our next section focuses on query performance, where we investigate various scenarios in preparation for your real-world situations.

This chapter is neither exhaustive nor prescriptive in approach but offers a starting point in determining performance bottlenecks and ways to investigate and remediate them. Your real-world experience will vary from the scenarios presented.

Naturally, there is always much more to discover. I trust this brief chapter provides sufficient information and context and delivers motivation for your further investigations.

Clustering

We know each micro-partition holds between 50 MB and 500 MB of data organized in hybrid columnar format, optimizing compression according to the column data type. But how does Snowflake determine the order in which data is organized in micro-partitions? In the absence of explicit cluster key information, Snowflake's only logic is to

A. Carruthers, *Building the Snowflake Data Cloud*, https://doi.org/10.1007/978-1-4842-8593-0_11

use the order in which data is stored in the source file to create the requested file size. In other words, data order is not changed, analyzed, or modified during load.

For many low-volume data sets, clustering by key value is not particularly relevant. For example, in those scenarios where all data for an object fit into a single micro-partition, no micro-partition pruning is possible. But for those large data sets spanning several or many micro-partitions with appropriate clustering keys declared, micro-partition pruning is possible, assuming query predicates match the cluster key.

This section focuses on cluster keys, recognizing that performance is also a function of CPU and memory. We later revisit warehouses.

What Is a Clustering Key?

A clustering key comprises a subset of attributes or expressions on a table, declared in the least selective to most selective order, with the express intent of co-locating data and designed to match the commonly used query predicates. Superficially a clustering key is "like" an index. But the similarity soon breaks down because only a single clustering key can be declared for a table. Unlike an index, all attributes are stored along with the clustered attributes. Clustering key documentation is at `https://docs.snowflake.com/en/user-guide/tables-clustering-keys.html`.

Cluster Key Operations

Declaring cluster keys uses serverless compute. The proof is to switch roles to ACCOUNTADMIN, click the Account tab, where usage includes AUTOMATIC_CLUSTERING consumption costs as shown in Figure 11-1.

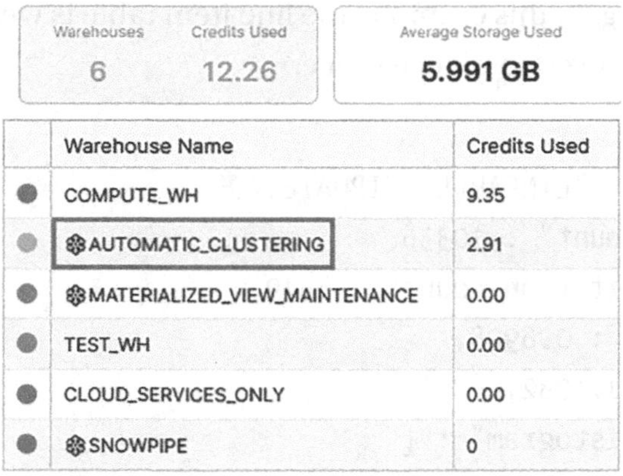

Figure 11-1. *Serverless compute costs*

Despite ALTER TABLE completing without delay, work continues in the background. To determine whether reclustering is complete, check the average_depth attribute in the result set for system$clustering_information, and when the average_depth value remains constant, reclustering is complete.

Default Clustering

In a departure from our immediate investigation to introduce clustering, we utilize the Snowflake supplied database SNOWFLAKE_SAMPLE_DATA as this contains tables with high data volumes set up for precisely this kind of investigation and look at clustering information.

```
USE ROLE      sysadmin;
USE DATABASE  snowflake_sample_data;
USE WAREHOUSE COMPUTE_WH;
USE SCHEMA    snowflake_sample_data.tpch_sf1000;
```

In case you are curious, the line item table contains 5999989709 rows, and because the metadata repository contains the rowcount, no warehouse is used, and the results are returned instantaneously.

```
SELECT system$clustering_information ( 'lineitem' );
```

The returned JSON is shown next. Note average_depth, where lower numbers indicate better clustering, and total_constant_partition_count, where higher numbers indicate better clustering. In this example, the line item table is well organized, noting some result set rows removed to reduce space.

```
{
  "cluster_by_keys" : "LINEAR(L_SHIPDATE)",
  "total_partition_count" : 10336,
  "total_constant_partition_count" : 8349,
  "average_overlaps" : 0.6908,
  "average_depth" : 1.4082,
  "partition_depth_histogram" : {
    "00000" : 0,
    "00001" : 8310,
    "00002" : 599,
    "00003" : 844,
    "00004" : 417,
    "00005" : 149,
    "00006" : 17,
...
  }
}
```

Take a few moments to examine partition_depth_histogram, which tells us the disposition of the data. The preceding example confirms the table is well clustered, showing high partition counts at level 00001 to low partition counts at level 00006.

Now we know how to interpret the clustering information result set. Let's now look at a less well-optimized table, partsupp.

```
SELECT system$clustering_information ( 'partsupp' );
```

Note that average_depth is higher, and total_constant_partition_count is zero, indicating the partsupp table is poorly organized. And the "notes" provide another clue. The clustering key should lead to low cardinality attributes and not high cardinality attributes; the least selective attribute first and the most selective attribute last when declaring the cluster key. Note that some result set rows were removed to reduce space.

```
{
  "cluster_by_keys" : "LINEAR(PS_SUPPKEY)",
```

```
  "notes" : "Clustering key columns contain high cardinality key PS_SUPPKEY
which might result in expensive re-clustering. Consider reducing the
cardinality of clustering keys. Please refer to https://docs.snowflake.net/
manuals/user-guide/tables-clustering-keys.html for more information.",
  "total_partition_count" : 2315,
  "total_constant_partition_count" : 0,
  "average_overlaps" : 1.8721,
  "average_depth" : 2.0043,
  "partition_depth_histogram" : {
    "00000" : 0,
    "00001" : 4,
    "00002" : 2303,
    "00003" : 2,
    "00004" : 6,
...
  }
}
```

Just as we did before, take a few moments to examine the partition_depth_
histogram, which tells us the disposition of the data. The preceding example confirms
the table is poorly clustered, showing low partition counts to high partition counts and
then low partition counts.

In contrast, the next query fails with this error message: "Invalid clustering keys or
table NATION is not clustered," which occurs because the table is not clustered. There
are only 25 rows.

```
SELECT system$clustering_information ( 'nation' );
```

Clustering depth documentation is at https://docs.snowflake.com/en/user-
guide/tables-clustering-micropartitions.html#clustering-depth.

Identify Cluster Key

In general, and as your real-world testing confirms, some general rules apply when
declaring cluster keys.

- Fewer attributes in the key are better. In preference, only one or two
  attributes

- Cluster key attribute order should be the lowest cardinality first to the highest cardinality last

- Apply expressions to reduce the cardinality of keys, for example, DATE_TRUNC()

Putting a high cardinality attribute before a low cardinality attribute reduces the effectiveness of clustering on the second attribute.

As each real-world scenario differs, the methodology described in the following subsections identifies cluster keys.

Step 1. Identify Access Paths

By profiling your queries, identify the most often used access paths to the object.

To do this using the Snowflake user interface, navigate to the History tab. Using the available filters, click the query ID corresponding to the SQL text of interest, then click the Profile tab.

For our investigation, we continue our use of supplied sample data.

```
SELECT * FROM partsupp WHERE ps_supplycost > 100;
```

To illustrate the absence of pruning, and to use our admittedly poor example query, Figure 11-2 shows the partitions scanned equal the partition's total, indicating no partition pruning has occurred.

Pruning
| Partitions scanned | 2,315 |
| Partitions total | 2,315 |

Figure 11-2. *No partition pruning*

We are not using the cluster key as a filter column in query predicates, or the table is not clustered.

Check the clustering information to identify the cluster key attributes.

```
SELECT system$clustering_information ( 'partsupp' );
```

The returned JSON confirms our sample query does not use the cluster key (ps_suppkey).

...

```
"cluster_by_keys" : "LINEAR(PS_SUPPKEY)"
...
```

Step 2. Gather Baseline Metrics

Using the History tab, we can sort by any column displayed. We should check for queries for long runtimes and gather information to inform our decision-making. Figure 11-3 shows where our query spent the most time for single query execution.

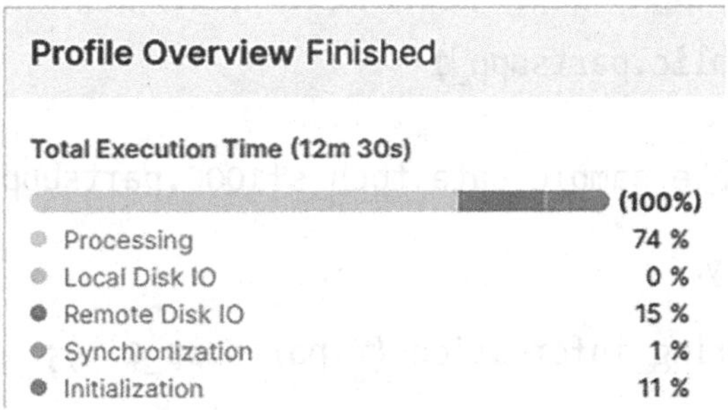

Profile Overview Finished

Total Execution Time (12m 30s)

	(100%)
● Processing	74 %
● Local Disk IO	0 %
● Remote Disk IO	15 %
● Synchronization	1 %
● Initialization	11 %

Figure 11-3. Query profile overview

Step 3. Assign Cluster Key

With the preceding information in hand, we should try another key aligned to predicates used in poorly performing queries identified earlier and retest. For this test, we need real data to work with, created in the next section.

However, reclustering uses serverless compute, and operations must complete before results can be checked and repeating earlier comments. To determine whether reclustering is complete, check the average_depth attribute in the result set for system$clustering_information, and when the average_depth value remains constant, reclustering is complete.

Step 4. Retest

As with every change, our results must be tested and proven good.

As the next section demonstrates, we should make a single change at a time and be methodical about how we approach testing.

Setup Test Data

For the rest of this chapter, we need to make changes to tables. As we cannot manipulate the shared database tables, we will create our own test objects.

```
USE ROLE      sysadmin;
USE DATABASE  test;
USE WAREHOUSE COMPUTE_WH;
USE SCHEMA    public;

CREATE TABLE test.public.partsupp_1
AS
SELECT * FROM snowflake_sample_data.tpch_sf1000.partsupp;
```

Check the cluster key.

```
SELECT system$clustering_information ( 'partsupp_1' );
```

Unsurprisingly, there is no cluster key, so let's declare one using an expression.

```
ALTER TABLE partsupp_1 CLUSTER BY (ps_supplycost > 100);
```

After confirming reclustering has been completed, recheck clustering where you see a problem. Note that some result set rows were removed to reduce space.

Better clustering is indicated by

average_depth = low

total_constant_partition_count = high

```
{
  "cluster_by_keys" : "LINEAR(ps_supplycost > 100)",
  "total_partition_count" : 1825,
  "total_constant_partition_count" : 1768,
  "average_overlaps" : 2.0055,
  "average_depth" : 3.0055,
  "partition_depth_histogram" : {
    "00000" : 0,
    "00001" : 1764,
...
    "00064" : 61
  }
}
```

In this example, we have partially achieved our desired effect.

average_depth desired = low, actual = medium (3.0055)

total_constant_partition_count desired = high, actual = high (1768)

Also, note that most micro-partitions are in a single bucket, which is not what we wish to achieve. Let's apply a new cluster key based upon our presumption both ps_ partkey and ps_suppkey are suitable candidates but first find out the cardinality.

```
SELECT COUNT(DISTINCT ps_partkey) count_ps_partkey,
       COUNT(DISTINCT ps_suppkey) count_ps_suppkey
FROM   partsupp_1;
```

ps_partkey = 200000000

ps_suppkey = 10000000

Following our general rules, we apply the least selective attribute first, noting neither attribute is low cardinality, and this is a test for illustration purposes only.

```
ALTER TABLE partsupp_1 CLUSTER BY (ps_suppkey, ps_partkey);
```

After confirming reclustering is complete, recheck clustering.

```
SELECT system$clustering_information ( 'partsupp_1' );
```

Superficially, the result looks good, a fairly even distribution in the partition_depth_ history.

```
{
  "cluster_by_keys" : "LINEAR(ps_suppkey, ps_partkey)",
  "notes" : "Clustering key columns contain high cardinality key PS_
  SUPPKEY, PS_PARTKEY which might result in expensive re-clustering.
  Consider reducing the cardinality of clustering keys. Please refer to
  https://docs.snowflake.net/manuals/user-guide/tables-clustering-keys.html
  for more information.",
  "total_partition_count" : 2022,
  "total_constant_partition_count" : 0,
  "average_overlaps" : 13.6973,
  "average_depth" : 8.7319,
  "partition_depth_histogram" : {
```

```
    "00000" : 0,
    "00001" : 0,
    "00002" : 0,
    "00003" : 24,
    "00004" : 224,
    "00005" : 501,
    "00006" : 478,
    "00007" : 346,
    "00008" : 135,
    "00009" : 51,
    "00010" : 2,
    "00011" : 2,
    "00012" : 0,
    "00013" : 0,
    "00014" : 0,
    "00015" : 14,
    "00016" : 22,
    "00032" : 139,
    "00064" : 84
  }
}
```

A closer look reveals the truth.

average_depth desired = low, actual = medium (8.7319)

total_constant_partition_count desired = high, actual = low (0)

But we can see notes providing useful information.

Try again using a single key. All remaining values are not keys. They are normal attributes.

```
ALTER TABLE partsupp_1 CLUSTER BY (ps_suppkey);
```

After confirming reclustering is complete, recheck clustering.

```
SELECT system$clustering_information ( 'partsupp_1' );
```

Better, but not perfect clustering, noting some result set rows removed to reduce space.

```
{
  "cluster_by_keys" : "LINEAR(ps_suppkey)",
  "notes" : "Clustering key columns contain high cardinality key PS_SUPPKEY
which might result in expensive re-clustering. Consider reducing the
cardinality of clustering keys. Please refer to https://docs.snowflake.net/
manuals/user-guide/tables-clustering-keys.html for more information.",
  "total_partition_count" : 1977,
  "total_constant_partition_count" : 0,
  "average_overlaps" : 4.3551,
  "average_depth" : 3.4401,
  "partition_depth_histogram" : {
    "00000" : 0,
    "00001" : 8,
    "00002" : 258,
    "00003" : 871,
    "00004" : 581,
    "00005" : 214,
    "00006" : 45,
...
  }
}
```

average_depth desired = low, actual = low (3.4401)

total_constant_partition_count desired = high, actual = low (0)

Our results are arguably better than those delivered for the source table where we began our investigation. In reality, they are only as good as the improvement in performance we experience in real-world usage. Caveat emptor. Under the circumstances, probably the best we can achieve.

Materialized Views

Having discussed materialized views in Chapter 10, we consider them worthy of a second mention.

Materialized views can be clustered using a different cluster key. As they are maintained behind the scenes using serverless compute, they can always be guaranteed to be consistent with the underlying table. Therefore, materialized views represent an alternative way to implement clustering more suited to different access paths than our primary use case, albeit incurring storage and maintenance costs.

Automatic Clustering

To be clear, not every table benefits from clustering. Only very large tables in the order of several terabytes and larger benefit from clustering. If data volumes and testing indicate adding cluster keys are beneficial, then consider enabling automatic clustering. Automatic clustering is where Snowflake utilizes serverless computers and manages all future maintenance on an as-needs basis. More information is at `https://docs.` `snowflake.com/en/user-guide/tables-auto-reclustering.html#automatic-` `clustering`.

Note   Automatic clustering is enabled by default when a table is reclustered.

Automatic clustering does not attempt to provide the best performance possible but instead aims to improve clustering depth to achieve both predictable and acceptable performance.

As we might expect, once enabled, we cannot influence automatic clustering behavior except by disabling it. There are no other levers available to apply.

To determine if automatic clustering is enabled on a table, look for the automatic_ clustering attribute that is either ON or OFF.

```
SHOW TABLES LIKE 'partsupp_1';
```

To disable automatic clustering.

```
ALTER TABLE partsupp_1 SUSPEND RECLUSTER;
```

And to re-enable automatic clustering.

```
ALTER TABLE partsupp_1 RESUME RECLUSTER;
```

For roles with MONITOR USAGE privilege, the following query identifies credit usage on a per-object basis noting full view contents are only visible to ACCOUNTADMIN by default. Further information is at https://docs.snowflake.com/en/sql-reference/ functions/automatic_clustering_history.html#automatic-clustering-history.

```
SELECT *
FROM TABLE(information_schema.automatic_clustering_history
          ( date_range_start => dateadd ( H, -24, current_timestamp )));
```

The corresponding view can be found in Account Usage. Note that the latency may be up to 3 hours. More information is at https://docs.snowflake.com/en/ sql-reference/account-usage/automatic_clustering_history.html#automatic-clustering-history-view.

Factors affecting credit consumption include the number and cardinality of key(s), how many micro-partitions are involved in reclustering, the size of the reclustered table, and the frequency and pattern of DML operations.

Clustering Recommendations

As previously stated, clustering is not for every table. I recommend starting with tables of 1 TB or more. For those smaller tables, you may see some benefits to adding a cluster key but proceed on a case-by-case basis.

Query predicates (WHERE, JOIN, GROUP BY columns) are a good indicator of appropriate cluster keys to add. We also have materialized views available to provide alternate clustering for the same table, and the optimizer may select from materialized views in preference to referenced tables.

In general, these principles apply.

- Begin with one or two cluster key attributes and aim to keep the total number of cluster keys small, in preference, less than three.

- Always order cluster key attributes from least selective to highest selective— low cardinality to high cardinality.

- Monitor table activity as DML operations can skew micro-partition content over time.

- Consider enabling automatic clustering, but note resource monitors do not currently track credit consumption.

Query Performance

Having looked at clustering, let's now look at how to identify poorly performing queries and discover remediation techniques. Documentation is at https://docs.snowflake.com/en/user-guide/ui-query-profile.html#query-operator-details.

Warehouse Considerations

As discussed in Chapter 3, three broad principles apply to scaling warehouses, as shown in Figure 11-4.

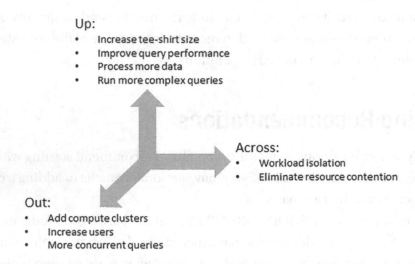

Figure 11-4. *Warehouse scaling options*

Our starting point is to identify poorly performing queries. For this, we revert to the Snowflake supplied database SNOWFLAKE_SAMPLE_DATA as this contains tables with high data volumes set up for precisely this kind of investigation.

```
USE ROLE       sysadmin;
USE DATABASE   snowflake_sample_data;
USE WAREHOUSE  compute_wh;
USE SCHEMA     snowflake_sample_data.tpcds_sf10tcl;
```

Note   Snowflake retains cached results for 24 hours and a maximum of 31 days subject to the same SQL re-running before the purge.

To establish a baseline for subsequent runs of the same query, we must ensure cached results are not reused; for this, we set use_cached_result to FALSE. More information is at https://docs.snowflake.com/en/user-guide/querying-persisted-results.html#using-persisted-query-results.

```
ALTER SESSION SET use_cached_result = FALSE;
```

Our warehouse compute_wh is X-Small, and we now set it to auto-suspend after 60 seconds.

```
ALTER WAREHOUSE compute_wh SET auto_suspend = 60;
```

We can check warehouse settings from the Snowflake user interface Warehouses tab.

Query Runtime and Memory Spills

This section addresses scaling up from Figure 11-4.

For this investigation, we resize our existing warehouse compute_wh, don't forget to resize it back to X-Small at the end of this exercise. We do not recommend resizing as a general practice.

To understand memory spills to disk, we will run the same SQL iteration using different size warehouses and query profile capture metrics.

```
SELECT i_product_name,
       SUM(cs_list_price)  OVER (PARTITION BY cs_order_number ORDER BY
       i_product_name ROWS BETWEEN UNBOUNDED PRECEDING AND CURRENT ROW)
       list_running_sum
FROM   catalog_sales, date_dim, item
WHERE  cs_sold_date_sk = d_date_sk
AND    cs_item_sk      = i_item_sk
AND    d_year IN (2000)
AND    d_moy  IN (1,2,3,4,5,6)
LIMIT 100;
```

Navigate to the History tab. Refresh the screen if necessary, noting an X-Small size. Then click Query ID for the most recent SQL statement invocation. Total Duration is 3 minutes and 37 seconds in my case. Click Profile. In Profile Overview (see Figure 11-5), note that micro-partition pruning results in just over 6% of the partitions selected.

Spilling shows where intermediate query results are written to a disk overflowing from memory. As expected, we have 0% scanned from the cache.

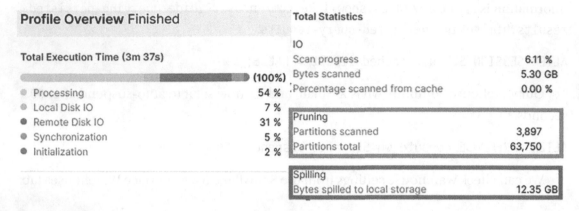

Figure 11-5. *Sample query profile overview 1*

For our next run, we resize to small, double the X-Small T-shirt size.

```
ALTER WAREHOUSE compute_wh SET warehouse_size = 'SMALL';
```

And re-run the same query, then repeat the same steps. Navigate to the History tab; refresh the screen if necessary. Note that the size is Small, and then click Query ID for the most recent SQL statement invocation. Total Duration is now 1 minute, 44 seconds in my case. Click Profile. In Profile Overview (see Figure 11-6), spilling is much lower. A new item, Network, shows the processing time when waiting for the network data transfer. As expected, we have 0% scanned from the cache, and all other values remain constant.

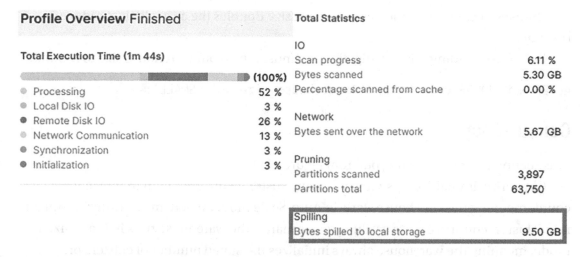

Figure 11-6. *Sample query profile overview 2*

For our next run, we resize to Medium, which is four times the X-Small T-shirt size.

```
ALTER WAREHOUSE compute_wh SET warehouse_size = 'MEDIUM';
```

And re-run the same query, and then repeat the same steps. Navigate to the History tab; refresh the screen if necessary. Note that the size is Medium, and then click Query ID for the most recent SQL statement invocation. Total Duration is now 59.1 seconds in my case. Click Profile. In Profile Overview, spilling has been reduced to 6.47 GB.

Our last run is with a Large size warehouse, which is eight times the X-Small T-shirt size.

```
ALTER WAREHOUSE compute_wh SET warehouse_size = 'LARGE';
```

Re-run the same query and then repeat the same steps. Navigate to the History tab; refresh the screen if necessary. Note that the size is Large, and then click Query ID for the most recent SQL statement invocation. Total Duration is now 29.3 seconds in my case. Click Profile. In Profile Overview, note the absence of spilling.

We can conclude for this sample query that either a Medium or a Large warehouse are reasonable compromises depending upon performance criteria. For your information, I re-ran using the X-Large warehouse (16 times the X-Small T-shirt size) with a reduction in runtime to 14 seconds.

One final consideration relating to cost. At the beginning of this section, we set auto-suspend at 60 seconds. Snowflake's minimum warehouse runtime change is 60 seconds per X-Small cluster. Therefore, we must consider the cost implications of running larger warehouses because each increase in T-shirt size doubles the cost, albeit at reduced runtime.

Before concluding this investigation, we must return our warehouse size to X-Small.

```
ALTER WAREHOUSE compute_wh SET warehouse_size = 'X-SMALL';
```

Concurrency

This section addresses scaling out from Figure 11-4.

Clustering is enabled by setting the max_cluster_count greater than min_cluster_count, meaning the warehouse starts in Auto-Scale mode. Where max_cluster_count and min_cluster_count are set to a value greater than 1, the warehouse runs in Maximized mode, meaning the warehouse always initializes the stated number of clusters on startup.

Scaling out determines when and how Snowflake turns on/off additional clusters in the warehouse, up to the max_cluster_count according to the query demand, maximizing query responsiveness and concurrency. There is no additional cost to implementing Auto-Scale mode as a single cluster is initially instantiated on demand. As demand rises, more clusters are instantiated, then shut down as demand falls. In contrast, Maximize mode always instantiates the stated number of clusters on startup and does not shut clusters down as demand falls. Therefore the initial cost is higher and may be acceptable for consistently repeatable performance. There are no hard and fast rules, simply guidelines according to the importance of response times, concurrency demands, and performance. We pay according to requirements.

Note   Snowflake implements a hard limit of 10 concurrent clusters, with each cluster capable of handling 8 concurrent requests.

To understand concurrency, first implement Auto-Scale mode by setting max_cluster_count greater than min_cluster_count. In this example, a maximum of 4 concurrent X-Small clusters can handle a total of 32 concurrent queries.

```
ALTER WAREHOUSE compute_wh
SET   warehouse_size    = 'X-SMALL',
      max_cluster_count = 4
      min_cluster_count = 1
```

Although the creation of multiple parallel execution threads is well beyond the scope of this chapter, you can get useful information on the effects of concurrent execution, SQL queueing, and clustering at `https://community.snowflake.com/s/article/Putting-Snowflake-s-Automatic-Concurrency-Scaling-to-the-Test`. I do not endorse third-party products. I simply offer this link to add value to your own investigations.

Assuming your investigations are complete, reset compute_wh to default clustering setting.

```
ALTER WAREHOUSE compute_wh
SET   warehouse_size    = 'X-SMALL',
      max_cluster_count = 1
      min_cluster_count = 1
      scaling_policy    = STANDARD;
```

Workload Segregation

This section addresses scaling across from Figure 11-4.

You have seen the effect of scaling up by increasing cluster size and improving query response times and discussed the effect of adding clusters to scale out. This section addresses scaling across and explains why we should consider adding warehouses of the same size but with multiple declarations.

To a point, adding more discretely named warehouses also addresses clustering concerns, but the aim is more subtle. When we scale across, we segment our billing and cross-charging consumption by warehouse usage, where each department's requirements for cost effectiveness and response times may differ. We also present a greater probability of result cache reuse as SQL queries issued in a department may have been run before the warehouse quiesces. A further benefit is the reduction in resource contention as warehouse usage is restricted to a subset of the user base and, therefore, more tunable to a narrower workload than might otherwise be possible.

The case for scaling across is less clear than for scaling up and out, but in larger organizations becomes clearer and is one left for your further investigations.

Summary

This chapter began by diving into clustering. You learned how to define a cluster key and determine better clustering as indicated by average_depth = low and total_constant_partition_count = high with a descending number of partitions across a shallow depth.

We then identified costs associated with automatic clustering. We determined how to disable and enable automatic clustering noting the aim is to improve clustering depth to achieve both predictable and acceptable performance.

Ending our discussion on clustering by providing recommendations, we discussed query performance and spilling data from memory to disk, commenting on the cost implications of increasing warehouse size.

Implementing concurrency falls outside the available space, though I hope you find the discussion useful. Finally, the comments on workload segregation may not resonate with smaller organizations.

Having looked at query performance, let's next look at data management.

## CHAPTER 12

Data Management

Data management is an established discipline that is gaining momentum in solving a multitude of challenges with ensuring proper ownership of data.

Good data architecture practices focus on managing the structure of environments where data is stored from an enterprise perspective, covering not only data stores but also related components, services, and metadata stores.

Data access controls are also considered in data management, ensuring only the right people are entitled to make decisions. Most pertinent to our discussion in this chapter is identifying, collating, organizing, and prioritizing problems for technologists to provide solutions. Most sections implement Snowflake constructs to assist our journey, offering worked examples to assist data management.

No single chapter can do justice to data management as a subject with a limited page count. Instead, I focus on tooling to assist our journey as we examine each subject area.

Metadata

Metadata has a wide definition and may mean different things to different people. In a broad sense, the word *metadata* is contextually bound according to the observer's viewpoint. In this book, metadata refers to contextual information about other data.

The absence of metadata for any subject data set results in "data littering." It is the deliberate act of creating and distributing data with missing or inadequate metadata annotation, thus rendering the data unusable junk for future applications. It is better described at `https://tdan.com/data-speaks-for-itself-data-littering/29122`. Thank you to my friend Diane Schmidt, PhD for making me aware of the phenomenon.

Much of what we describe in this chapter aims to enrich the metadata content associated with our data. We will revisit object tagging and Snowflake data classification tooling. Another way to improve our metadata is by implementing a metadata management tool that can centrally manage, curate, and enrich metadata for our

© Andrew Carruthers 2022
A. Carruthers, *Building the Snowflake Data Cloud*, https://doi.org/10.1007/978-1-4842-8593-0_12

organization. Next, I propose interaction between our metadata management tool and Snowflake, resulting in automated integration.

Recent benchmarking exercises, reinforced by direct feedback from regulators, have established that financial institutions do not have an industry best practice model for data management. To move toward having a better data management process, a series of shortfalls in the current model must be addressed.

The natural organic growth of any financial organization through product innovation and business development, combined with strategic mergers and acquisitions, results in a fragmented business, technology, and data landscape, which is common across many large organizations. As a direct result of this, financial institutions now incur significant costs.

- The complex, overly manual processes that are required to keep their landscape functioning

- Increasing project costs as amending or adding to the landscape becomes more complex and difficult

- Increasing risks due to discrepancies, duplication, inconsistent interpretation, and inconsistent use of data as it flows through organization systems and processes, ultimately resulting in weakness in management decision-making processes

- Increased scrutiny and punitive directives from regulators, directly and indirectly resulting from an inability to demonstrate good data management practices

The management of metadata is the frame of reference giving data its context and meaning. Having proper metadata management for data governance is the key to success. Effectively governed metadata provides a view into the flow of data, the ability to perform an impact analysis, a common business vocabulary and accountability for its terms and definitions, and finally, an audit trail for compliance. Metadata management becomes an important capability enabling IT to oversee changes while delivering trusted, secure data in a complex data integration environment.

Without effective metadata management, we cannot say we know very much about our data, and the problem compounds as data volumes grow, eventually passing the diode in our upward trajectory, beyond which we cannot reverse and correct problems without starting again. With this in mind, if your organization has not yet adopted a metadata management tool as part of its data management strategy, please investigate this soon.

Data Custodian

As discussed in Chapter 2, the subject of data ownership is emotive. For our technical environments, we prefer *data custodians* because this title better reflects the role of an individual in safeguarding data artifacts. Regardless of the title, the role of a data custodian is to maintain secure environments in which our data resides. The role of a data steward is to manage the data itself, not the environment. With this distinction in mind, and recognizing our focus is Snowflake, we focus on data custodian aspects, not data stewards.

Our organizations operate in increasingly more regulated environments. We spend much time navigating regulation and governance, protecting our technical infrastructure from unexpected, unwanted, or inappropriate change. We need to find ways to reduce risks inherent with development which may lead to a compromised delivery.

While a full discussion of the data custodian role is beyond the scope of this book, one such way to reduce risk is by adopting pattern-based delivery utilizing a low-code framework to generate Snowflake objects. Let's begin by providing an overview of a low-code environment, as shown in Figure 12-1.

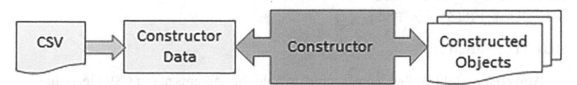

Figure 12-1. *Code generation overview*

Reading Figure 12-1 from left to right, we create a comma-separated variable (CSV) file containing our data pipeline metadata. The CSV file is loaded into a table. A stored procedure references the metadata; a suite of constructed objects is generated for deployment.

The constructor can be extended into our continuous integration pipeline and auto-deploy the constructed objects according to need. Once we have a constructor, the possibilities are (almost) endless. As Snowflake's capability expands, it may also include historization, validation, enrichment, data masking, object tagging, role creation, automation, and other deliverables. All we need is the imagination to engineer our future.

Turning our attention to implementation, we would typically start with a new role, database, and schema dedicated to generating our patterns. Trusting you will forgive the lack of setup script due to space constraints, let's reuse the test database from previous chapters.

As you might expect, we set up our context.

```
USE ROLE       sysadmin;
USE DATABASE   TEST;
USE WAREHOUSE  COMPUTE_WH;
USE SCHEMA     public;
```

Create a file format for CSV files.

```
CREATE OR REPLACE FILE FORMAT TEST.public.test_csv_format
TYPE                 = CSV
FIELD_DELIMITER      = ','
SKIP_HEADER          = 1
NULL_IF              = ( 'NULL', 'null' )
EMPTY_FIELD_AS_NULL  = TRUE
SKIP_BLANK_LINES     = TRUE;

SHOW file formats;
```

And create a table called "constructor" to hold the contents of our CSV file, noting that I dual-purposed some of the attributes explained.

```
CREATE OR REPLACE TABLE constructor
(
dataset              VARCHAR(255),
position             NUMBER,
label                VARCHAR(255),
source_attribute     VARCHAR(255),
source_datatype      VARCHAR(255),
source_precision     VARCHAR(255),
target_attribute     VARCHAR(255),
target_datatype      VARCHAR(255),
target_precision     VARCHAR(255)
);
```

Let's assume we want to implement a data pipeline to ingest data and historize using the Slowly Changing Dimension 2 pattern (SCD2) for a feed called EMPLOYEE.

Edit a CSV file to contain the following data shown in Figure 12-2. This is also available as the ch12_constructor.csv file.

DATASET	POSITION	LABEL	SOURCE_ATTRIBUTE	SOURCE_DATATYPE	SOURCE_PRECISION	TARGET_ATTRIBUTE	TARGET_DATATYPE	TARGET_PRECISION
EMPLOYEE	1	STAGING_TABLE	STG_EMPLOYEE	TEST.public				
EMPLOYEE	2							
EMPLOYEE	3	PRIMARY_KEY	preferred_name	STRING	255	preferred_name	STRING	255
EMPLOYEE	4	ATTRIBUTE	surname_preferred	STRING	255	surname_preferred	STRING	255
EMPLOYEE	5	ATTRIBUTE	forename_preferred	STRING	255	forename_preferred	STRING	255
EMPLOYEE	6	ATTRIBUTE	gender	STRING	255	gender	STRING	255
EMPLOYEE	7	ATTRIBUTE	national_insurance_number	STRING	255	national_insurance_number	STRING	255
EMPLOYEE	8	ATTRIBUTE	social_security_number	STRING	255	social_security_number	STRING	255
EMPLOYEE	9	ATTRIBUTE	postcode	STRING	255	postcode	STRING	255
EMPLOYEE	10	ATTRIBUTE	zip_code	STRING	255	zip_code	STRING	255
EMPLOYEE	11	DATE_MATCH_KEY	stg_last_updated	STRING	255	stg_last_updated	STRING	255

Figure 12-2. *Sample CSV file content*

Navigate to Databases ➤ Test, click the whitespace for the row labeled CONSTRUCTOR, and then click Load Data. Leaving the warehouse set to COMPUTE_ WH, click Next, and then Select Files. In File Explorer, navigate to the directory where ch12_constructor.csv resides, select the file, and click Open. Return to the Load Data dialog and click Next. Select the TEST_CSV_FORMAT file format, and then click Load. You should see a dialog box, as shown in Figure 12-3. Click OK.

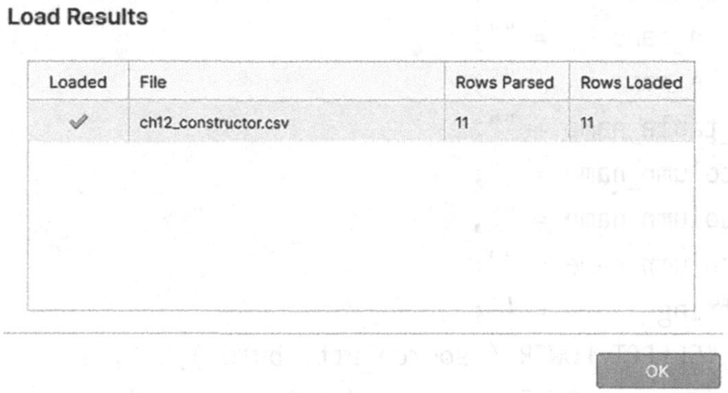

Load Results

Loaded	File	Rows Parsed	Rows Loaded
✓	ch12_constructor.csv	11	11

OK

Figure 12-3. *Load results*

Returning to the worksheet, we should find the header record has been removed.

```
SELECT * FROM constructor ORDER BY position ASC;
```

This statement illustrates a "rule" built into the constructor stored procedure, always ORDER BY position, which must be unique in the data set. Failure to implement this rule results in code execution failure.

With our data loaded, let's create our sp_constructor stored procedure, where we leverage the data loaded into the constructor table. In this example, I have left debug inline, so invocation returns the generated SQL statements, and once again, my apologies for removing whitespace.

```
CREATE OR REPLACE PROCEDURE sp_construct( P_DATASET         STRING,
                                          P_AUTODEPLOY      STRING )
RETURNS STRING
LANGUAGE javascript
EXECUTE AS CALLER
AS
$$
    var sql_stmt            = "";
    var stmt                = "";
    var recset              = "";
    var result              = "";
    var stg_table_name      = "";
    var stg_column_name     = "";
    var strm_table_name     = "";
    var seq_stg_table_name  = "";
    var stg_pk_column_name  = "";
    var stg_at_column_name  = "";
    var stg_dt_column_name  = "";
    var debug_string        = '';
    sql_stmt  = "SELECT LOWER ( source_attribute ),\n";
    sql_stmt += "       LOWER ( source_datatype  ),\n";
    sql_stmt += "       label\n";
    sql_stmt += "FROM   constructor\n";
    sql_stmt += "WHERE  label IN ( 'STAGING_TABLE' )\n";
    sql_stmt += "AND    dataset = :1;\n\n";
    debug_string += sql_stmt;
    stmt = snowflake.createStatement ({ sqlText:sql_stmt, binds:[P_
    DATASET] });
```

```
try
{
   recset = stmt.execute();
   while(recset.next())
   {
       if( recset.getColumnValue(3) == "STAGING_TABLE")
       {
           stg_table_name     = recset.getColumnValue(2) + "." + recset.
                                getColumnValue(1);
           strm_table_name    = recset.getColumnValue(2) + "." + "strm_" +
                                recset.getColumnValue(1);
           seq_stg_table_name = recset.getColumnValue(2) + "." + "seq_"  +
                                recset.getColumnValue(1) + "_id";
       }
   }
}
catch { result = sql_stmt; }
sql_stmt  = "SELECT LOWER ( source_attribute ),\n";
sql_stmt += "        LOWER ( source_datatype  ),\n";
sql_stmt += "        LOWER ( source_precision ),\n";
sql_stmt += "        LOWER ( target_attribute ),\n";
sql_stmt += "        LOWER ( target_datatype  ),\n";
sql_stmt += "        LOWER ( target_precision ),\n";
sql_stmt += "        label\n";
sql_stmt += "FROM    constructor\n";
sql_stmt += "WHERE   dataset = :1\n";
sql_stmt += "AND     label IN ( 'PRIMARY_KEY', 'ATTRIBUTE', 'DATE_MASTER_
KEY' )\n"
sql_stmt += "ORDER BY TO_NUMBER ( position ) ASC;\n\n";
debug_string += sql_stmt;
stmt = snowflake.createStatement ({ sqlText:sql_stmt, binds:[P_
DATASET] });
try
{
    recset = stmt.execute();
```

```
    while(recset.next())
    {
        stg_column_name += recset.getColumnValue(1) + " " + recset.
        getColumnValue(2) + "(" + recset.getColumnValue(3) + "),\n"

        if( recset.getColumnValue(7) == "PRIMARY_KEY")
        {
            stg_pk_column_name += recset.getColumnValue(1) + ",\n"
        }
        else if( recset.getColumnValue(7) == "ATTRIBUTE")
        {
            stg_at_column_name += recset.getColumnValue(1) + ",\n"
        }
        else if( recset.getColumnValue(7) == "DATE_MASTER_KEY")
        {
            stg_dt_column_name += recset.getColumnValue(1)
        }
    }
}
catch { result = sql_stmt; }
stg_column_name = stg_column_name.substring(0, stg_column_name.
length -2)
sql_stmt  = "CREATE OR REPLACE SEQUENCE " + seq_stg_table_name + " START
= 1 INCREMENT = 1;\n\n";
debug_string += sql_stmt;
if( P_AUTODEPLOY == "TRUE" )
{
    try
    {
        stmt = snowflake.createStatement ({ sqlText:sql_stmt });
        recset = stmt.execute();
    }
    catch { result = sql_stmt; }
}
sql_stmt  = "CREATE OR REPLACE TABLE " + stg_table_name + "\n";
sql_stmt += "(\n"
```

```
sql_stmt += stg_column_name + "\n"
sql_stmt += ")\n";
sql_stmt += "COPY GRANTS;\n\n";
debug_string += sql_stmt;
if( P_AUTODEPLOY == "TRUE" )
{
   try
   {
      stmt = snowflake.createStatement ({ sqlText:sql_stmt });
      recset = stmt.execute();
   }
   catch { result = sql_stmt; }
}
sql_stmt  = "CREATE OR REPLACE STREAM " + strm_table_name + " ON TABLE "
+ stg_table_name + ";\n\n";
debug_string += sql_stmt;
if( P_AUTODEPLOY == "TRUE" )
{
   try
   {
      stmt = snowflake.createStatement ({ sqlText:sql_stmt });
      recset = stmt.execute();
   }
   catch { result = sql_stmt; }
}
return debug_string;
//   return result;
$$;
```

The first parameter corresponds to the data set just loaded into the constructor table.
The second parameter, FALSE, prevents the generated code from being automatically
deployed.

```
CALL sp_construct ( 'EMPLOYEE',
                    'FALSE' );
```

In the next example, we automatically deploy the generated code. Note your organization may not allow automated code generation and deployment. The example illustrates the "art of the possible."

```
CALL sp_construct ( 'EMPLOYEE',
                    'TRUE' );
```

Now prove objects were successfully generated.

```
SELECT seq_stg_employee_id.NEXTVAL;
```

```
SELECT * FROM strm_stg_employee;
```

This example shows how to implement code generation. The code is easily extended so long as a few basic rules are followed. I leave it for your further investigation to extend the constructor capability.

Data Privacy

While data privacy is not a part of data management, as a related topic and potential consumer of object tagging, I trust you will forgive the digression. I feel inclusion is important because data privacy is complex. Each jurisdiction typically has its own regulations making a thorough discourse of data privacy impossible in the limited space available. Since regulations change over time, we focus on general principles regardless of jurisdiction.

Each of us is responsible for protecting the personal data we access, not only in Snowflake but any time we move data to our desktops or anywhere else in private and public domains. Prevention is better than cure. Anticipating privacy challenges, putting effective controls, and tooling in place to assist later data discovery are valuable techniques in reducing time and effort when resolving investigations.

We must assume privacy is our default position, protecting our data. Snowflake already provides RBAC, RLS, and other techniques to ensure privacy is embedded and integrated into our system design. Commensurate with protecting our data is the premise of retaining data only for as long as necessary, meaning we must implement a data retention policy. Not every data item persists forever, and there are many scenarios where data must be securely removed from our systems. Throughout the data lifecycle, we must respect user privacy, maintain strong controls, and conform to our regulatory responsibilities.

Supporting our regulatory obligations requires a data privacy policy for a specific jurisdiction, and from the policy, we develop processes, then procedures that implement each process. Each procedure is a discrete series of steps that may be either automated or manual and sometimes a combination of both.

Most organizations have either a team or nominated individuals responsible for data privacy. It is usual for all proposed data movement between systems to be assessed for data privacy and for existing data sets to ensure compliance with the policy. Chapter 9 explained how to implement object tagging. I recommend each Snowflake system change also contain object tagging proposals. For existing data sets, a review of implemented object tags should be conducted to ensure consistency and conformance noting the requirement to maintain uniqueness. With appropriate object tagging in place, we equip our business stakeholders with the capability to identify objects and attributes by searching for specific tags, effectively a metadata search without needing to know each database, schema, and attribute.

For completeness, our data privacy object tagging may reside in a metadata management tool where object tags can be declared, managed, and maintained centrally for distribution to all systems in our organization, not just Snowflake. We may choose to implement an object tagging feed from our metadata management tool into Snowflake. This is discussed next using data privacy tags in our sample code.

Object Tagging Automation

Chapter 9 explained how to implement object tagging. Let's now expand upon the object tagging demonstration, also setting up our later investigation into data classification.

Depending upon our approach to data management, our organizations may use a metadata management tool from where tags may be deployed to classify our data. We may also have periodic uploads to various repositories, including Snowflake, for deployment from a centrally maintained location. Leaving aside considerations of tag uniqueness covered in Chapter 9, and assuming tags have been landed into a table, let's consider an approach for automating tag deployment.

Set up Test Data

This section sets up our execution context and then creates objects to support our test case in two parts. First, we create a sample target employee table to which we apply tags. Next, we create a simple metadata model for declaring object tags.

The following sets up our context.

```
USE ROLE sysadmin;
USE DATABASE  TEST;
USE WAREHOUSE COMPUTE_WH;
USE SCHEMA    public;
```

Create a table to apply tags on noting some attributes created for later use.

```
CREATE OR REPLACE TABLE employee
(
preferred_name             VARCHAR(255),
surname_preferred          VARCHAR(255),
forename_preferred         VARCHAR(255),
gender                     VARCHAR(255),
national_insurance_number  VARCHAR(255),
social_security_number     VARCHAR(255),
postcode                   VARCHAR(255),
zip_code                   VARCHAR(255),
salary                     NUMBER
);
```

Let's create a data model to contain our tags and relationships to schema objects as shown in Figure 12-4, denormalizing into the v_tag_object itself wrapped by v_tags for later use.

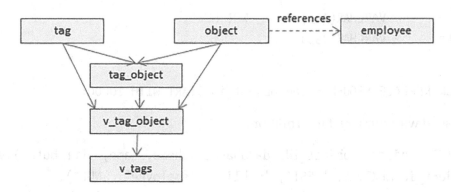

Figure 12-4. *Sample object tagging metadata model*

Now create a tag metadata model starting with the tag table and sequence.

```
CREATE OR REPLACE TABLE tag
(
tag_id  NUMBER PRIMARY KEY NOT NULL,
name    VARCHAR(255)       NOT NULL,
comment VARCHAR(2000)
);
```

```
CREATE OR REPLACE SEQUENCE seq_tag_id START WITH 10000;
```

Create sample tags.

```
INSERT INTO tag ( tag_id, name, comment ) VALUES
(seq_tag_id.NEXTVAL, 'PII',          'Personally Identifiable
                                      Information'),
(seq_tag_id.NEXTVAL, 'PII_S_Name',   'Personally Identifiable
                                      Information -> Sensitive -> Name'),
(seq_tag_id.NEXTVAL, 'PII_N_Gender', 'Personally Identifiable
                                      Information -> Non-Sensitive ->
                                      Gender');
```

Now create a table holding target object information.

```
CREATE OR REPLACE TABLE object
(
object_id       NUMBER PRIMARY KEY NOT NULL,
database        VARCHAR(255)       NOT NULL,
schema          VARCHAR(255)       NOT NULL,
```

```
name              VARCHAR(255)        NOT NULL,
attribute         VARCHAR(255)
);

CREATE OR REPLACE SEQUENCE seq_object_id START WITH 10000;
```

And seed with our target information.

```
INSERT INTO object ( object_id, database, schema, name, attribute ) VALUES
(seq_object_id.NEXTVAL, 'TEST', 'public', 'employee',  NULL),
(seq_object_id.NEXTVAL, 'TEST', 'public', 'employee',  'preferred_name'),
(seq_object_id.NEXTVAL, 'TEST', 'public', 'employee',  'surname_
preferred'),
(seq_object_id.NEXTVAL, 'TEST', 'public', 'employee',  'forename_
preferred'),
(seq_object_id.NEXTVAL, 'TEST', 'public', 'employee',  'gender');

CREATE OR REPLACE TABLE tag_object
(
tag_object_id NUMBER    NOT NULL,
tag_id        NUMBER    NOT NULL REFERENCES tag    (tag_id),
object_id     NUMBER    NOT NULL REFERENCES object (object_id) --
Foreign key
);

CREATE OR REPLACE SEQUENCE seq_tag_object_id START WITH 10000;
```

Populate the intersection table to map tags to target object and some attributes.

```
INSERT INTO tag_object ( tag_object_id, tag_id, object_id )
WITH
c AS (SELECT tag_id FROM tag WHERE name = 'PII'),
o AS (SELECT object_id FROM object WHERE database = 'TEST' AND schema =
'public' AND name = 'employee' AND attribute IS NULL)
SELECT seq_tag_object_id.NEXTVAL, c.tag_id, o.object_id
FROM   c, o;

INSERT INTO tag_object ( tag_object_id, tag_id, object_id )
WITH
```

```
c AS (SELECT tag_id FROM tag WHERE name = 'PII_S_Name'),
o AS (SELECT object_id FROM object WHERE database = 'TEST' AND schema =
'public' AND name = 'employee' AND attribute = 'preferred_name')
SELECT seq_tag_object_id.NEXTVAL, c.tag_id, o.object_id
FROM    c, o;

INSERT INTO tag_object ( tag_object_id, tag_id, object_id )
WITH
c AS (SELECT tag_id FROM tag WHERE name = 'PII_S_Name'),
o AS (SELECT object_id FROM object WHERE database = 'TEST' AND schema =
'public' AND name = 'employee' AND attribute = 'surname_preferred')
SELECT seq_tag_object_id.NEXTVAL, c.tag_id, o.object_id
FROM    c, o;

INSERT INTO tag_object ( tag_object_id, tag_id, object_id )
WITH
c AS (SELECT tag_id FROM tag WHERE name = 'PII_S_Name'),
o AS (SELECT object_id FROM object WHERE database = 'TEST' AND schema =
'public' AND name = 'employee' AND attribute = 'forename_preferred')
SELECT seq_tag_object_id.NEXTVAL, c.tag_id, o.object_id
FROM    c, o;

INSERT INTO tag_object ( tag_object_id, tag_id, object_id )
WITH
c AS (SELECT tag_id FROM tag WHERE name = 'PII_N_Gender'),
o AS (SELECT object_id FROM object WHERE database = 'TEST' AND schema =
'public' AND name = 'employee' AND attribute = 'gender')
SELECT seq_tag_object_id.NEXTVAL, c.tag_id, o.object_id
FROM    c, o;
```

And create a view to denormalize the model noting this view may be replaced by a table delivered by our metadata management tool.

```
CREATE OR REPLACE VIEW v_tag_object
AS
SELECT cto.tag_object_id,
       ct.tag_id,
       o.object_id,
```

```
        ct.name         AS tag_name,
        ct.comment      AS tag_comment,
        o.database,
        o.schema,
        o.name          AS object_name,
        o.attribute
FROM    tag_object      cto,
        tag             ct,
        object          o
WHERE   cto.tag_id      = ct.tag_id
AND     cto.object_id   = o.object_id;
```

Using the preceding view, let's now create a helper view to generate on-the-fly SQL statements. We encapsulate this view in a stored procedure.

```
CREATE OR REPLACE VIEW v_tags COPY GRANTS
AS
SELECT 'CREATE OR REPLACE TAG '||tag_name||' COMMENT = '''||tag_
comment||''';' AS create_tag_stmt,
        CASE
            WHEN attribute IS NULL THEN
                'ALTER TABLE '||database||'.'||schema||'.'||object_name||' SET
                TAG '||tag_name||' = '''||tag_comment||''';'
            ELSE
                'ALTER TABLE '||database||'.'||schema||'.'||object_name||'
                MODIFY COLUMN '||attribute||' SET TAG '||tag_name||' =
                '''||tag_comment||''';'
        END AS apply_tag_stmt,
        tag_name,
        tag_comment,
        database,
        schema,
        object_name,
        attribute
FROM    v_tag_object;
```

And now check the results from our view.

```
SELECT * FROM v_tags;
```

Apply Tags

With apologies for removing whitespace to save page count, our next sp_apply_object_ tag stored procedure utilizes a v_tags view to generate and apply tags. I have left the return value as debug_string to expose the inner workings for your testing. Note that the code is supplied for illustration purposes and overwrites tags for each run, simply representing the contents of view v_tags. You may wish to refactor for real-world use.

```
CREATE OR REPLACE PROCEDURE sp_apply_object_tag() RETURNS STRING
LANGUAGE javascript
EXECUTE AS CALLER
AS
$$
    var sql_stmt        = "";
    var stmt            = "";
    var create_tag_stmt = "";
    var apply_tag_stmt  = "";
    var result          = "";
    var retval          = "";
    var debug_string    = "";

    sql_stmt  = "SELECT DISTINCT create_tag_stmt\n"
    sql_stmt += "FROM    v_tags\n"
    sql_stmt += "ORDER BY create_tag_stmt ASC;\n\n"

    stmt = snowflake.createStatement ({ sqlText:sql_stmt });
    debug_string += sql_stmt;
    try
    {
        result = stmt.execute();
        while(result.next())
        {
            create_tag_stmt = result.getColumnValue(1);
```

```
            stmt = snowflake.createStatement ({ sqlText:create_tag_stmt });
            debug_string += create_tag_stmt + "\n";

            try
            {
                retval = stmt.execute();
                retval = "Success";
            }
            catch (err) { retval = create_tag_stmt + "\nCode: " + err.code +
            "\nState: " + err.state + "\nMessage: " + err.message + "\nStack
            Trace: " + err.stackTraceTxt }
        }
    }
    catch (err) { retval = sql_stmt + "Code: " + err.code + "\nState: "
+ err.state + "\nMessage: " + err.message + "\nStack Trace: " + err.
stackTraceTxt }

    sql_stmt  = "\nSELECT apply_tag_stmt\n"
    sql_stmt += "FROM    v_tags\n"
    sql_stmt += "ORDER BY create_tag_stmt ASC;\n\n"

    stmt = snowflake.createStatement ({ sqlText:sql_stmt });
    debug_string += sql_stmt;
    try
    {
        result = stmt.execute();
        while(result.next())
        {
            apply_tag_stmt = result.getColumnValue(1);
            stmt = snowflake.createStatement ({ sqlText:apply_tag_stmt });
            debug_string += apply_tag_stmt + "\n";

            try
            {
                retval = stmt.execute();
                retval = "Success";
            }
```

```
    catch (err) { retval = apply_tag_stmt + "\nCode: " + err.code +
    "\nState: " + err.state + "\nMessage: " + err.message + "\nStack
    Trace: " + err.stackTraceTxt }
    retval = "Success";
  }
}
catch (err) { retval = sql_stmt + "Code: " + err.code + "\nState: "
+ err.state + "\nMessage: " + err.message + "\nStack Trace: " + err.
stackTraceTxt }
return debug_string;
//   return retval;
$$;
```

We now invoke sp_apply_object_tag.

```
CALL sp_apply_object_tag();
```

Checking Tag Usage

Snowflake provides several ways to identify tag usage with these SQL statements returning the tag comment. While informative, these commands are not very useful as the tag, object/attribute, and object type must be declared.

```
SELECT system$get_tag('PII', 'employee', 'TABLE');
```

```
SELECT system$get_tag('PII_N_GENDER', 'employee.gender', 'COLUMN');
```

These SQL statements are more useful.

```
SELECT *
FROM   TABLE ( information_schema.tag_references_all_columns ( 'employee',
'TABLE' ));
```

```
SELECT *
FROM   TABLE ( information_schema.tag_references_all_columns ( 'employee',
'TABLE' ))
WHERE   tag_name IN ( 'PII', 'PII_N_GENDER' );
```

We can also interrogate Snowflake's Account Usage store noting latency of up to 2 hours.

```
USE ROLE accountadmin;

SELECT *
FROM    snowflake.account_usage.tags
WHERE   deleted IS NULL
ORDER BY tag_name;
```

And reset the role to sysadmin.

```
USE ROLE sysadmin;
```

Documentation is at https://docs.snowflake.com/en/user-guide/object-tagging.html#step-4-track-the-tags.

Automating Tag Deployment

With our object tagging model in place, this section demonstrates how to automate the deployment of object tags using a stream and task.

Create a stream to capture changes in tag assignment to objects and attributes.

```
CREATE OR REPLACE STREAM strm_tag_object ON TABLE tag_object;
```

Create a task to automatically apply tags as data changes in the source.

```
CREATE OR REPLACE TASK task_apply_object_tag
WAREHOUSE = COMPUTE_WH
SCHEDULE  = '1 minute'
WHEN system$stream_has_data ( 'strm_tag_object' )
AS
CALL sp_apply_object_tag();
```

Set Task to run.

```
ALTER TASK task_apply_object_tag RESUME;
```

Monitor and Suspend Task

With our task deployed and set to resume, we must check the task is executing correctly before suspending it to preserve credits. The following commands illustrate how to do this.

```
SELECT timestampdiff ( second, current_timestamp, scheduled_time ) as
next_run,
       scheduled_time,
       current_timestamp,
       name,
       state
FROM   TABLE ( information_schema.task_history())
WHERE  state = 'SCHEDULED'
ORDER BY completed_time DESC;
```

Once your testing is complete, suspend the Task.

```
ALTER TASK task_apply_object_tag SUSPEND;
```

Next, we continue data classification and reuse the objects defined in this section; therefore, cleanup is shown later.

Data Classification

Data classification is a Snowflake-supplied emerging capability currently in public preview. Unlike object tagging, which categorizes objects and attributes according to user-defined tags, data classification addresses the challenge of categorizing according to the actual data content of attributes.

Overview

Figure 12-5 provides an overview of data classification showing interactions between Snowflake supplied Categories and Semantics, our target application tables, and resultant JSON output from which classifications emerge.

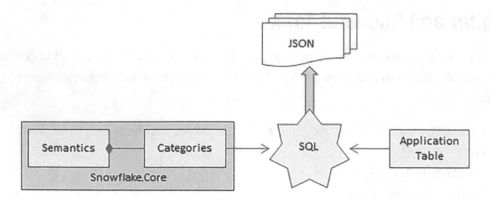

Figure 12-5. *Data classification schematic*

At the time of writing, category and semantic scope are quite limited but sufficient to demonstrate ambition to expand further. Tag mapping documentation is at `https://docs.snowflake.com/en/user-guide/governance-classify-sql.html#tag-mappings`. We explore the boundaries and hint at future capability but do not cover every aspect of data classification. For completeness, documentation is at `https://docs.snowflake.com/en/user-guide/governance-classify.html#data-classification`.

Identify Data Classifications

We assume the code developed in previous sections in this chapter has been deployed as the subsequent examples rely upon these objects.

We begin by programmatically identifying Snowflake supplied tag categories and comments for account and data classification schema (snowflake.core).

```
USE ROLE    accountadmin;
USE SCHEMA snowflake.core;

SHOW tags IN ACCOUNT;

SHOW tags IN SCHEMA snowflake.core;
```

Note for both SHOW commands, the owner is NULL and converting the SHOW command result set into a SQL query.

```
SELECT "name", "comment" FROM TABLE ( RESULT_SCAN ( last_query_id()));
```

There is no out-of-the-box way to fetch the list of tag categories and their associated values. But, with some further thought, the capability to do so facilitates a two-way feed between Snowflake and our metadata management tool.

Noting data classification operates on real data and not tags, we reuse our employee table defined and prove our assertion.

```
USE ROLE sysadmin;
USE DATABASE TEST;
```

Prove data classification works on data, not existing tags, and ensure tags are created against the employee table and attributes.

```
SELECT *
FROM   TABLE ( information_schema.tag_references_all_columns ( 'employee',
'TABLE' ));
```

Assuming tags are returned from the preceding query and no data in the employee table, our next SQL statement returns an empty JSON document.

```
SELECT extract_semantic_categories( 'TEST.public.employee' );
```

Create two test records in employee table.

```
INSERT INTO employee ( preferred_name, surname_preferred, forename_
preferred, gender, national_insurance_number, social_security_number,
postcode, zip_code, salary ) VALUES
('John Doe', 'Doe', 'John', 'Male', 'FA123456Z', '123-45-6789', 'DN13 7ZZ',
99950, 100000 ),
('Jane Doe', 'Doe', 'Jane', 'Female', 'XS123456Y', '234-56-7890', 'YP22
9HG', 99949, 120000 );
```

Check employee records.

```
SELECT * FROM employee;
```

Now use supplied function extract_semantic_categories to automatically examine our employee data using Snowflake supplied categories and semantics.

```
SELECT extract_semantic_categories( 'TEST.public.employee' );
```

Our test data was deliberately set up to investigate data classification boundaries known at the time of writing; therefore, the returned JSON document has expected gaps to save space. The following JSON document has been edited to remove redundant or duplicate sections.

```
{
  "EMPLOYEE_ID": {
    "extra_info": {
      "alternates": [],
      "probability": "0.80"
    },
    "privacy_category": "QUASI_IDENTIFIER",
    "semantic_category": "US_POSTAL_CODE"
  },
  "FORENAME_PREFERRED": {
    "extra_info": {
      "alternates": [],
      "probability": "0.99"
    },
    "privacy_category": "IDENTIFIER",
    "semantic_category": "NAME"
  },
  "GENDER": {
    "extra_info": {
      "alternates": [],
      "probability": "1.00"
    },
    "privacy_category": "QUASI_IDENTIFIER",
    "semantic_category": "GENDER"
  },
  "NATIONAL_INSURANCE_NUMBER": {
    "extra_info": {
      "alternates": []
    }
  },
...
```

```
"SALARY": {
  "extra_info": {
    "alternates": [],
    "probability": "1.00"
  },
  "privacy_category": "SENSITIVE",
  "semantic_category": "SALARY"
},
"SOCIAL_SECURITY_NUMBER": {
  "extra_info": {
    "alternates": [],
    "probability": "1.00"
  },
  "privacy_category": "IDENTIFIER",
  "semantic_category": "US_SSN"
},
...
"ZIP_CODE": {
  "extra_info": {
    "alternates": [
      {
        "privacy_category": "QUASI_IDENTIFIER",
        "probability": "0.80",
        "semantic_category": "US_POSTAL_CODE"
      }
    ]
  }
}
}
```

The preceding JSON shows all three privacy categories—IDENTIFIER, QUASI_IDENTIFIER, and SENSITIVE—along with their associated semantic categories. The probability attribute represents a confidence factor for each derivation. as we infer from the results, alternates are proposed where one or more options exist. We also see that UK-specific national_insurance_number is not recognized; it is nominally the same as US social_security_number, which is recognized. Of concern is the misidentification of employee_id as a QUASI_IDENTIFIER of type us_postal_code; note the 80% probability assigned.

Flattening Data Classifications

Now that you understand the returned JSON document, let's turn the result set into something more useful.

```
SELECT  f.key,
        f.value:"privacy_category"::VARCHAR              AS privacy_
                                                         category,
        f.value:"semantic_category"::VARCHAR             AS semantic_
                                                         category,
        f.value:"extra_info":"probability"::NUMBER(10,2) AS probability
FROM    TABLE(FLATTEN(extract_semantic_categories('TEST.public.
employee')::VARIANT)) AS f
WHERE   privacy_category IS NOT NULL
ORDER BY privacy_category, semantic_category, probability;
```

Tip   If you struggle to navigate JSON, add f.* to display all available attributes.

Figure 12-6 identifies data classifications from the preceding query.

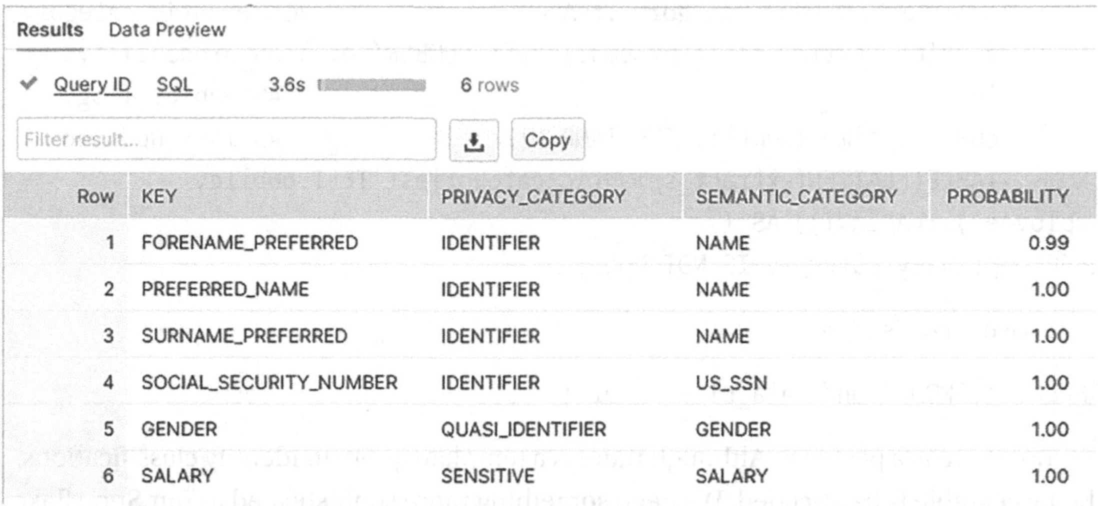

| Results | Data Preview |

✓ Query ID   SQL   3.6s ▓▓▓▓▓▓   6 rows

| Filter result... | | ⬇ | Copy |

Row	KEY	PRIVACY_CATEGORY	SEMANTIC_CATEGORY	PROBABILITY
1	FORENAME_PREFERRED	IDENTIFIER	NAME	0.99
2	PREFERRED_NAME	IDENTIFIER	NAME	1.00
3	SURNAME_PREFERRED	IDENTIFIER	NAME	1.00
4	SOCIAL_SECURITY_NUMBER	IDENTIFIER	US_SSN	1.00
5	GENDER	QUASI_IDENTIFIER	GENDER	1.00
6	SALARY	SENSITIVE	SALARY	1.00

Figure 12-6. *Snowflake-derived data classifications*

We might consider inserting directly into a table noting deploy_flag, which is covered later.

```
CREATE OR REPLACE TABLE found_data_classifiers COPY GRANTS
(
path_to_object    VARCHAR(2000) NOT NULL,
attribute         VARCHAR(2000) NOT NULL,
privacy_category  VARCHAR(2000) NOT NULL,
semantic_category VARCHAR(2000) NOT NULL,
probability       NUMBER        NOT NULL,
deploy_flag       VARCHAR(1)    DEFAULT 'N' NOT NULL,
last_updated      TIMESTAMP_NTZ DEFAULT current_timestamp()::TIMESTAMP_NTZ
                  NOT NULL
);
```

And now insert records.

```
INSERT INTO found_data_classifiers
SELECT 'TEST.public.employee'                        AS path_to_object,
       f.key                                         AS attribute,
       f.value:"privacy_category"::VARCHAR           AS privacy_
                                                        category,
```

```
        f.value:"semantic_category"::VARCHAR              AS semantic_category,
        f.value:"extra_info":"probability"::NUMBER(10,2) AS probability,
        'N'                                              AS deploy_flag,
        current_timestamp()::TIMESTAMP_NTZ               AS last_updated
FROM    TABLE(FLATTEN(extract_semantic_categories('TEST.public.
employee')::VARIANT)) AS f
WHERE   privacy_category IS NOT NULL;
```

Prove records exist.

```
SELECT * FROM found_data_classifiers;
```

But there is a problem. Although there is a template query to identify classifications, the target table is hard-coded. We need something more sophisticated as our Snowflake applications may have tens of thousands of objects.

Semi-Automating Data Classifications

The next step is to semi-automate data classification by implementing a stored procedure to analyze each schema on demand.

```
CREATE OR REPLACE PROCEDURE sp_classify_schema( P_DATABASE   STRING,
                                                P_SCHEMA     STRING )
RETURNS STRING
LANGUAGE javascript
EXECUTE AS CALLER
AS
$$
   var sql_stmt        = "";
   var stmt            = "";
   var outer_recset    = "";
   var inner_recset    = "";
   var result          = "";
   var debug_string    = '';
   sql_stmt  = "SELECT LOWER ( table_catalog )||'.'||\n";
   sql_stmt += "        LOWER ( table_schema  )||'.'||\n";
   sql_stmt += "        LOWER ( table_name    )\n";
```

```
sql_stmt += "FROM    information_schema.tables\n";
sql_stmt += "WHERE   table_catalog = UPPER ( :1 )\n"
sql_stmt += "AND     table_schema  = UPPER ( :2 )\n"
sql_stmt += "AND     table_type    IN ( 'BASE TABLE', 'MATERIALIZED
VIEW' )\n"
sql_stmt += "AND     table_owner IS NOT NULL;\n\n";
debug_string += sql_stmt;
stmt = snowflake.createStatement ({ sqlText:sql_stmt, binds:[P_DATABASE,
P_SCHEMA] });
try
{
   outer_recset = stmt.execute();
   while(outer_recset.next())
   {
      sql_stmt  = `INSERT INTO found_data_classifiers\n`
      sql_stmt += `SELECT '` + outer_recset.getColumnValue(1) +
        AS path_to_object,\n`
      sql_stmt += `         f.key
      AS attribute,\n`
      sql_stmt += `         f.value:"privacy_
      category"::VARCHAR               AS privacy_category,\n`,
      sql_stmt += `         f.value:"semantic_category"::VARCHAR
      AS semantic_category,\n`,
      sql_stmt += `         f.value:"extra_info":"probability"::NUMB
      ER(10,2) AS probability,\n`
      sql_stmt += `         'N'
      AS deploy_flag,\n`
      sql_stmt += `       current_timestamp()::TIMESTAMP_NTZ
      AS last_updated\n`
      sql_stmt += `FROM    TABLE(FLATTEN(extract_semantic_categories(`` +
      outer_recset.getColumnValue(1) + ``')::VARIANT)) AS f\n`
      sql_stmt += `WHERE   privacy_category IS NOT NULL;\n\n`
      debug_string += sql_stmt;

      stmt = snowflake.createStatement ({ sqlText:sql_stmt });
      try
```

```
        {
            inner_recset = stmt.execute();
        }
        catch { result = sql_stmt; }
      }
    }
    catch { result = sql_stmt; }
    return debug_string;
//   return result;
$$;
```

Remove existing data from previous runs.

```
TRUNCATE TABLE found_data_classifiers;
```

Call our new stored procedure recognizing this may take a few minutes to run, and the output is SQL statements for debugging.

```
CALL sp_classify_schema ( 'TEST',
                          'PUBLIC' );
```

When the stored procedure completes, check the output records, noting that some false positives occur due to the Snowflake internal data classification code attempting to match primary and surrogate keys to declared tags. Your results may vary according to the tables in the TEST.public schema.

```
SELECT * FROM found_data_classifiers;
```

This output provides a starting point for object tag creation, and while not perfect, we expect Snowflake data classification to be enhanced over time.

Apply Data Classifications

Snowflake provides the means to automatically apply identified data classifiers to our data, for which documentation is at https://docs.snowflake.com/en/sql-reference/stored-procedures/associate_semantic_category_tags.html#associate-semantic-category-tags.

> **Note**   I suggest automatically applying identified data classifiers is not yet mature enough to implement directly due to false positives.

Before automatically applying data classifiers, establish a baseline by running this SQL and export results to a CSV file.

```
SELECT *
FROM   TABLE ( information_schema.tag_references_all_columns ( 'employee',
'TABLE' ));
```

Entitle sysadmin role to use Snowflake database, then revert to sysadmin.

```
USE ROLE securityadmin;
```

```
GRANT IMPORTED PRIVILEGES ON DATABASE snowflake TO ROLE sysadmin;
```

```
USE ROLE sysadmin;
```

This SQL statement automatically assigns Snowflake data classifiers to our employee table, noting that this generates false positives.

```
CALL associate_semantic_category_tags
    ('TEST.public.employee',
     extract_semantic_categories('TEST.public.employee'));
```

Note the returned values.

```
Applied tag semantic_category to 7 columns.
Applied tag privacy_category to 7 columns.
```

Re-run our baseline query to provide a before and after comparison.

```
SELECT *
FROM   TABLE ( information_schema.tag_references_all_columns ( 'employee',
'TABLE' ));
```

Figure 12-7 shows the original result set in green, new data classifiers in blue, and false positives in red.

TAG DATABASE	TAG SCHEMA	TAG NAME	TAG VALUE	LEVEL	OBJECT DATABASE	OBJECT SCHEMA	OBJECT NAME	DOMAIN	COLUMN NAME
TEST	PUBLIC	PII	Personally Identifiable Information	TABLE	TEST	PUBLIC	EMPLOYEE	COLUMN	EMPLOYEE_ID
TEST	PUBLIC	PII	Personally Identifiable Information	TABLE	TEST	PUBLIC	EMPLOYEE	COLUMN	PREFERRED_NAME
TEST	PUBLIC	PII	Personally Identifiable Information	TABLE	TEST	PUBLIC	EMPLOYEE	COLUMN	SURNAME_PREFERRED
TEST	PUBLIC	PII	Personally Identifiable Information	TABLE	TEST	PUBLIC	EMPLOYEE	COLUMN	FORENAME_PREFERRED
TEST	PUBLIC	PII	Personally Identifiable Information	TABLE	TEST	PUBLIC	EMPLOYEE	COLUMN	GENDER
TEST	PUBLIC	PII	Personally Identifiable Information	TABLE	TEST	PUBLIC	EMPLOYEE	COLUMN	NATIONAL_INSURANCE_NUMBER
TEST	PUBLIC	PII	Personally Identifiable Information	TABLE	TEST	PUBLIC	EMPLOYEE	COLUMN	SOCIAL_SECURITY_NUMBER
TEST	PUBLIC	PII	Personally Identifiable Information	TABLE	TEST	PUBLIC	EMPLOYEE	COLUMN	POSTCODE
TEST	PUBLIC	PII	Personally Identifiable Information	TABLE	TEST	PUBLIC	EMPLOYEE	COLUMN	ZIP_CODE
TEST	PUBLIC	PII	Personally Identifiable Information	TABLE	TEST	PUBLIC	EMPLOYEE	COLUMN	SALARY
TEST	PUBLIC	PII_N_GENDER	Personally Identifiable Information -> Non-Sensitive -> Gender	COLUMN	TEST	PUBLIC	EMPLOYEE	COLUMN	GENDER
TEST	PUBLIC	PII_S_NAME	Personally Identifiable Information -> Sensitive -> Name	COLUMN	TEST	PUBLIC	EMPLOYEE	COLUMN	PREFERRED_NAME
TEST	PUBLIC	PII_S_NAME	Personally Identifiable Information -> Sensitive -> Name	COLUMN	TEST	PUBLIC	EMPLOYEE	COLUMN	SURNAME_PREFERRED
TEST	PUBLIC	PII_S_NAME	Personally Identifiable Information -> Sensitive -> Name	COLUMN	TEST	PUBLIC	EMPLOYEE	COLUMN	FORENAME_PREFERRED
SNOWFLAKE	CORE	PRIVACY_CATEGORY	IDENTIFIER	COLUMN	TEST	PUBLIC	EMPLOYEE	COLUMN	PREFERRED_NAME
SNOWFLAKE	CORE	PRIVACY_CATEGORY	IDENTIFIER	COLUMN	TEST	PUBLIC	EMPLOYEE	COLUMN	SURNAME_PREFERRED
SNOWFLAKE	CORE	PRIVACY_CATEGORY	IDENTIFIER	COLUMN	TEST	PUBLIC	EMPLOYEE	COLUMN	FORENAME_PREFERRED
SNOWFLAKE	CORE	PRIVACY_CATEGORY	IDENTIFIER	COLUMN	TEST	PUBLIC	EMPLOYEE	COLUMN	SOCIAL_SECURITY_NUMBER
SNOWFLAKE	CORE	PRIVACY_CATEGORY	QUASI_IDENTIFIER	COLUMN	TEST	PUBLIC	EMPLOYEE	COLUMN	EMPLOYEE_ID
SNOWFLAKE	CORE	PRIVACY_CATEGORY	QUASI_IDENTIFIER	COLUMN	TEST	PUBLIC	EMPLOYEE	COLUMN	GENDER
SNOWFLAKE	CORE	PRIVACY_CATEGORY	SENSITIVE	COLUMN	TEST	PUBLIC	EMPLOYEE	COLUMN	SALARY
SNOWFLAKE	CORE	SEMANTIC_CATEGORY	GENDER	COLUMN	TEST	PUBLIC	EMPLOYEE	COLUMN	GENDER
SNOWFLAKE	CORE	SEMANTIC_CATEGORY	NAME	COLUMN	TEST	PUBLIC	EMPLOYEE	COLUMN	PREFERRED_NAME
SNOWFLAKE	CORE	SEMANTIC_CATEGORY	NAME	COLUMN	TEST	PUBLIC	EMPLOYEE	COLUMN	SURNAME_PREFERRED
SNOWFLAKE	CORE	SEMANTIC_CATEGORY	NAME	COLUMN	TEST	PUBLIC	EMPLOYEE	COLUMN	FORENAME_PREFERRED
SNOWFLAKE	CORE	SEMANTIC_CATEGORY	SALARY	COLUMN	TEST	PUBLIC	EMPLOYEE	COLUMN	SALARY
SNOWFLAKE	CORE	SEMANTIC_CATEGORY	US_POSTAL_CODE	COLUMN	TEST	PUBLIC	EMPLOYEE	COLUMN	EMPLOYEE_ID
SNOWFLAKE	CORE	SEMANTIC_CATEGORY	US_SSN	COLUMN	TEST	PUBLIC	EMPLOYEE	COLUMN	SOCIAL_SECURITY_NUMBER

Figure 12-7. Snowflake-derived data classifications

Convert to JSON

We are missing the ability to review and either select or reject tags before application. In fairness, the documentation at `https://docs.snowflake.com/en/sql-reference/ stored-procedures/associate_semantic_category_tags.html#examples` does offer suggestions. But the implementation syntax is cryptic and command-line driven, whereas we want to make life as easy as possible for our business colleagues.

Instead, I suggest a more rounded, user-friendly proposal using the found_data_ classifiers table populated earlier.

```
SELECT * FROM found_data_classifiers;
```

With this output, we can identify false positives and leave the deploy_flag set to N, preventing object tag creation. We could also update table contents interactively through the user interface or the SnowSQL command line. An alternative approach would be to download found_data_classifiers contents, manipulate them using Excel, then implement a CSV file ingestion pattern to overwrite found_data_classifiers data with the new values.

For our example, we use DML to interactively modify the found_data_classifiers contents.

```
UPDATE found_data_classifiers
SET    deploy_flag   = 'Y'
WHERE  path_to_object = 'test.public.employee'
```

```
AND     attribute     != 'EMPLOYEE_ID';
```

The associate_semantic_category_tags function expects a VARIANT input; therefore, we must convert found_data_classifiers attributes into a VARIANT data type for consumption and associated identifiers.

There may be easier ways to do this but remember this is also a tutorial. Let's create a view to illustrate how to manually convert from strings to JSON.

```
CREATE OR REPLACE VIEW v_found_data_classifiers AS
SELECT path_to_object,
       '  "'||attribute||'": {'||
       '    "extra_info": {'||
       '      "alternates": [],'||
       '      "probability": "'||probability||'"'||
       '      },'||
       '    "privacy_category": "'||privacy_category||'",'||
       '    "semantic_category": "'||semantic_category||'"'||
       '    },'                                          AS json_inner,
       deploy_flag
FROM   found_data_classifiers;
```

Now aggregate and convert to JSON.

```
CREATE OR REPLACE VIEW v_found_data_classifiers_list AS
SELECT path_to_object,
       LISTAGG ( json_inner )                            AS
       json_list,
       '{'||SUBSTR ( json_list, 1, LENGTH ( json_list ) -1 )||'}' AS
        json_source,
       to_json(parse_json ( json_source ))::VARIANT      AS
       json_string,
       deploy_flag
FROM   v_found_data_classifiers
GROUP BY path_to_object,
         deploy_flag;
```

During testing and to aid debugging, I used `https://jsonformatter.org` to check that the JSON format is correct. I cut and pasted the JSON output from the json_source attribute in the preceding view before attempting to convert to the VARIANT data type.

Note   You may need to comment out json_string conversion to VARIANT while debugging.

To assist your debugging, these queries may help. Attempt to convert from string to variant.

```
SELECT *
FROM   v_found_data_classifiers_list;

SELECT to_json(parse_json ( json_string ))
FROM   v_found_data_classifiers
WHERE  deploy_flag    = 'Y'
AND    path_to_object = 'test.public.employee';
```

Deploy Selected Tags

We can now deploy the tags, excluding those that generate false positives.

```
CALL associate_semantic_category_tags
    ('TEST.public.employee',
     (SELECT json_string
      FROM   v_found_data_classifiers_list
      WHERE  deploy_flag = 'Y'
      AND    path_to_object = 'test.public.employee'));
```

Noting the number of tags applied.

```
Applied tag semantic_category to 6 columns.
Applied tag privacy_category to 6 columns.
```

Check the outcome.

```
SELECT *
```

```
FROM   TABLE ( information_schema.tag_references_all_columns ( 'employee',
'TABLE' ));
```

Data Lineage

Data lineage is neatly described at `https://en.wikipedia.org/wiki/Data_lineage`.
In short, data lineage is the mechanism by which we track the lifecycle of data as it
flows through our systems from ingestion through to consumption and every process,
transformation, or manipulation along the way. If we can automatically trace every data
processing step, we can achieve several valuable business objectives. In reality, tracing
full data lineage is much more complex and a very difficult objective for a variety of
reasons, some of which we touch upon next.

At the macro level, imagine a situation where data is bought, ingested, manipulated,
cleansed, transformed, and enriched before the sale. And now, overlay every process
in your organization and look for the common systems and tooling enabling your
organization to function. Some natural outcomes would be to identify common systems
through which a high percentage of your organization's life-blood (data) flows, along
with key technologies which are at the end of life and require upgrade or replacement.
Such capability would also identify systems critical to data consumption and revenue
generation, providing an audit trail of change over time. We might consider how a *digital
twin* can assist our organizations in mapping system interactions and identifying risk
concentration and critical components. A useful reference is at `https://en.wikipedia.
org/wiki/Digital_twin`.

Snowflake system development is one such component in our data lineage at the
micro level. But everything we do must be consistent with the aim of achieving full
data lineage. External tooling helps identify data lineage but typically does not cope
with dynamic SQL, for example, encapsulated in stored procedures and struggling
to interpret parameterized code. Snowflake is one component in our data lifecycle,
from ingestion to consumption. We may already have an accepted architecture and
considerable code-base built over several years, which works very well but is not best
suited to exposing data lineage.

What can we do to facilitate data lineage in Snowflake?

Following on, and reliant upon objects created in previous sections, let's investigate
Snowflake supplied capability to identify objects by applied tags. We may have

search capability in our metadata management tool, but we know catalogs and target environments can lose synchronization in real-world use.

From our previous investigation, you know how to identify tags associated with an object.

```
SELECT *
FROM   TABLE ( information_schema.tag_references_all_columns ( 'employee',
'TABLE' ));
```

But what if we wish to search by tag.

Fortunately, Snowflake provides the means. Note the use of the TABLE function as there is latency when using Snowflake's Account Usage store and the tag name must be in uppercase.

```
SELECT *
FROM   TABLE ( snowflake.account_usage.tag_references_with_lineage ( 'TEST.
PUBLIC.PII_S_NAME' ));
```

Further information is at https://docs.snowflake.com/en/sql-reference/ functions/tag_references_with_lineage.html#tag-references-with-lineage.

Data Consumption

One of the biggest challenges our non-technical colleagues face is identifying the data consumers of the data sets under their stewardship. In effect, they want to know who can see what, and, most importantly, how to prevent access to unauthorized users. As technical practitioners, it is incumbent upon us to remove the veil of secrecy and expose how data sets are treated to appropriately authorized colleagues. To better understand the business context, we must take a step back and look at the provenance of the issue.

Remembering we are all on a journey, and everyone finds themselves somewhere on the timeline from starting their journey through to full awareness and understanding, we use a business-focused lens to explain the challenge in simple terms. Each custodian of a discrete data set in an organization applies safeguards, including knowing why access to the data set is required, who will have access to the data set, and the purpose of the data set.

When the data custodian is satisfied their safeguards are met, we should be granted access to their data sets, but only via approved tooling and interfaces and for the purpose expressly articulated. From the data custodian perspective, they have lost control once data has left their source system. They can no longer mandate their safeguards because the data now resides in a different system under new custodians and stewardship. Furthermore, if a change to data access scope is required, how is the original data custodian informed and allowed to ensure their safeguards are met?

You know from Chapter 5 that Snowflake provides robust, secure role-based entitlements to source objects. As the technical platform owner, we are responsible for retaining total data isolation for each source system up to the point of data consumption, which is covered later. From a technical perspective, we have absolute confidence in our ability to highly segregate our data sets. Furthermore, should make every effort to inform, educate and win over our business colleagues with technical solutions proposed to meet their data storage requirements. Essentially, we must conduct a "hearts and minds" campaign while allowing our business colleagues to air their concerns and address any issues raised. But this approach only goes part way to solving the core issue of "who can see what" and brings us back to the original question.

Listing the available controls gives us a starting point in articulating evidence and may include ensuring Snowflake user provision is only via Active Directory (or equivalent) provisioning tools. RBAC roles constrain access to specific data sets; sensitive role assignment is monitored, and alerts are raised automatically. And periodic role membership recertification is performed. All these controls and more are retrospective. They all work after a change has been applied and do not proactively prevent inadvertent access to data. In other words, by the time we have implemented a change in Snowflake, there is an explicit assumption all safeguards have been satisfied and supporting evidence provided.

But what if this is not the case?

The best answer lies with implementing a metadata management tool where entitlement grants to roles and assignments to AD groups are managed externally to Snowflake. Adopting this approach requires a centralized team to ensure consistent applications of each data steward policy, staff continuity, and automated tooling to deploy changes.

Summary

We began this chapter with an overview of data management before discussing the importance of metadata and the costs of not managing our organizations' metadata.

We then looked at the role of the data custodian and showcased how to automate the repeated construction of a design pattern. I leave you to extend into fully fledged data pipelines using examples found elsewhere in this book.

Data privacy is not typically part of data management but should be a key consumer of object tagging, where we spend much of this chapter.

With the assumption object tags are maintained in an external metadata management tool, I explained how an automated object tagging process could be implemented, offering a template solution for further customization.

Our investigation into data classification shows the potential for object tag identification from data values in our current data sets and provides a degree of automation. Recognizing data classification is an emerging Snowflake capability, we developed tooling to build capability and mitigate against perceived limitations.

A brief foray into data lineage showcasing the use of tags to identify objects followed before lastly discussing how to proactively protect data sets contained in Snowflake and prevent inadvertent access before the point of deploying changes into Snowflake.

Having discussed data management, let's look at another pillar, information, or data modeling.

CHAPTER 13

Data Modeling

This chapter briefly discusses data modeling, providing an overview of Third Normal Form (3NF), data vault modeling (e.g., Data Vault 2.0), and dimensional modeling techniques. Many books offer complete explanations better suited to dive deep into theory and practice. Instead, I provide an overview of the main data modeling options to both allow identification and inform why certain techniques are preferred at different points in a data lifecycle.

If you are familiar with data modeling, feel free to skip this chapter, which is intended to be a simple explanation for newcomers. However, I contend that understanding data modeling is important when explaining the why, how, and what of data ingestion through to data consumption.

This chapter is a very brief guide to only get you started. A full explanation of database normalization is at `https://en.wikipedia.org/wiki/Database_normalization`.

Third Normal Form (3NF)

Whether Snowflake is the target database of choice or not, 3NF is optimized for database writes. Typically, we insert, update, or delete from a single table at a time. I do not discuss transaction boundaries here but draw your attention to Snowflake micro-partitions, discussed in Chapter 3, and earlier comments on not using Snowflake for heavy OLTP workloads. The salient point is that 3NF is not a read-optimized pattern because we must join several tables together to derive full context. Each table holds information about a single "thing." Object joins can be expensive operations, and the data model must be understood to ensure desired meaning can be derived. We must also ensure uniqueness is enforced in our 3NF model and prevent duplicates, which may consist of programmatically derived unique constraints and date attributes for temporal data.

© Andrew Carruthers 2022
A. Carruthers, *Building the Snowflake Data Cloud*, https://doi.org/10.1007/978-1-4842-8593-0_13

Database normalization is the process by which the core database design is distilled into a suite of objects containing a single expression of like or grouped values with keys providing relationships. Joining objects via keys informs the model, resulting in a logical and meaningful representation of the business context. 3NF has been the optimal and most encountered implementation pattern for database structures for decades; it is best understood and widely accepted. Now that we know what we are trying to achieve, how do we implement a 3NF data model?

Consider Figure 13-1, which represents part of a sample physical data model where there are parties (companies), contacts (people), and agreements (documents).

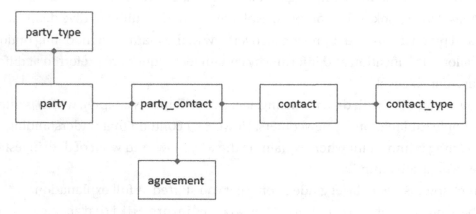

Figure 13-1. *Sample 3NF model*

Figure 13-1 depicts a typical scenario where the tables suffixed "_type" hold values used to restrict or constrain values held in the parent tables. While Snowflake does not enforce constraints apart from NOT NULL, we contend declaring constraints allows relationships to be discovered by tooling, easing our business colleagues understanding of our data models.

As you dive deeper into data modeling, you better understand how relationships are the "glue" binding our model together using natural keys (i.e., the physical labels that provide uniqueness quickly becomes unmanageable). Instead, we use surrogate keys expressed as sequences. You may have seen sequences used in previous chapters. Let's discuss the reasoning. Figure 13-2 illustrates how sequences improve the readability of the data model. The first image shows natural keys in the party_contact intersection table and agreement table, whereas the second image shows the addition of surrogate keys generated from sequences, one per table.

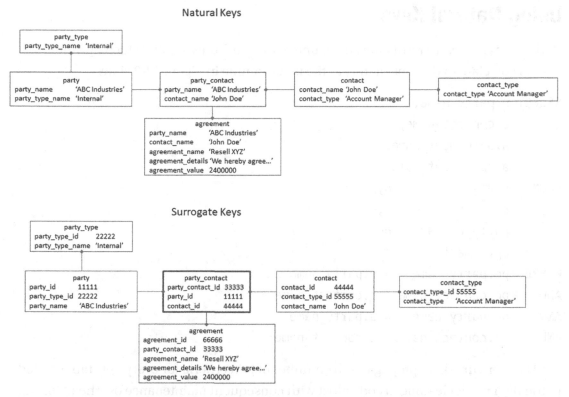

Figure 13-2. *Natural keys and surrogate keys*

The importance of using surrogate keys may not be readily apparent from the example, because adding sequences also adds an attribute to each table. But imagine the situation without surrogate keys where party_name changes. For example, a merger or acquisition may predicate the legal entity being renamed. In such a scenario, the party_ name change would break the referential integrity between party and party_contact and all downstream tables after that. Better to make the data model agnostic of name changes and use immutable surrogate keys to join tables rather than mutable real-world labels. Surrogate keys provide flexibility and are adaptable, whereas natural keys are inflexible and do not adapt to change cascading through our models. What we also find in practice is most, if not all, SQL optimizers prefer numbers to character equivalence checks. Therefore, the sequence-generated surrogate keys offer a simple way to improve performance.

Let's look at how sample SQL statements might look, assuming we have created appropriate tables, sequences, and records not shown here.

Using Natural Keys

With natural keys, we must know the name of the attributes to join, which may not always correlate to the table name, as the first example in Figure 13-2 shows.

```
SELECT  p.party_name,
        c.contact_name,
        a.agreement_name,
        a.agreement_value
FROM    party           p,
        contact         c,
        party_contact   pc,
        agreement       a
WHERE   pc.party_name   = p.party_name
AND     pc.contact_name = c.contact_name
AND     pc.party_name   = a.party_name
AND     pc.contact_name = a.contact_name;
```

Using natural keys propagates the number of join criteria for every new table added to the data model leading to code bloat with consequent maintenance overhead for our support team and when adding new features.

Using Surrogate Keys

With surrogate keys, we may reasonably assume most (but not all) data models correlate to the table name, as the second example in Figure 13-2 shows.

```
SELECT  p.party_name,
        c.contact_name,
        a.agreement_name,
        a.agreement_value
FROM    party               p,
        contact             c,
        party_contact       pc,
        agreement           a
WHERE   pc.party_id         = p.party_id
AND     pc.contact_id       = c.contact_id
AND     pc.party_contact_id = a.party_contact_id;
```

Using surrogate keys adds readability to our code by reducing the number of join conditions. Not only is the data model more understandable, but adding a new relationship is relatively easy. Adding constraints is also simpler as we do not need to list all the constituent attributes but instead rely upon surrogate keys alone. Where surrogate keys are used, the number of core join conditions is always one less than the number of objects joined.

3NF Assumptions

When working with 3NF data models, there is an implicit assumption that reference data satisfies any declared referential integrity. While Snowflake does not enforce referential integrity, we might reasonably assume our ingested data is correct at the point of delivery and, to our earlier point of not rejecting any invalid records, might optionally set missing or invalid attributes to defaults while validating and reporting invalid records back to the source.

3NF Summary

Joining tables together allows information to be derived from the discrete data sets but implies a high degree of familiarity with the data model. We may also see challenges with identifying data sets for date ranges. Each component table may have its own temporal attributes, particularly where SCD2 has been implemented, implying that further domain knowledge is required. We would extend the number of SQL predicates to explicitly include the date ranges required recognizing the code quickly becomes bloated. However, as previously stated, 3NF is a storage (write) pattern, not a retrieval (read) pattern. We are most concerned with writing data into Snowflake as fast as possible for data ingestion. We later discuss retrieval patterns.

Adding attributes at a higher level of grain than currently supported in a data model is generally very easy. For example, adding a geographical region where a country exists on the party table would require a new reference table for the region and an intersection entity between the country table (not shown) and the region. After which, all reporting referring to a country may need amending to add a filter for the region. Similarly, adding a lower grain of detail to a 3NF data model is generally doable with a little effort and some disruption but typically not nearly as invasive as for dimensional modeling described next.

Denormalization

No discussion on normalization is complete without mentioning denormalization, the process by which we join our discrete 3NF tables to provide meaningful business objects for reporting. While this topic deserves more considered treatment than we can provide here, I offer a few suggestions.

We prefer to represent all data in our data warehouse (see `https://en.wikipedia.org/wiki/Data_warehouse`) and not just those that match our reference data. We contend that all data, regardless of quality, is valid, and our business colleagues are responsible for their data quality. We must not exclude records by filtering out "bad" data but instead notify source systems of records failing validation.

Our success is entirely dependent upon adoption by our users. We must enable an infrastructure that enables results driving business value. In other words, we must retain our business users' perspective and provide what they want, how they want it, and not deliver what we think they want. From this position, we may enrich our data and suggest summaries, aggregates, and pre-filtered data sets tuned to our customer needs.

We must make our data sets easy to discover and self-serve. Adding object tags, as discussed in Chapter 12, is one very powerful way to enable our business community. Lastly, we must not forget data security and adhere to the policy.

Dimensional Modeling

In contrast to 3NF optimized for transactional speed of writing data into an RDBMS, dimensional models are concerned with the speed of reading data from our data warehouse. Two different paradigms with interrelated structures and complementary approaches. Ralph Kimball developed dimensional modeling; more information is at `https://en.wikipedia.org/wiki/Dimensional_modeling`. For further reading, I recognize Bill Inmon as the father of data warehousing. More information is at `https://en.wikipedia.org/wiki/Bill_Inmon`.

Dimensional modeling is often referred to as a star schema reflecting the visual representation of the data model. Re-purposing our earlier 3NF data model into a star schema depicted in Figure 13-3 illustrates how entities are conflated into dimensions containing textual attributes, with a new fact entity containing numeric data with foreign keys to dimensions. Putting aside temporal considerations for dimension entries, we arrive at a business-friendly data model where our business users can intuitively understand and self-serve with minimal intervention or explanation from technical staff.

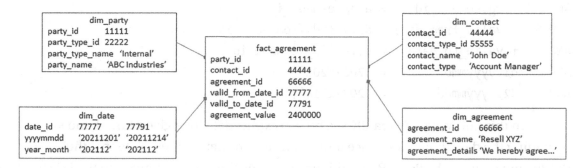

Figure 13-3. *Sample dimension model*

Date Dimension

The prevalence of a date dimension used twice next for date banding queries is more prominent in dimensional modeling. We might create a date dimension using Julian Day range from before our earliest system record to some point in the far distance, easily created using SQL.

Sample Query

Dimension keys typically correlate to the dimension name as our example from Figure 13-3 shows.

```
SELECT  p.party_name,
        c.contact_name,
        a.agreement_name,
        a.agreement_value
FROM    fact_agreement    f,
        dim_party         p,
        dim_contact       c,
        dim_agreement     a,
        dim_date          d1,
        dim_date          d2
```

```
WHERE   f.party_id        = p.party_id
AND     f.contact_id      = c.contact_id
AND     f.agreement_id    = a.agreement_id
AND     f.valid_from_date = d1.valid_from_date_id
AND     f.valid_to_date   = d2.valid_to_date_id
AND     d1.yyyymmdd       = '20211201'
AND     d2.yyyymmdd       = '20211214';
```

From this query, we can readily see how to extract data for a given time-band by joining the date dimension twice and setting appropriate values. Note that the dimension contains VARCHAR definitions, removing the need to convert to the DATE data type to make usage easy for our business consumers. We might add more predicates to filter on specific criteria or summarize data.

Dimensional Modeling Assumptions

When working with star schemas, an implicit assumption dimension data satisfies all fact data. While Snowflake does not enforce referential integrity, we might reasonably assume our ingested data is correct at the point of delivery and, to our earlier point of not rejecting any invalid records, might optionally set missing or invalid attributes to dimension default values.

Dimensional Modeling Summary

Dimensions are generally quite small, sometimes as large as a few thousands of records, changing slowly. Facts are generally huge, in the order of billions of records, with new data sets added frequently.

One can readily observe fact table contents are easily aggregated and summarized, further speeding up periodic reporting over defined time buckets. Moreover, the same dimensions can be used to report against summarized and aggregated data. With our date dimension attributes extended to months, quarters, and years, the natural calendar relationships enable time bucket–based reporting. We might extend our party dimension to include industry classifications, sectors, and geographical locations, and our agreement dimension includes similar attributes. With the ability to extend dimensions by adding textual search attributes without impacting our fact tables, we should take the opportunity to consult with our business colleagues for their input. Closer working relationships are often achievable, leading to more successful business outcomes. Considering that we may have poor quality data, each dimension must contain a default value against which all non-matching facts are assigned to enable complete reporting. However, we must not apply 3NF to our dimensions and subset the contents into what is paradoxically called a "snowflake" schema, as doing so may significantly impact performance.

Given our knowledge of Snowflake micro-partitions, the potential for querying huge data volumes, and knowledge of the dimensions commonly used for lookups, we must consider appropriate cluster keys and materialized views for summarization, aggregation, and alternative lookup paths.

Adding attributes at a higher level of grain than currently supported is generally very easy; for example, adding a geographical region where a country exists as the new dimension attribute self-evidently rolls up. But what happens when the grain of the dimension is too high to support evolving business requirements? For example, what are the options if we need to decompose an attribute in an existing dimension to a finer grain of detail? Our first option is to restructure the original fact table to decompose existing facts into the same finer grain along with a re-created dimension table with extended attributes and increased surrogate key ranges. Our second option is to create a new fact table with the new data and dimension table. Neither option is easy to implement. The first option is fraught with danger, whereas the second option would be expected to reconcile with the original fact table.

Data Vault 2.0

Data vault modeling breaks down the relationship between elements of an entity and its primary key into hubs and satellites joined by link tables. Data vault is a distinct modeling technique that takes time to master. If done badly, expect a poor outcome.

Note   A data vault is not intended as an end-user query repository, and such usage must be avoided.

A raw data vault is designed to represent unchanged data as it was loaded from the source. We always represent data as supplied regardless of how complete or imperfect on receipt, and our work up to this point in this book explains how to historize using the

SCD2 pattern. A raw data vault represents a "system of record". It may be better expressed as connecting directly to staging tables as the historization comes from the timestamp of the latest record and not through the SCD2 pattern implementing valid_to and valid_ from timestamps.

For existing 3NF schemas with a history built up, we should reflect the history of data vault modeling using the valid_from timestamp as the timestamp of the latest record. The key is that it's not Data Vault 2.0 versus Kimball; it's Data Vault 2.0 underpinning Kimball and avoiding the expensive rework of a "direct to Kimball" approach.

Now that you understand where data vault modeling fits, let's address the distinct advantages that data vault modeling offers over 3NF and dimensional modeling.

- Ability to bulk load data in parallel

- Efficient loading and storing data from disparate sources

- Flexibility to add new hubs, links, and satellites without disrupting the existing model

- Track and audit data with ease

- Scalable, adaptable, and consistent architecture

However, the key disadvantage of data vault modeling is the expansion in the number of objects used in the model, rendering a data vault unsuitable for end-user reporting, which must be avoided. The data vault methodology never recommends direct access to the data vault structures. The proposal is *always* to build virtual or materialized star schema data marts over the data vault for familiar and optimized reporting.

Flexibility comes with complexity, requiring a mindset shift from traditional data modeling. And data vault modeling is not a good candidate for single source systems and/or relatively static data.

Re-purposing our earlier 3NF data model into a data vault format depicted in Figure 13-4 illustrates how entities are expanded into hubs containing business keys, satellites with textual attributes, and a link table containing the relationship between hubs.

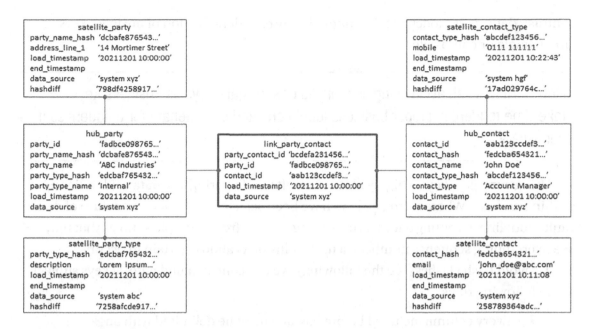

Figure 13-4. Sample data vault model

Success Factors

A data vault is not a silver bullet. Poor understanding and execution of core data vault concepts and principles result in poor implementation and general dissatisfaction. Data vaults have been well described in various media. This section seeks to inform basic principles rather than dive into technicalities; it is broken down into a series of headings explaining each principle.

Experience Counts

Without adherence to the data vault methodology, projects fail badly. Having at least one seasoned data vault practitioner on the team is highly recommended to ensure best practice is observed. While costly, this is a risk mitigation strategy. I have been on a failed data vault project. I speak from experience.

Business Keys, Not Surrogate Keys

Every practical 3NF system uses surrogate keys instead of natural keys to represent relationships making both code cleaner and physical data models easier to read. In contrast, data vault modeling relies upon the correct identification of business keys, either single or composite.

Note   At all cost, avoid using sequences or surrogate keys as business keys, take time to identify proper business identifiers, and concatenate for uniqueness if necessary

---

Using surrogate keys is highly likely to result in refactoring as a data vault relies upon hashing for keys. Where surrogate keys have been selected and given the nature of data vault modeling accepting data sets into a single entity from multiple sources that may use overlapping sequence numbers, a hash collision is almost certain to occur.

To avoid hash clashes (see the following), we use concatenated natural keys subject to the following rules.

- Every column included before hashing must be delimited with an uncommon field like double-pipe (||) between columns.

- All null values are converted to blank before hashing. Null values break the hash. For surrogate hash keys, these are substituted with zero keys ('-1').

- Column order must be consistent, and new columns added to the satellite table must be added to the end of the record hash.

- Leading and trailing blanks of every column must be removed before hashing.

- For surrogate hash keys only, business keys must be up-cased to ensure passive integration across source applications of business keys loading to a common hub. Alternate solutions may be applicable if variable business key treatments are needed.

- All business keys must be cast to text. This is a standard treatment because business keys are not metrics. This is done before hashing.

- We may optionally include a *business key collision code* (BKCC) to differentiate keys from disparate sources where overlap would cause hash clashes.

Hashing

Data vault extensively uses hashing, a one-way algorithm translating input value into a fixed length 128-bit value which can be stored in a 32-character length field. Hashes are not used for cryptographic functions but simply to generate a manageable key. A far more comprehensive explanation of hashing is at `https://en.wikipedia.org/wiki/Hash_function`. For our purposes, we use the Snowflake functionality documented at `https://docs.snowflake.com/en/sql-reference/functions/md5.html`.

Hash keys are generated from all non-primary key values with a fixed field separator and stored in a field called hashdiff. Comparing hashdiff for existing data with hashdiff for new or modified records identifies whether any non-primary key value has changed. However, the comparison must only be against the most recent record and not the historical record; otherwise, records reverted to earlier values may not be detected as the original hashdiff is calculated and no delta detected.

Hash Clash

Hash clash (also known as *hash collision*) occurs when two different input values resolve to the same hash value. With MD5, hash clash can become more likely when over 5.06Bn records exist in a table. Therefore, I suggest the probability is so small as to be insignificant; more information is at `https://en.wikipedia.org/wiki/Hash_collision`. If hash clash is perceived to be an issue, alternatives exist, such as SHA1/SHA2/SHA3, noting hash generation time is longer than for MD5. Various strategies can be implemented to test for hash collision, some of which are at `https://dwa.guide/2017/10/20/2017-10-20-implementing-a-hash-key-collision-strategy/`.

Rehashing

Recognizing data vault modeling as an insert-only model, there may be the temptation to "knife and fork" data without re-creating the associated hash key and subsequent propagation to hubs and links, rendering the explicit relationship between business key and hash broken. We must exclude all manual updates in our systems and rely upon correctly supplied data submission only to retain data fidelity. If data is wrong at the point of submission, our systems regard the data as valid for the full lifecycle of the data. Correction by resubmission is the preferred option as reporting may have relied upon the bad data.

Business Ontology

Business ontology is a consistent representation of data and data relationships by providing a structured way to interpret corporate data as a consolidation of taxonomies. A taxonomy is a clearly defined hierarchical structure for categorizing information.

Note   Without a business ontology, we are destined to fail.

Expect Change

Data models change over time, and data vault modeling facilitates change by adopting a flexible architecture. Incorporating new features and capabilities in data vault modeling is relatively easy compared to 3NF and dimensional models.

Data Representation

Loading new data sets into an existing model is straightforward as data vault modeling expects each record to be identified back to the source by default. Typically our 3NF systems segregate by data source where each schema or data model corresponds to a single data source.

Data Vault Core Components

Noting earlier comments on identifying natural business keys rather than surrogate keys, let's look at guiding principles for designing each major data vault table type using Figure 13-5 as an example.

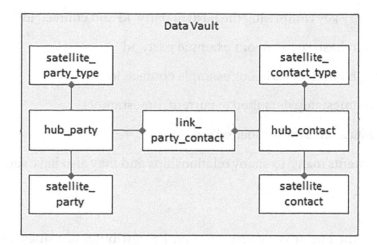

Figure 13-5. Sample data vault pattern

Hub

A hub is a central entity containing business keys enabling all related satellites and links to interconnect.

When creating hub entries, avoid using sequences or surrogate keys as business keys, take time to identify proper business identifiers, and concatenate for uniqueness if necessary.

Hub tables typically contain these attributes.

- A primary key comprising the hash of the business key(s) (Preferably, they should be one or more textual identifiers separated by a fixed delimiter, not sequence or surrogate key.)

- Alternate keys from satellites, in our example: party_name_hash/ party_name and party_type_hash/party_type_name

- A load timestamp defaulted to current_timestamp()

- The data source for the loaded data

A hub should have at least one satellite table.

Link

The intersection between the party hub and contact hub is the link table. A link table supports many-to-many relationships and typically contains these attributes.

- A primary key comprising the hash of party_id and contact_id

- A foreign hash key #1 in our example party_id

- A foreign hash key #2 in our example contact_id

- A load timestamp defaulted to current_timestamp()

- The data source for the loaded data

A link implements many-to-many relationships and may also have satellites

Satellite

Satellites contain the business context (i.e., non-key attributes of a single hub or link). For example, party and party_type are both satellites of the party hub. We might also separate volatile attributes (i.e., those which change frequently) from non-volatile attributes (i.e., those which are static or change infrequently) into separate satellite tables. Separating high and low-frequency data attributes can assist with the speed of data loading and significantly reduce the space that historical data consumes.

Normally, a new satellite table would be designed for each data source, and another common consideration is data classification. This approach also enables data to be split apart based on classification or sensitivity. This makes it easier to handle special security considerations by physically separating data elements.

A satellite can only link to a single hub with load_timestamp used to determine the most recent record.

Satellites always have these attributes.

- A composite primary key consisting of the parent hub (or link) primary key and load timestamp

- Non-key attributes

- A load timestamp defaulted to current_timestamp()

- The data source for the loaded data

- The hash_diff key (hash of all attributes that we want to track for SCD2 changes)

Other Data Vault Tables

Other data vault table types exist for reference data, bridge, and point-in-time tables. Largely, they may be used for edge cases, not core data vault delivery, and may be encountered occasionally.

Data Vault Summary

Data vault modeling must not be used for reporting but instead may be an intermediary step between the data source (or 3NF) and star schema, as Figure 13-6 shows.

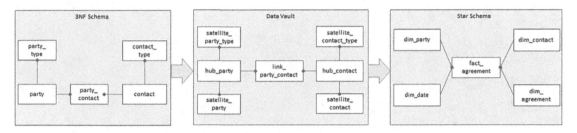

Figure 13-6. *Sample data model usage*

Summary

This chapter began with a simplified discussion of Third Normal Form (3NF) and pseudocode examples showing the benefits of using surrogate keys, particularly as real-world labels change and data models expand, change, or adapt over time. We also saw how 3NF models are relatively easy to change compared to star schemas. Our 3NF discussion concluded with denormalization, the term used to describe how we bring a 3NF data model back into business usable reporting structures.

Moving on to dimensional modeling, we proposed how our 3NF model might be re-imagined as a dimensional model. I explained how dimensions are created, and attributes may be extended to include a more business-friendly context while recognizing star schemas are difficult to change once implemented.

Finally, you learned the advantages of implementing data vault modeling as an intermediate step between 3NF and star schema became apparent. Data vault modeling delivers successful business outcomes by enabling rapid dimensional model deployment with corresponding object tagging and embedded data ownership, facilitating later self-service.

I suggest these comparisons among the three modeling approaches discussed.

- 3NF: Faster database writes, moderately flexible approach

- Data vault modeling: Parallel writes, hybrid flexible approach

- Dimensional modeling: Faster reads, rigid approach

Having discussed data management, let's look at the next chapter.

## CHAPTER 14

Snowflake Data Cloud by Example

This chapter discusses implementing two paradigms: Snowflake Data Exchange and Snowflake Marketplace. Both are underpinned by Snowflake Secure Data Sharing, which we delve into first. You know from Chapter 1 that Snowflake Data Exchange and Snowflake Marketplace are similar in concept but serve different target audiences. Secure Direct Data Share underpins both Snowflake Data Exchange and Snowflake Marketplace. Chapter 3 explained how data shares are implemented using micro-partitions. Let's investigate use cases via hands-on practical examples.

For Secure Direct Data Share, we continue our investigation using command-line scripts; but for Snowflake Data Exchange and Snowflake Marketplace, there is no approved command-line approach for configuration. Instead, and in line with the Snowflake-approved approach, we use Snowsight. Furthermore, to participate in Snowflake Marketplace, there is a formal engagement process to endure with a commitment to producing specific data sets, not to be undertaken lightly. More information is at `https://other-docs.snowflake.com/en/marketplace/becoming-a-provider.html#step-1-submit-a-request-to-join-the-snowflake-data-marketplace`.

Two separate Snowflake accounts are required for this chapter. For ease of testing, please ensure both are in the same cloud service provider and region. The data share examples will not work otherwise! Using a single email address, multiple trial Snowflake accounts may be provisioned at `https://signup.snowflake.com`.

© Andrew Carruthers 2022
A. Carruthers, *Building the Snowflake Data Cloud*, https://doi.org/10.1007/978-1-4842-8593-0_14

Snowflake is constantly evolving, and screenshots may be enhanced or reorganized between writing and publication. However, the general content and navigation are sure to remain relevant. Be aware that some details may change. Due to space constraints, I have not explored every option available for every feature but instead focused on what is believed to be the core capabilities resulting in end-to-end data provision and consumption.

Note   Terminology and product positioning have changed since inception. Please refer to the documentation for the latest information.

For an overview of the three product offerings discussed in this chapter, please also refer to the documentation at `https://docs.snowflake.com/en/user-guide/data-sharing-product-offerings.html#overview-of-the-product-offerings-for-secure-data-sharing`.

This chapter is what I would have liked to have at the outset of my Snowflake journey, a blueprint to get started, investigate further, and build out into a comprehensive, secure, business-focused solution.

By the end of this chapter, you will be equipped to implement pragmatic solutions immediately useful to your organization's internal and external facing needs.

A bold claim. Let's get started!

Overview

Let's start by reviewing our objectives in the context of Snowflake's platform, as shown in Figure 14-1.

SNOWFLAKE'S PLATFORM

Figure 14-1. *Snowflake platform*

Earlier chapters explained how to ingest data into Snowflake by presenting data for usage using a common pattern: data ingestion on the left and data consumption on the right. Recognizing we have not addressed data science and data applications, both are better served by other authors and media, not the least due to space constraints in this book.

Our focus now shifts to data sharing, the final part of our journey to interacting with internal and external consumers. In support, we should know where Snowflake has established its presence, as Figure 14-2 shows.

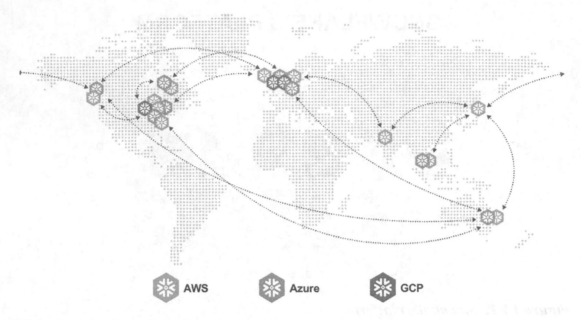

Figure 14-2. *Snowflake locations*

The latest Snowflake location information is at `https://docs.snowflake.com/en/user-guide/intro-regions.html`.

We must be aware of Snowflake's presence for various reasons, not the least of which are costs when moving data between cloud service providers (CSPs) and regions, see documentation at `https://docs.snowflake.com/en/user-guide/billing-data-transfer.html#data-transfer-billing-use-cases`.

Let's consider the purpose of Secure Direct Data Share, Data Exchange, and Data Marketplace. Figure 14-3 compares options and provides context for this chapter.

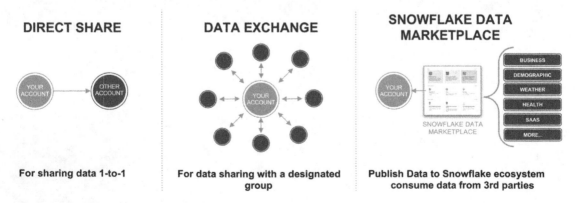

Figure 14-3. *Snowflake data interchange options*

Secure Direct Data Share

If the receiving Snowflake account is in a different region or hosted in a different CSP than the primary, then you must replicate first before sharing, as Figure 14-4 illustrates.

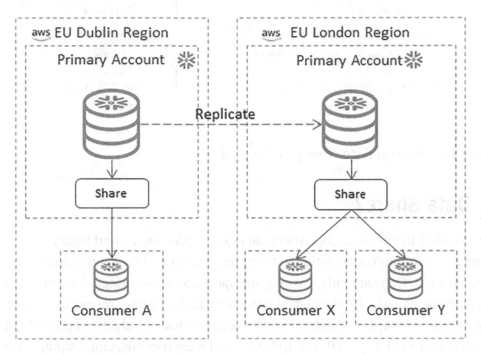

Figure 14-4. *Replicate before share*

In the same CSP and region, either Secure Direct Data Share or replication may be used, as illustrated in Figure 14-5.

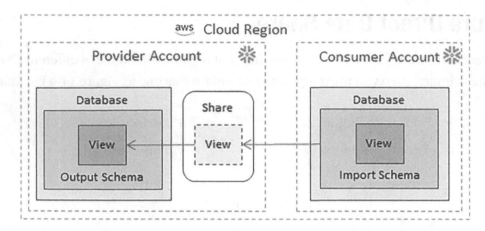

Figure 14-5. *Shared data conceptual model*

Why Data Share?

We mostly work in environments where data silos predominate, and legacy thinking pervades our conversations, stifling innovation and preventing true data democratization. Consequently, we may only have access to curated data sets via limited access paths or, worse still, partial data sets from which to derive meaningful answers. Without wishing to repeat Chapter 2, we are largely stuck with the legacy paradigm born in the 1980s of FTP/SFTP, API, ETL/ELT, and most recently, cloud storage. Yes, it is surprising to see the latest emergent technology considered a legacy. But the stark reality is that many organizations have wrapped up cloud storage in legacy constructs because these are both familiar and well understood.

I recognize the irony. Most of this book has been concerned with ingesting data via legacy techniques. But the reality on the ground (at the time of writing) is that most organizations are not yet willing to move away from their comfort zone and embrace the new data-sharing paradigm.

I hope you are among those willing to consider a trial and then adopt what is considered the future of data interchange. Snowflake has envisioned a new paradigm, Secure Direct Data Share, where service on demand is seamless and transparent, and data access is available out of the box. Secure Direct Data Share is a built-in, highly secure, core Snowflake capability reliant on cloud storage to share data with internal business colleagues, partners, and Snowflake customers.

Note   The aim is to streamline data access by implementing frictionless interfaces while retaining data security and system integrity regardless of location.

Suitable for all applications, Secure Direct Data Share explicitly controls customers' access to shared data. It is a great way to securely and safely monetize your data while always retaining full control.

Data sharing removes barriers, regardless of cloud, region, workload, or organizational domain, and gives instant access through a single copy of data. The number of data copies is set to rise from 9:1 to 10:1. Reusing a single copy of data should cause everyone to sit up and pay attention. Data sharing is a true game-changer.

Data sharing enables business continuity by eliminating disruptions and delivering a seamless customer experience, with no more processing chains or single points of failure from source to destination. Utilizing Snowflake's global data mesh guarantees consistent, reliable delivery with full security embedded.

Incorporating object tagging gains the distributed benefit of our organization's perspective on data categorization applied consistently across all uses of the subject data. Cross-domain governance is more easily managed centrally and consistently rather than distributed or fragmented across geographically disparate teams; one consistent implementation pattern and multiple aspects satisfy all use cases.

How to Data Share

Shares are a first-order object whose declaration and management are typically performed using the ACCOUNTADMIN role, the use of which should be tightly restricted. Snowflake provides instructions on devolving entitlement to lower privileged roles at `https://docs.snowflake.com/en/user-guide/security-access-privileges-shares.html#enabling-non-accountadmin-roles-to-perform-data-sharing-tasks`. In our view, devolving entitlement should not be done lightly as the law of unintended consequences quickly follows `https://en.wikipedia.org/wiki/Unintended_consequences`. I recommend that share entitlement is not devolved.

Let's set our context by reusing our TEST database and using the ACCOUNTADMIN role.

```
USE ROLE      accountadmin;
USE DATABASE  TEST;
```

```
USE WAREHOUSE COMPUTE_WH;
USE SCHEMA     public;
```

Let's create btsdc_share.

```
CREATE OR REPLACE SHARE btsdc_share;
```

With our share in place, let's assign entitlement enabling the share to access objects.

```
USE ROLE securityadmin;
```

```
GRANT USAGE ON DATABASE TEST              TO SHARE btsdc_share;
GRANT USAGE ON SCHEMA    TEST.test_owner TO SHARE btsdc_share;
```

To identify shares.

```
SHOW shares;
```

Set context, switch role to test_object_role.

```
USE ROLE       test_object_role;
USE DATABASE   TEST;
USE SCHEMA     TEST.test_owner;
```

Create a test table.

```
CREATE OR REPLACE TABLE csv_test
(
id      NUMBER,
label   VARCHAR(30)
);
```

Now grant SELECT privilege on test table to share, noting test_object_role does not have entitlement to use share btsdc_share; therefore, we must use securityadmin.

```
USE ROLE securityadmin;
```

```
GRANT SELECT ON TEST.test_owner.csv_test TO SHARE btsdc_share;
```

We can see entitlement granted to a share.

```
SHOW GRANTS TO SHARE btsdc_share;
```

Attempting to enable an account outside the CSP or region or an invalid account identifier results in this error.

```
SQL compilation error: Following accounts cannot be added to this share:
<account_name>.
```

To enable an account in the same CSP and region to access our share, we execute the following commands, recognizing each share may be accessed by many accounts, not just the one used in our example.

```
USE ROLE accountadmin;

ALTER SHARE btsdc_share SET
ACCOUNTS = <your_consumer_account>,
COMMENT  = 'Test share to Account';
```

And to remove an account from a share.

```
ALTER SHARE btsdc_share REMOVE ACCOUNT = <your_consumer_account>;
```

Replacing <your_consumer_account> with a consumer account in the same CSP and region.

To identify your Snowflake account, `select current_account();`.

We have created a share, assigned entitlement, and enabled another account in the same CSP and region to import the share. To do so, we must switch to the consumer Snowflake account and prepare a database.

```
USE ROLE accountadmin;
```

In the next command, replace <your_provider_account> with the name of the Snowflake provider account used to create the share.

```
CREATE OR REPLACE DATABASE btsdc_import_share
FROM SHARE <your_provider_account>.btsdc_share
COMMENT = 'Imported btsdc_share';
```

As a temporary measure and to prove imported shared objects exist, grant to SYSADMIN. This step illustrates an important point. Role-based access control (RBAC) is not carried forward with the share. They must be rebuilt in the recipient account. I strongly recommend giving due consideration to provisioning roles and entitlement for imported shares, which are not further explained here in the interests of brevity.

```
USE ROLE securityadmin;

GRANT IMPORTED PRIVILEGES ON DATABASE btsdc_import_share TO ROLE sysadmin;
```

We switch to the SYSADMIN role. Ensure imported objects can be seen in the share.

```
USE ROLE sysadmin;

SHOW SCHEMAS IN DATABASE btsdc_import_share;

SHOW TABLES IN SCHEMA btsdc_import_share.test_owner;
```

Let's assume we need to share another table, switch back to our provider account, set our context, and add another table.

```
USE ROLE       test_object_role;
USE DATABASE   TEST;
USE SCHEMA     TEST.test_owner;

CREATE OR REPLACE TABLE csv_test_2
(
id      NUMBER,
label   VARCHAR(30)
);

USE ROLE securityadmin;

GRANT SELECT ON TABLE TEST.test_owner.csv_test_2 TO SHARE btsdc_share;

SHOW GRANTS TO SHARE btsdc_share;
```

Now switch back to our consumer account and check whether the table has been shared.

```
SHOW TABLES IN SCHEMA btsdc_import_share.test_owner;
```

We should see the new table csv_test_2 in our imported share. Naturally, all other share consumers can also see the new table.

Further information on consuming from shares is at https://docs.snowflake.com/en/user-guide/data-share-consumers.html#data-consumers.

Data Share Summary

Consumers may create a single database per share. An inbound database must be created before the share can be queried, and created databases can be renamed and dropped. As stated, roles and entitlement are not currently carried forward with the share. They must be rebuilt in the recipient account. However, row-level security, column-level security, external tokenization, and data masking policies are honored.

Managing the share is trivial. Adding and removing objects from the share does not require much effort, and all share consumers also receive the same shared objects.

Most importantly, no data is copied. Metadata alone enables the secure sharing of data in underlying storage. Since no data is copied from the provider account, the Snowflake consumer account is not charged for storage.

Only the most recent micro-partitions are available in the share. The Time Travel feature is not supported for objects imported via a share, although it is supported for the original objects. Not all schema object types can be shared, and for those shareable object types, their secure counterparts must be used to preserve strict control of access to data. Further information is at `https://docs.snowflake.com/en/user-guide/data-sharing-provider.html#working-with-shares`.

Reader Accounts

Reader accounts are provisioned for organizations who are not yet Snowflake customers enabling data query but not upload, update, or delete of data. A key advantage of reader accounts is reduced data integration cost with direct access to live, ready-to-query data.

For Snowflake customers, the compute cost is paid for by the consuming account. For non-Snowflake consumers using a reader account, consumption is paid for by the provider.

We do not provide an in-depth study of reader accounts but instead refer to Snowflake documentation at `https://docs.snowflake.com/en/user-guide/data-sharing-reader-create.html#managing-reader-accounts`.

Tracking Usage

Before investigating data access, we must be mindful of latency when accessing Snowflake's Account Usage views. For query_history, latency can be up to 45 minutes, and for access_history, up to 3 hours. Latency is not a major consideration for tracking usage as the information is not of critical importance. More information is at `https://docs.snowflake.com/en/sql-reference/account-usage.html#account-usage-views`.

Access History

At the time of writing, the access_history view is a public preview and subject to change; therefore, the following code sample is indicative only. Documentation is at `https://docs.snowflake.com/en/user-guide/access-history.html#access-history`.

It is not currently possible to track consumer activity through the Snowflake Account Usage store. Each account tracks its own usage, for which these SQL statements may prove useful.

First, enable access to the Account Usage store.

```
USE ROLE securityadmin;

GRANT IMPORTED PRIVILEGES ON DATABASE snowflake TO ROLE test_object_role;
```

Revert to the test_object_role.

```
USE ROLE       test_object_role;
USE WAREHOUSE TEST_WH;
USE DATABASE  TEST;
USE SCHEMA     TEST.test_owner;
```

Create some test data.

```
INSERT INTO csv_test VALUES
(1, 'aaa'),
(2, 'bbb'),
(3, 'ccc');
```

Create a view v_access_history. You may wish to modify this to suit your needs.

```
CREATE OR REPLACE SECURE VIEW v_access_history AS
SELECT qh.query_text,
```

```
        qh.user_name||' -> '||qh.role_name||' -> '||qh.warehouse_name
        AS user_info,
        qh.database_name||'.'||qh.schema_name||' -> '||qh.query_type||' ->
        '||qh.execution_status AS object_query,
        ah.query_start_time,
        ah.direct_objects_accessed,
        ah.base_objects_accessed,
        ah.objects_modified,
        ah.query_id
FROM    snowflake.account_usage.access_history ah,
        snowflake.account_usage.query_history  qh
WHERE   ah.query_id                           = qh.query_id
ORDER BY ah.query_start_time DESC;
```

If a Data Sharing provider account shares objects to a Data Sharing consumer accounts through a share, then for:

- Data Sharing provider accounts: The queries and logs on the shared objects executed in the provider account are not visible to Data Sharing consumer accounts.

- Data Sharing consumer accounts: The queries on the data share executed in the consumer account are logged and only visible to the consumer account, not the Data Sharing provider account. The base tables accessed by the data share are not logged.

Data Sharing Usage

Snowflake is developing a suite of monitoring views where we can track shared data usage. Current monitoring capability is quite limited and does not allow identifying the rows accessed by any specific query. The latency may be up to two days.

Please also refer to Snowflake documentation at https://docs.snowflake.com/en/sql-reference/data-sharing-usage.html.

Both classic and Snowsight browsers provide access to the Data Sharing Usage views. The classic view is shown in Figure 14-6.

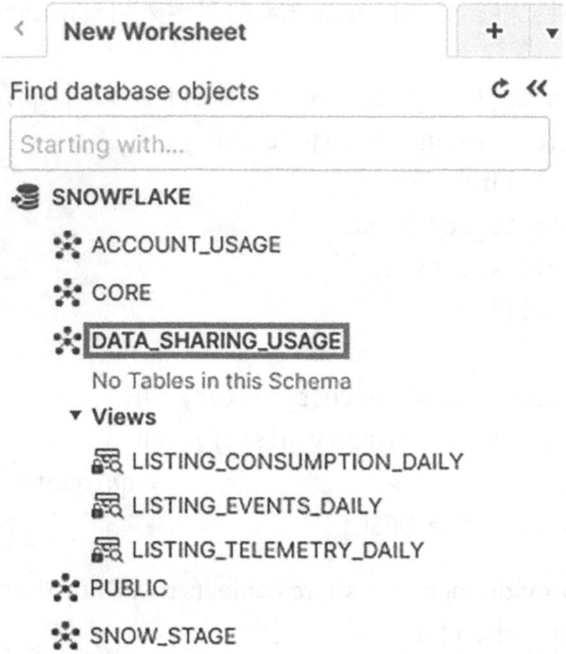

Figure 14-6. Data sharing usage views

I recommend periodic checks of Data Sharing Usage views. The following are sample queries.

```
USE ROLE       test_object_role;
USE WAREHOUSE TEST_WH;
USE DATABASE   snowflake;
USE SCHEMA     snowflake.data_sharing_usage;
```

We expect zero rows unless shares have been consumed.

```
SELECT *
FROM   snowflake.data_sharing_usage.listing_consumption_daily;
```

```
SELECT *
FROM   snowflake.data_sharing_usage.listing_events_daily;
```

Intriguingly, during testing, history has been retained for prior implementation of Data Exchange.

```
SELECT *
FROM   snowflake.data_sharing_usage.listing_telemetry_daily;
```

Centralized Cost Monitoring

This section explores the potential for implementing a common cost monitoring pattern using data share and local tables.

Note   It is assumed that all code in this chapter has been run to this point.

Conceptual Model

Building upon the code developed in Chapter 6, let's propose an extension to (almost) plug-and-play capability. Figure 14-7 illustrates a conceptual model of how a centralized cost monitoring share-based implementation may operate.

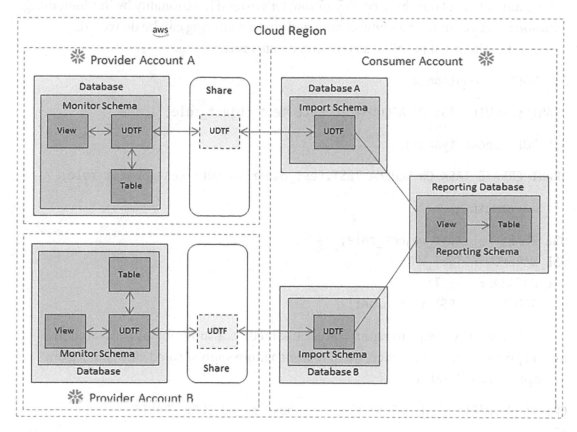

Figure 14-7. *Conceptual model*

Figure 14-7 shows two provider accounts (A & B) with a common suite of baselined reporting components deployed in a monitor schema. Each view overlays the provider Account Usage store or information schema with tabular SQL user-defined functions (UDTFs), one per view. Each provider account shares its UDTFs, and the consumer account imports each share into a separate database, one per provider.

Taking advantage of UDTF capability to return a result set, scheduling via a task in the consumer account calls each UDTF inserting data into its own tables, one for each UDTF.

In the reporting database, result sets from each provider account reporting component are joined using views for common reporting.

Cost Monitoring Setup

Taking advantage of our share configuration, let's extend functionality by implementing components to demonstrate how centralized cost monitoring can be delivered.

Grant entitlement to both executes and create tasks.

```
USE ROLE accountadmin;

GRANT EXECUTE TASK ON ACCOUNT TO ROLE test_object_role;

USE ROLE securityadmin;

GRANT CREATE TASK ON SCHEMA TEST.test_owner TO ROLE test_object_role;
```

Set context.

```
USE ROLE        test_object_role;
USE WAREHOUSE  TEST_WH;
USE DATABASE   TEST;
USE SCHEMA      TEST.test_owner;
```

Let's assume we want to report warehouse credit consumption by the warehouse for the previous day. In practice, we may want a more sophisticated view, but for our example, this will suffice.

```
CREATE OR REPLACE SECURE VIEW v_warehouse_spend COPY GRANTS
AS
SELECT wmh.warehouse_name,
       SUM ( wmh.credits_used )            AS credits_used,
```

```
         EXTRACT ( 'YEAR',  wmh.start_time )||
         EXTRACT ( 'MONTH', wmh.start_time )||
         EXTRACT ( 'DAY',   wmh.start_time ) AS spend_date
FROM     snowflake.account_usage.warehouse_metering_history wmh
WHERE    TO_DATE ( spend_date, 'YYYYMMDD' ) = current_date() -1
GROUP BY wmh.warehouse_name,
         spend_date
ORDER BY spend_date DESC, wmh.warehouse_name ASC;
```

Prove v_warehouse_spend returns results.

```
SELECT * FROM v_warehouse_spend;
```

We should see results from our Snowflake activity. We may also see unexpected rows from serverless compute.

We cannot add v_warehouse_spend directly to our share; otherwise, we receive this error: "SQL compilation error: A view or function being shared cannot reference objects from other databases." The error is caused by referencing the Snowflake database but using the TEST database for sharing. Instead, we must use a task to extract data into a table in the TEST database and then share the table.

Create a table named warehouse_spend_hist to hold the view output.

```
CREATE OR REPLACE TABLE warehouse_spend_hist
(
warehouse_name VARCHAR(255),
credits_used   NUMBER(18,6),
spend_date     VARCHAR(10),
last_updated   TIMESTAMP_NTZ DEFAULT current_timestamp()::TIMESTAMP_NTZ
               NOT NULL
);
```

Now create a task named task_load_warehouse_spend_hist, which runs once daily to load the warehouse_spend_hist table with the previous day's spend data. The schedule has been set to 1 minute for testing purposes only with a once per day (at 01:00) execution schedule commented out for later use. CRON was used to demonstrate an alternative method of scheduling.

```
CREATE OR REPLACE TASK task_load_warehouse_spend_hist
```

```
WAREHOUSE = TEST_WH
SCHEDULE  = 'USING CRON * * * * * UTC'
--SCHEDULE  = 'USING CRON 0 1 * * * UTC'
AS
INSERT INTO warehouse_spend_hist
SELECT warehouse_name,
       credits_used,
       spend_date,
       current_timestamp()::TIMESTAMP_NTZ
FROM   v_warehouse_spend;
```

Ensure task is registered.

```
SHOW tasks;
```

Enable task_load_warehouse_spend_hist but be sure to SUSPEND after the first run.

```
ALTER TASK task_load_warehouse_spend_hist RESUME;
```

Check next_run to ensure that the task is running.

```
SELECT timestampdiff ( second, current_timestamp, scheduled_time ) as
next_run,
       scheduled_time,
       current_timestamp,
       name,
       state
FROM   TABLE ( information_schema.task_history())
WHERE  state = 'SCHEDULED'
ORDER BY completed_time DESC;
```

When the run is complete, suspend the task.

```
ALTER TASK task_load_warehouse_spend_hist SUSPEND;
```

Note   Failure to suspend the task results in duplicate data.

When testing is complete, reset the task schedule to run once daily, and resume the task.

We could add warehouse_spend_hist to our share, but instead, to expand our knowledge, I prefer to create a fn_get_warehouse_spend UDTF. We must create UDTF as a SECURE function. For more information, see the documentation at `https://docs.snowflake.com/en/sql-reference/udf-secure.html#secure-udfs`.

Note   There is no COPY GRANTS clause for functions. Entitlement must be redone if a function is redefined.

```
CREATE OR REPLACE SECURE FUNCTION fn_get_warehouse_spend()
RETURNS TABLE ( warehouse_name VARCHAR,
                credits_used   NUMBER(18,6),
                spend_date     VARCHAR,
                last_updated   TIMESTAMP )
AS
$$
   SELECT warehouse_name,
          credits_used,
          spend_date,
          last_updated
   FROM   warehouse_spend_hist
$$
;
```

And prove secure function fn_get_warehouse_spend() returns data.

```
SELECT warehouse_name,
       credits_used,
       spend_date ,
       last_updated
FROM   TABLE ( fn_get_warehouse_spend());
```

We should see the result set from our earlier SELECT from warehouse_spend_hist noting the credits_used attribute has been coerced to NUMBER (18,6).

We are now ready to add the fn_get_warehouse_spend UDTF to our share.

```
USE ROLE securityadmin;

GRANT USAGE ON FUNCTION TEST.test_owner.fn_get_warehouse_spend() TO SHARE
btsdc_share;
```

Prove the fn_get_warehouse_spend UDTF has been added to btsdc_share.

```
SHOW GRANTS TO SHARE btsdc_share;
```

Reporting Database Setup

Switch to the Snowflake consumer account and create a new reporting database, report, with the report_owner schema and entitlement to use btsdc_import_share.

```
USE ROLE sysadmin;

CREATE OR REPLACE DATABASE report DATA_RETENTION_TIME_IN_DAYS = 90;
CREATE OR REPLACE SCHEMA   report.report_owner;

USE ROLE securityadmin;

CREATE OR REPLACE ROLE report_owner_role;

GRANT ROLE report_owner_role TO ROLE securityadmin;

GRANT USAGE    ON DATABASE  report                 TO ROLE report_owner_role;
GRANT USAGE    ON WAREHOUSE compute_wh             TO ROLE report_owner_role;
GRANT OPERATE  ON WAREHOUSE compute_wh             TO ROLE report_owner_role;
GRANT USAGE    ON SCHEMA    report.report_owner TO ROLE report_owner_role;

GRANT CREATE TABLE              ON SCHEMA report.report_owner TO ROLE
                                   report_owner_role;
GRANT CREATE VIEW               ON SCHEMA report.report_owner TO ROLE
                                   report_owner_role;
GRANT CREATE SEQUENCE           ON SCHEMA report.report_owner TO ROLE
                                   report_owner_role;
GRANT CREATE STREAM             ON SCHEMA report.report_owner TO ROLE
                                   report_owner_role;
GRANT CREATE MATERIALIZED VIEW ON SCHEMA report.report_owner TO ROLE
                                   report_owner_role;
```

```
GRANT IMPORTED PRIVILEGES ON DATABASE btsdc_import_share TO ROLE report_
owner_role;
```

Set the context to the new reporting database.

```
USE ROLE      report_owner_role;
USE DATABASE  report;
USE WAREHOUSE compute_wh;
USE SCHEMA    report.report_owner;
```

Prove we can access the fn_get_warehouse_spend UDTF. For later extension, add UNION with SELECT from subsequent data sources.

```
SELECT 'Provider Account A' AS source_account,
       warehouse_name,
       credits_used,
       spend_date,
       last_updated                       AS source_last_updated,
       current_timestamp()::TIMESTAMP_NTZ AS insert_timestamp
FROM   TABLE ( btsdc_import_share.test_owner.fn_get_warehouse_spend());
```

Create target table to persist source account data.

```
CREATE OR REPLACE TABLE source_warehouse_spend_hist
(
source_account      VARCHAR(255),
warehouse_name      VARCHAR(255),
credits_used        NUMBER(18,6),
spend_date          VARCHAR(10),
source_last_updated TIMESTAMP_NTZ,
insert_timestamp    TIMESTAMP_NTZ DEFAULT current_timestamp()::TIMESTAMP_
                    NTZ NOT NULL
);
```

Create a wrapper view for share and UDTF.

```
CREATE OR REPLACE SECURE VIEW v_source_warehouse_spend COPY GRANTS
AS
SELECT 'Provider Account A' AS source_account,
```

```
        warehouse_name,
        credits_used,
        spend_date,
        last_updated                          AS source_last_updated,
        current_timestamp()::TIMESTAMP_NTZ AS insert_timestamp
FROM    TABLE ( btsdc_import_share.test_owner.fn_get_warehouse_spend());
```

Insert reporting data.

```
INSERT INTO source_warehouse_spend_hist
SELECT * FROM v_source_warehouse_spend;
```

I leave it for you to create a task to schedule data ingestion from the source. Note the dependency upon data extracted into the originating table.

Replication

Replication has been discussed throughout the book. Now, let's work through setting up replication.

Note   We do not discuss replication in the context of full disaster recovery (DR) implementation; we simply use DR as a target account.

This example uses two accounts: primary and disaster recovery (DR). It replicates a database from Primary to DR. Figure 14-8 illustrates the scenario.

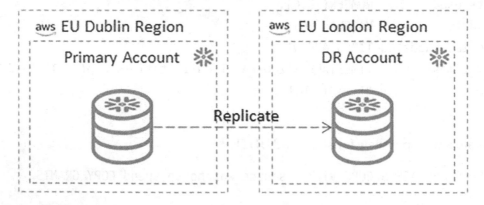

Figure 14-8.   *Database replication from primary to DR*

While we can configure replication in the Snowflake user interface, we contend deeper understanding is gained from working through command line invocation recognizing any automation we develop works through scripts and not using the interface. In other words, our replication strategy depends on more than a single user interface-driven strategy, probably with complex dependencies including failover, failback, and client redirect.

Not every object type is currently available for replication. Please refer to the documentation at `https://docs.snowflake.com/en/user-guide/database-replication-intro.html#replicated-database-objects`. Replication has further limitations. For example, RBAC is not replicated; therefore, the receiving account must implement RBAC locally; databases with external tables cannot be replicated; databases created from shares cannot be replicated.

This example illustrates simple database replication in support of data shares and is not a full treatise on failover and failback, which falls beyond the scope of this book. Also, note there are cost implications to replicating data across accounts, CSPs, and regions.

Using the primary account, we must identify accounts in our organization enabled for replication.

```
USE ROLE accountadmin;

SHOW REPLICATION ACCOUNTS;
```

We need two (or more) accounts enabled for replication. Your organization may prefer each account to be enabled individually, in which case a Snowflake support ticket should be raised. Response times are usually very short. Otherwise, Snowflake support can assign accounts the ORGADMIN role, as documented at `https://docs.snowflake.com/en/user-guide/organizations-manage-accounts.html#enabling-accounts-for-replication`.

Once ORGADMIN has been enabled for your account, then self-service to enable replication is possible.

```
USE ROLE orgadmin;

SHOW ORGANIZATION ACCOUNTS;

SELECT current_account();
```

In the next SQL statement, replace <your_account> with the response from the preceding SQL statement.

```
SELECT system$global_account_set_parameter ( '<your_account>', 'ENABLE_
ACCOUNT_DATABASE_REPLICATION', 'true' );
```

Re-check replication is enabled.

```
USE ROLE accountadmin;
```

```
SHOW REPLICATION ACCOUNTS;
```

With replication enabled in the primary account, you can now enable replication to the DR account by replacing <your_database> with your chosen database to replicate recalling limitations mentioned. Also, replace <your_target_account> with your chosen target account to receive a replica. Documentation is at `https://docs.snowflake.com/en/user-guide/database-replication-config.html#step-2-promoting-a-local-database-to-serve-as-a-primary-database`.

```
ALTER DATABASE <your_database> ENABLE REPLICATION TO ACCOUNTS <your_target_
account>;
```

While strictly speaking beyond the scope of our immediate aims, we might also enable failover for our primary database.

```
ALTER DATABASE <your_database> ENABLE FAILOVER TO ACCOUNTS <your_target_
account>;
```

Next, log in to the second account and view databases available for import.

```
SHOW REPLICATION DATABASES;
```

You should see <your_database> in the list of available databases for import. Note that is_primary is "true".

Let's create the replica. Replace <your_database> with your chosen database but respect the convention to retain naming conventions back to the source.

```
CREATE DATABASE <your_database>
AS REPLICA OF <your_account>.<your_database>
DATA_RETENTION_TIME_IN_DAYS = 90;
```

Check that our new database has been imported correctly by running this SQL statement.

```
SHOW REPLICATION DATABASES;
```

And refreshing the left-hand pane of our browser where our new database is listed.

With our database replicated, we have two remaining tasks: create and apply a security model and then share desired objects, which I leave you to complete.

Finally, we may wish to remove our replicated database.

```
DROP DATABASE <your_database>;
```

Data Exchange

Now that we understand data sharing, let's investigate Data Exchange, the internal nexus for colleagues, partners, and customers to collaborate, where our citizen scientists discover and self-serve available data sets. Please refer to the Snowflake documentation at `https://docs.snowflake.com/en/user-guide/data-exchange.html#data-exchange`.

Utilizing Snowflake's built-in capabilities, Data Exchange is its internal place where data is published to consumers—the people and organizations in and outside the publishing organization boundary.

Data Exchange is limited to 20 accounts for each listing across all CSPs and regions. Consideration must be given to creating a data share topology that does not replicate data silos but meets business needs and facilitates the delivery of successful business outcomes. Throughout this section, there is repeated reference to working with our business colleagues when developing our data exchanges. Only through collaboration will organizations benefit from the power and flexibility that Data Exchange offers. But equally, if done badly, it results in frustration and disillusionment with Data Exchange as a capability.

Request Data Exchange

Data Exchange is not enabled by default. Please contact Snowflake support to enable Data Exchange for your account. Your support ticket should contain the following.

- Business case, team, and/or participants in the Data Exchange: An example might be, "We are developing content for distribution throughout our organization and wish to run a proof-of-concept using Data Exchange."

- Data Exchange name (without spaces): XYZ_CORP_DX

- Public Data Exchange name: XYZ Corporation Data Exchange

- Your administrative account URL: for example, https://<account>. eu-west-2.aws.snowflakecomputing.com

- Your account locator: Run SELECT current_account();

Enabling Data Exchange can take 48 hours.

Snowsight

This section moves from the classic console to use Snowsight, the future-state Snowflake console, because there is no approved command-line approach to implementing Data Exchange features.

If using AWS PrivateLink or Azure Private Link, you may need to create a CNAME record as described in the documentation at https://other-docs.snowflake.com/en/ marketplace/intro.html#introduction-to-the-snowflake-data-marketplace.

Figure 14-9 highlights the button to click.

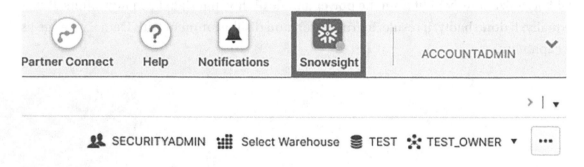

Figure 14-9. *Enable Snowsight*

A new browser window and login dialog appear alongside our existing classic console. Log in, accept the prompt to import tabs, and take a moment to familiarize yourself with the new look and feel.

On the day of writing, Snowflake released Snowsight as generally available, prompting migration to the new user interface with a button on the left-hand side of the browser to revert to the classic console. You may need to click the Home button before progressing to investigate Data Exchange, as shown in Figure 14-10.

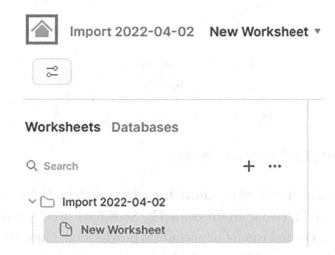

Figure 14-10. *Home button*

Accessing Data Exchange

When ready, change your role to ACCOUNTADMIN by clicking your login name. Then click Home ➤ Data ➤ Private Sharing, as shown in Figure 14-11.

Figure 14-11. *Data Exchange navigation*

We are now presented with all private data exchanges shared with our account, as shown in Figure 14-12. Two shares provided by Snowflake overlay the imported Snowflake databases created when our account is provisioned. The corresponding data exchange entry is also removed if a sample database has been deleted.

I leave the investigation of data from direct shares for your investigation and focus on XYZ Corporation Data Exchange, our fictional data exchange ready for configuration.

Note   To administer the private exchange, use the ACCOUNTADMIN role.

We may have multiple data exchanges for our organization, which appear alongside our initial data exchange, as shown in Figure 14-12.

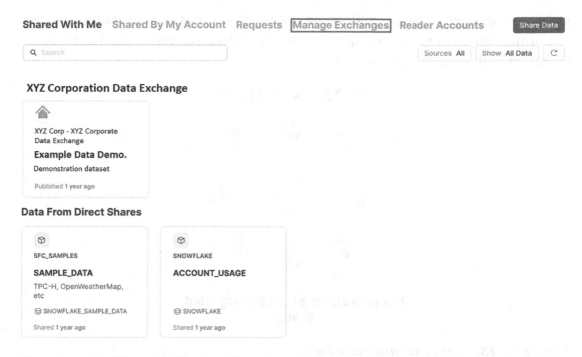

Figure 14-12. *Data Exchange Shared With Me*

Tony Robbins said, "Complexity is the enemy of execution." I strongly recommend a simple approach to data exchange is adopted by provisioning the minimal number of data exchanges necessary to meet business objectives while retaining meaningful data isolation boundaries.

Managing Data Exchange

In contrast to data share, accounts may be in any region or CSP. Beware of egress charges, however.

Note   All Data Exchange members may publish and consume the data sets of other members, but only the owner account can administer members.

Before proceeding, a brief reminder of our intent is to configure one-to-many data sharing in a designated group from a single central Snowflake account, as shown in Figure 14-13. In the following discussion, there is a single publisher account with admin privilege to manage exchanges and several consumer accounts without admin privilege. The distinction is evident in the request made to Snowflake support to provision Data Exchange, as illustrated in Figure 4-12.

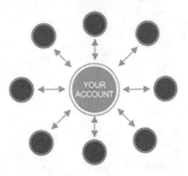

For data sharing with a designated group

Figure 14-13. *Data Exchange topology*

We can now manage our data exchange by adding and removing Snowflake accounts. Click Manage Exchanges ➤ XYZ Corporation Data Exchange, as shown in Figure 14-14.

Shared With Me Shared By My Account Requests **Manage Exchanges** Reader Accounts	Share Data ⌄
NAME	EXCHANGE ROLE
XYZ Corporation Data Exchange	Admin, Provider

Figure 14-14. *Manage exchanges*

Figure 14-15 shows two data exchange member accounts, in this example, those used to configure Secure Direct Data Share discussed previously in this chapter.

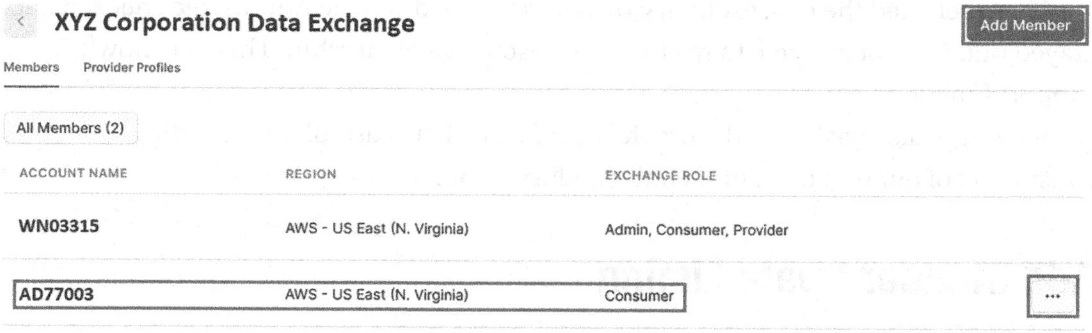

Figure 14-15. *Manage members and assign roles*

Adding a new member to our data exchange is simple. Click the Add Member button, populate with the Snowflake account URL, and select the role, as shown in Figure 14-16. By default, both Consumer and Provider are enabled, which may not be appropriate for your data exchange where one-way traffic may be preferred. Populate the account with a valid URL noting the user interface validates the entered URL, set the appropriate role, and then click the Add button. Accounts may be located in any CSP or region.

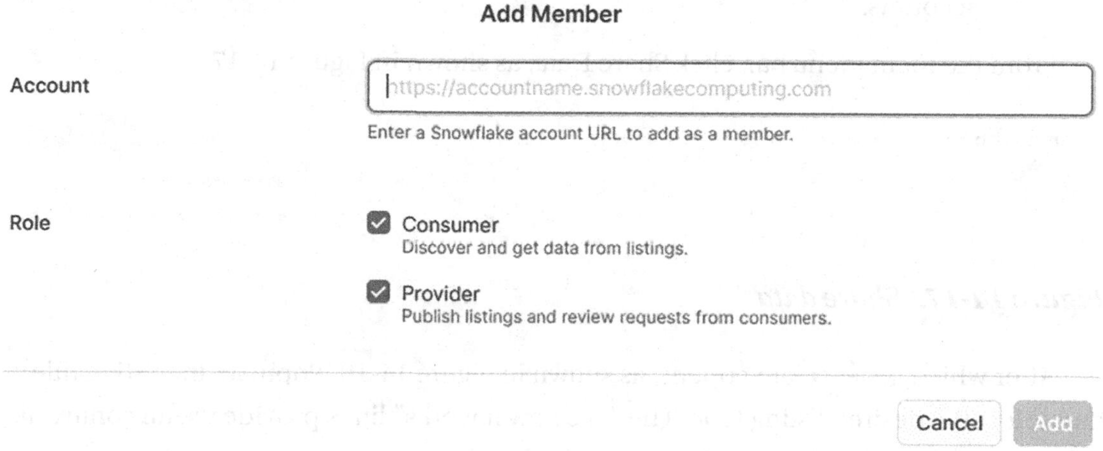

Figure 14-16. *Add account and role(s)*

Selecting an existing data exchange member enables their role to be changed, or clicking the three-dotted box allows removal from the data exchange.

If you selected the data exchange owner, you would see the Admin role checked and grayed out. It is not possible to reassign data exchange ownership. This is a Snowflake support function.

We might also wish to add a provider profile, which is useful for informing consumers of our organization's contact information.

Add Standard Data Listing

With data exchange members added to our data exchange, let's consider how to manage data sets in the data exchange. We have already encountered the two roles available, which are self-explanatory: providers and consumers.

Data exchange supports two types of listing.

- Standard: As soon as the data is published, all consumers have immediate access to the data.

- Personalized: Consumers must request access to the data for subsequent approval, or data is shared with a subset of consumer accounts.

From the main menu bar, click Share Data, as shown in Figure 14-17.

Figure 14-17. *Share data*

After which, a new dialog opens, as shown in Figure 14-18. Populate the listing title and select the desired listing type. The "See how it works" links provide useful contextual information. I recommend you investigate both links.

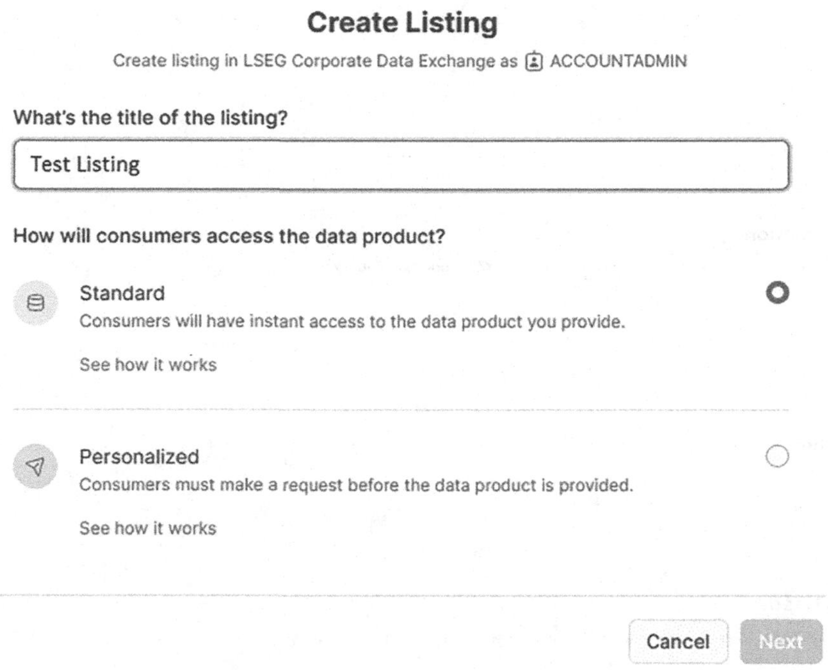

Create Listing

Create listing in LSEG Corporate Data Exchange as 🔒 ACCOUNTADMIN

What's the title of the listing?

Test Listing

How will consumers access the data product?

🗄	**Standard** Consumers will have instant access to the data product you provide. See how it works	●
◁	**Personalized** Consumers must make a request before the data product is provided. See how it works	○

Cancel   Next

Figure 14-18. *Create listing*

In our example, we use Test Listing as our title, select Standard, then click Next. The title is presented in preview form requiring further information, as shown in Figure 14-19. Note additional tabs for Consumer Requests and Settings for your further investigation.

Note   Once a listing has been published, changing the listing type is impossible.

Unpublishing a listing reverts to preview format, which is discussed later.

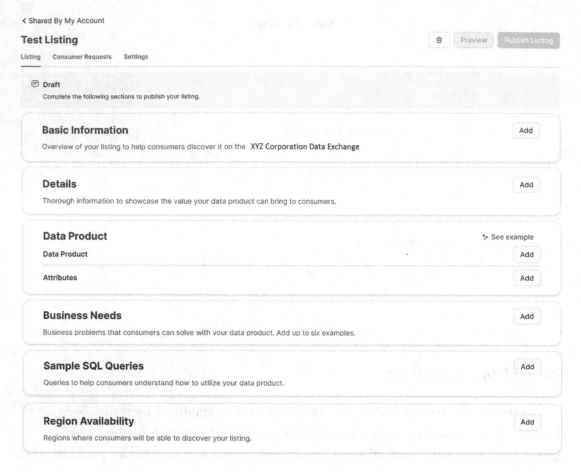

Figure 14-19. *Listing details*

I strongly recommend your business colleagues are actively involved in the creation of content to populate the listing details. Each listing is a showcase and shop front for your organization's internal colleagues and selected external customers; therefore, well-crafted descriptions and documentation representing approved messaging are essential.

Some callouts for further consideration.

- The terms of service requires your organization's legal team's input and approval before publication.

- Sample SQL Queries should sufficiently represent consumers to gain insight and pique their interest in using your data.

- Limit query runtime and data sets as the provider account pays for consumption costs.

- Protect your data model by provisioning access to views in preference to tables.

Although each entry is self-explanatory, there is a lot to consider when provisioning objects into our data exchange. We do not dwell on populating. I leave them for your further investigation.

When complete, click Publish Listing.

Manage Shared Data

In the pop-up dialog box shown in Figure 14-20, we may choose to manage our shared data by clicking Manage Shared Data.

Your listing is live

Your listing is visible to consumers in all
regions. You must attach a Secure
Share in each region in order for
consumers in those regions to gain
immediate access to your data.

Figure 14-20. *Manage shared data dialog*

According to the regions enabled in Figure 14-21, there are options to make our shared data available in consumer accounts. Note the requirement to log in to each account. Simply select the required region in the drop-down box where further options become available (not shown).

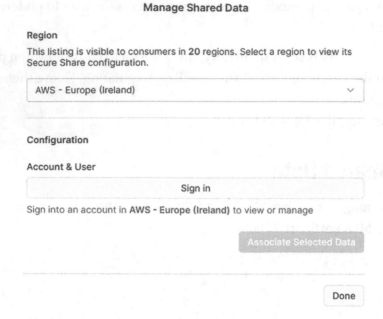

Figure 14-21. *Enable Data Share*

When complete, click Done. The listing banner changes to show two new buttons—View on Exchange and Live, as shown in Figure 14-22.

Figure 14-22. *Listing management options*

Clicking View on Exchanges provides a single page view of listings with summary information which I leave for your further investigation. Navigating to Home ➤ Data ➤ Private Sharing, as shown in Figure 14-12, displays a second listing alongside the original (not shown).

Unpublish a Data Listing

The second option, Live, allows you to unpublish a listing noting existing consumers retain access, as shown in Figure 14-23.

Unpublish Listing

Test Listing will no longer be visible in the XYZ
Corporation Data Exchange

Existing consumers of your data will continue to have
access to it.

Cancel   Unpublish

Figure 14-23. *Unpublish listing dialog*

Unpublished listings revert to preview status, as shown in Figure 14-19.

Accessing Data

Assuming we have shared our new data listing with our second snowflake account
region, we can view it in our second Snowflake account. Log in and navigate to Home
➤ Data ➤ Private Sharing ➤ Test Listing. Figure 14-24 illustrates the dialog box now
enabled in the consumer account.

Free

Unlimited queries

Query Data

View Database

Figure 14-24. *Consumer account accessing data*

The View Database screen (not shown) displays the Database Details tab by default. There is a second tab for Schemas. I leave these for your further investigation as the content is self-explanatory. Note the context menu navigates to Home ➤ Data ➤ Databases.

In common with Secure Direct Data Share, RBAC is not imported with shares. You may need to set your browser role to ACCOUNTADMIN or configure a new role with IMPORT SHARE privilege.

Click the Get Data button. Figure 14-25 shows the opportunity to assign a role, which should be pre-created.

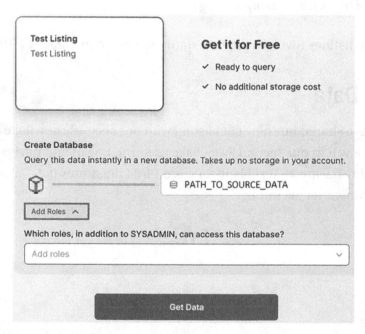

Figure 14-25. *Import share and assign RBAC*

If the chosen role does not have sufficient entitlement, the following error occurs.

Your selected role (SYSADMIN) cannot get data. Choose a role that has the IMPORT SHARE privilege to get this data or please contact your account administrator.

Insufficient privilege to accept DATA_EXCHANGE_LISTING terms.

Once RBAC entitlement considerations have been resolved, the dialog in Figure 14-26 is displayed.

Data is Ready to Query

🗄 PATH_TO_SOURCE_DATA

Query Data

Figure 14-26. *Query data*

A new browser tab opens pre-populated with the sample query and comments from the listing Usage Examples. Add warehouse to execute the sample query and execute.

Data Exchange Summary

The following are other menu options.

- Shared By My Account lists each created listing, whether published or not.

- Requests lists inbound and outbound data requests along with approval status.

- Reader Accounts allows providers to share data with consumers who are not already Snowflake customers without requiring the consumers to become Snowflake customers. See documentation for further information at `https://docs.snowflake.com/en/user-guide/data-sharing-reader-create.html`.

Having walked through hands-on examples of configuring and accessing objects provisioned via Data Exchange using a Standard Listing, I leave you to develop data access paths for your user community to interact with provisioned data sets.

Provisioning steps for Personalized listings are broadly the same as for Standard Listings, except the consumer must request access to the data set, and the provider must approve the request. These steps are self-explanatory.

Snowflake Marketplace

Now that you understand Data Exchange, the internal nexus for colleagues, partners, and customers to collaborate, let's investigate Snowflake Marketplace, the external nexus for individual and business collaboration. It is a worldwide environment where our citizen scientists discover and self-serve available data sets. Please refer to Snowflake documentation at `https://other-docs.snowflake.com/en/data-marketplace. html#snowflake-data-marketplace` and further summary information at `www. snowflake.com/data-marketplace/`.

At this point, you may be asking how Data Exchange and Snowflake Marketplace differ, so I offer these broad comparisons.

- Data Exchange is focused on closed groups with a maximum of 20 participant accounts. Snowflake Marketplace is global in scope with unlimited participation.

- Data Exchange is thematic in approach with either a single or limited topics. Snowflake Marketplace has a general approach where all themes and topics are available ubiquitously.

Both Data Exchange and Snowflake Marketplace are evolving rapidly where the boundaries and scope are converging and overlapping. You see an example later in this chapter.

In the same manner as Data Exchange and utilizing Secure Direct Data Sharing capability, organizations create listings on the Snowflake Marketplace, the single Internet location for seamless data interchange. The same security and real-time features remain though the scope of Snowflake Marketplace is external partners and consumers instead of internally focused, as is the case for Data Exchange.

Further information on joining the Snowflake Marketplace or requesting new data sets is at `https://other-docs.snowflake.com/en/marketplace/intro.html#how-do-i-request-new-data-or-data-providers-to-be-added-to-the-data-marketplace`.

Snowflake claims over 500 listings from more than 160 vendors representing a 76% increase in usage in the six months leading up to June 2021, along with new usage-based purchase options enabling consumers to transact entirely on the Snowflake Marketplace. For more information, see Snowflake news at `www.snowflake.com/news/snowflake-accelerates-data-collaboration-with-more-than-500-listings-in-snowflake-data-marketplace-and-announces-monetization-in-the-data-cloud/`.

Becoming a data provider involves populating an online form to enroll in the Snowflake Partner Network. For organizations, this step will likely involve your procurement, legal, and data governance teams as a minimum, along with a clearly articulated and approved proposal on organizational objectives.

Snowflake Marketplace offers standard, sample/purchase, and personalized data listings. This exploration implements a standard data listing while leaving the personalized data listing for you to investigate further.

When writing this book, Snowflake Marketplace was undergoing significant changes, many of which relate to navigation. Although I have attempted to reflect the latest information, further changes will probably render the navigation outdated, but the core data entry screens remain intact.

Setting up as a Provider

When ready, change the role to ACCOUNTADMIN by clicking your login name. Then click Home ➤ Data ➤ Provider Studio, as shown in Figure 14-27. The direct URL is `https://app.snowflake.com/provider-studio`.

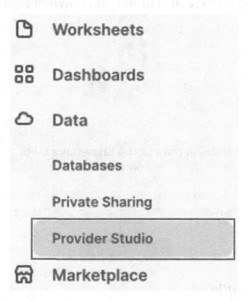

Figure 14-27. Snowflake marketplace navigation

If our account has not been entitled to participate in the Snowflake Marketplace, the message shown in Figure 14-28 appears, contact your Snowflake account executive, or fill out the form at `https://other-docs.snowflake.com/en/marketplace/becoming-a-provider.html#step-1-submit-a-request-to-join-the-snowflake-data-marketplace`.

Permission Required

To manage listings in Snowflake Data
Marketplace, select a role with the correct
permissions or contact your account
administrator.

***Figure 14-28.** Permission denied*

Assuming your account has been approved to participate in the Snowflake Marketplace, the next step is to create a profile, as shown in Figure 14-29.

**Publish data to the Snowflake Data
Marketplace**

Build a profile, manage listings and view
analytics in the provider studio. Learn more

Set Up Profile

***Figure 14-29.** Set up profile*

We are now required to populate our profile. Note the subset of fields shown in Figure 14-30.

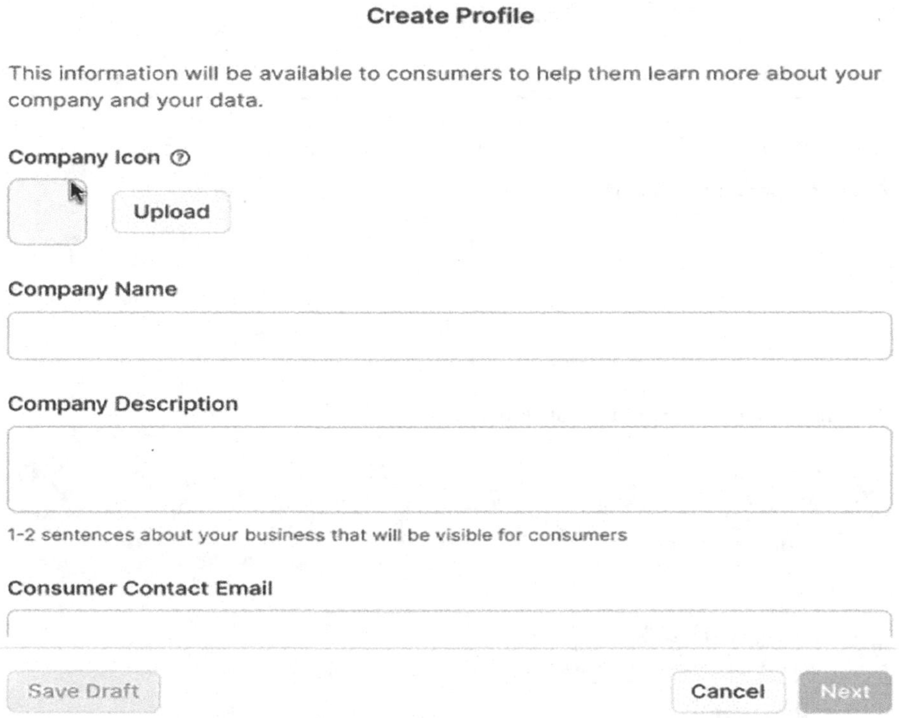

Create Profile

This information will be available to consumers to help them learn more about your company and your data.

Company Icon ⑦

[Upload]

Company Name

[   ]

Company Description

[   ]

1-2 sentences about your business that will be visible for consumers

Consumer Contact Email

[   ]

[Save Draft]   [Cancel] [Next]

Figure 14-30. *Set up profile*

Add Standard Data Listing

Snowflake Inc. requires content in Microsoft Word for its operations team to review and approve prior to creating the first listing. This is Snowflake Inc.'s standard practice.

For all listing types (see Figure 14-31), prospective consumers must register their interest by populating contact/company information.

There is an overlap with Data Exchange insofar as private listings may be made on Snowflake Marketplace, and long term, it is possible Data Exchange will be folded into Snowflake Marketplace.

Snowflake Marketplace listings attract a rebate on credits consumed by usage.

When the profile is populated, click Next. The screen shown in Figure 14-31 should appear.

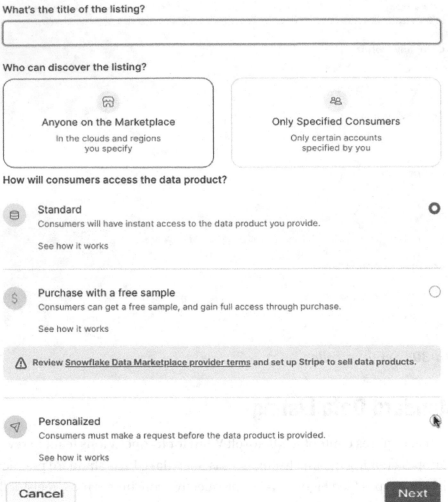

Figure 14-31. *Listing information*

Snowflake Marketplace supports three types of listing.

- Standard: As soon as the data is published, all consumers have immediate access to the data.

- Purchase with free sample: Upgrade to full data set upon payment.

- Personalized: Consumers must request access to the data for subsequent approval, or data is shared with a subset of consumer accounts.

Once the listing has been created (not shown but very similar to creating a Data Exchange listing), there is one further consideration.

There is one further consideration. Consumption analysis may indicate data set publication may best be focused on particular regions; therefore, I recommend analyzing available data before publishing to all regions.

Let's decide how our listing will be published, as shown in Figure 14-32. The options are automatically published upon the Snowflake operations team's approval. We might prefer to set up multiple listings and decide to coordinate release as part of our marketing strategy.

Figure 14-32. *Publish setting*

Accessing Snowflake Marketplace

We are now presented with the most recent data sets made available via the Snowflake Marketplace, as shown in Figure 14-33. Your content will differ.

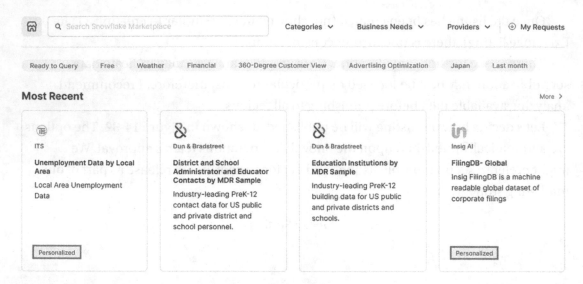

Figure 14-33. *Snowflake marketplace content*

Immediately we can see the distinct difference between internal Data Exchange and external Snowflake Marketplace. The two listings are standard and personalized as outlined.

Using Snowflake Marketplace

A closer reading of the Standard data set listing shown in Figure 14-33 illustrates the marketplace nature of published data. Both are sample data sets.

To gain experience with Snowflake Marketplace, click Home ➤ Marketplace or select Providers, as shown in Figure 14-34.

Figure 14-34. *Select provider*

Then scroll down and select Snowflake Inc., where the user interface refreshes to display the content shown in Figure 14-35.

Snowflake Inc.

Contact

Financial

Snowflake delivers the Data Cloud — a global network where thousands of organizations mobilize data with near-unlimited scale, concurrency, and performance. Inside the Data Cloud, organizations unite their siloed data, easily discover and securely share governed data, and execute diverse analytic workloads. Wherever data or users live, Snowflake delivers a single and seamless experience across multiple public clouds. Snowflake's platform is the engine that powers and provides access to the Data Cloud, creating a solution for data warehousing, data lakes, data engineering, data science, data ap... More

Snowflake Inc.

Quarterly Financial Statements

Snowflake Financial Data

Figure 14-35.   Snowflake Inc. financial data

Note content may also be searched by category and business need.

Content can also be referenced directly from the corresponding URL `https://app.snowflake.com/marketplace/listings/Snowflake%20Inc`. You see one or more published data sets.

Selecting a data set results in a detailed screen similar to the one seen when consuming Data Exchange content, which I consider sufficiently familiar to not require further explanation. from our work

Managing Snowflake Marketplace

Assuming all obligations have been met and our request to become an approved data provider (`https://spn.snowflake.com/s/become-a-partner`) to the Snowflake Marketplace is successful, we may now participate in the Snowflake Marketplace. Our participation may impose restrictions according to the type and nature of the data we provide.

Note   Listings must always comply with relevant laws and contractual obligations. We must have legal and contractual rights to share data.

In contrast to data share, but common with data exchange, be aware of egress charges. And like Data Exchange, there is no approved command line approach to implementing Snowflake Marketplace features.

Before proceeding, a brief reminder of our intent is to configure our Snowflake account to publish data into the Snowflake Marketplace and/or to consume data from third parties sharing data in the Snowflake Marketplace, as shown in Figure 14-36.

**Publish Data to Snowflake ecosystem
of consume data from 3rd parties**
Access data from 175+
providers across 18+ categories

Figure 14-36. *Snowflake marketplace topology*

Accessing Data

To access a data set, we might be required to validate our email address when selecting a listing, as shown in Figure 14-37.

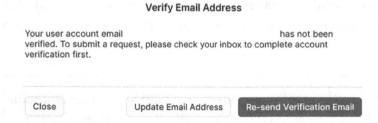

Figure 14-37. *Verify email*

Otherwise, data should immediately become available for download, as shown in Figure 14-38. The database to be created (in this example, FINANCIAL_STATEMENTS_ SHARE) can be overwritten; if done so, I recommend a naming convention be established to facilitate tracing back to the source.

The Add Roles dialog allows another existing role with IMPORT SHARE entitlement in your Snowflake account to access the imported database.

Then click Get Data.

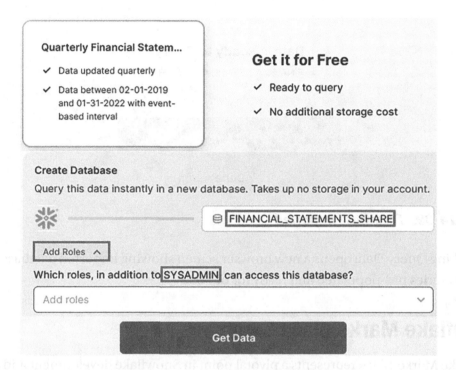

Figure 14-38. *Import database*

If all is well, our data is imported into a new database and made available for use; otherwise, you may see an error.

> ⚠ **Your selected role (SYSADMIN) cannot get data. Choose a role that has the IMPORT SHARE privilege to get this data or please contact your account administrator.**
>
> Insufficient privilege to accept DATA_EXCHANGE_LISTING terms.

Note   In Figure 14-38, the SYSADMIN role is set at the browser level, underneath the username.

Change role and retry, after which we should see a dialog shown in Figure 14-39.

Data is Ready to Query

🛢 FINANCIAL_STATEMENTS_SHARE

Query Data

Done

Figure 14-39. *Database imported*

Clicking Query Data opens a new browser screen showing imported databases with sample queries pre-populated and ready for our use.

Snowflake Marketplace Summary

Snowflake Marketplace represents a pivotal point in Snowflake development and delivery, facilitating citizen scientists to rapidly acquire desired data sets into a single account and providing an opportunity to monetize data sets. Participation is critical for Snowflake's success, and the ease of data set integration presents an almost frictionless opportunity for our organizations to derive immense value.

Automating integration with Snowflake Marketplace requires tooling outside of the user interface. Allowing programmatic interaction, along with improved consumption metrics, will encourage adoption. And Snowflake is signaling steps in this direction.

Summary

This chapter overviewed its objectives before diving into data sharing, explaining why by unpacking the advantages of adopting a new approach. We then developed a simple data share example using two separate Snowflake accounts, noting the use of the ACCOUNTADMIN role and caution over delegating entitlement to lower privileged roles. You began to see the power of sharing, where objects can be added and removed from a share with immediate visibility in the consumer account. Many accounts may consume a single share.

Using your newfound knowledge, we implemented a simple but readily extensible cost monitoring solution sharing daily Warehouse consumption costs from one account to another while introducing UDTFs and CRON scheduling for tasks. And for those who require replication before sharing, we walked through replicating a database between accounts.

Next, we focused on Snowflake Data Exchange, noting the use of Snowsight user interface for configuration, and walked through an example of how to create and manage a standard data listing noting that once published, we cannot change a listing type and unpublishing a listing does not remove existing consumer access.

Lastly, we investigated Snowflake Marketplace identifying how to access listings and publish our organizations' listings. The process is largely similar to Snowflake Data Exchange, with additional governance.

Index

A

Access control, 55, 63

Access model
- assign roles to user, 136
- create task, 144–146
- database, 133
- discretionary access control, 130
- end user/service user, 147
- object creation, 137, 138
- overview, 130
- owner schema build, 139, 140, 142
- RBAC, 132
- remove all objects, 148
- role grants, 135
- roles, 134, 135
- sample automation, 139
- sample object/role layout, 131
- schema, 134
- search path, 149
- SET declarations, 132
- staging data load, 142
- test stored procedure, 143
- troubleshooting, 148
- warehouse, 133

Account management, 129

Account Usage store
- access, 155
- best practices, 154
- database/warehouse/schema, 156
- declarations, 155
- definition, 153

- enabling, 159, 160
- functional requirements, 151
- role grants, 157
- roles, 157, 159
- streams/tasks, 152

Active Directory (AD), 95

Active Directory Federated Services (ADFS), 96

API Gateway, 322–324

API integration, 317, 324–326

Application programming interfaces (APIs), 42–44

Assumptions, 66, 74, 124, 231, 399, 402

Asynchronous monitoring
- account parameters, 170–172, 175
- network parameters, 176
- SCIM, 177, 178

Attack vectors, 97

Authentication, 11, 25, 39, 55, 63, 95, 204

AUTO_INGEST Snowpipe, 223
- configure AWS SQS, 224, 225
- notification channel, 224
- set up snowflake objects, 223, 224
- SQS integration, 223
- test, 225, 226

Automatic clustering, 74, 83, 348–349, 356

AWS Management Console, 189, 194, 196, 198, 201, 204, 225, 291, 294, 307, 308, 321, 322, 325, 327, 332

AWS PrivateLink, 191, 438

AWS storage integration, 90

Azure Private Link, 191, 438

Printed in the United States
by Baker & Taylor Publisher Services

Printed in the United States
by Baker & Taylor Publisher Services